BEYOND SPACETIME

One of the greatest challenges in fundamental physics is to reconcile quantum mechanics and general relativity in a theory of quantum gravity. A successful theory would have profound consequences for our understanding of space, time, and matter. This collection of essays written by eminent physicists and philosophers discusses these consequences and examines the most important conceptual questions among philosophers and physicists in their search for a quantum theory of gravity. Comprised of three parts, the book explores the emergence of classical spacetime; the nature of time; and important questions of the interpretation, metaphysics, and epistemology of quantum gravity. These essays will appeal to both physicists and philosophers of science working on problems in foundational physics, specifically that of quantum gravity.

NICK HUGGETT is LAS Distinguished Professor at the University of Illinois at Chicago. He has worked in the field of the philosophy of quantum gravity for over 20 years. He is coeditor of *Physics Meets Philosophy at the Planck Scale* (Cambridge University Press, 2001) and is Co-Director of the Beyond Spacetime research project, funded by the National Science Foundation, Foundational Questions Institute, John Templeton Foundation, and American Council of Learned Societies.

KEIZO MATSUBARA is an affiliated researcher at Uppsala University. His philosophical work primarily addresses string theory. He has held postdoctoral positions at the Rotman Institute of Philosophy at Western University and at the University of Illinois at Chicago.

CHRISTIAN WÜTHRICH is Associate Professor of Philosophy at the University of Geneva. He works in philosophy of physics, philosophy of science, and metaphysics. Starting with his doctoral research, the primary focus of his research has long been the philosophy of quantum gravity. He is Co-Director of the Beyond Spacetime research project, funded by the National Science Foundation, Foundational Questions Institute, John Templeton Foundation, and American Council of Learned Societies.

BEYOND SPACETIME

The Foundations of Quantum Gravity

Edited by

NICK HUGGETT
University of Illinois at Chicago

KEIZO MATSUBARA
Uppsala University

CHRISTIAN WÜTHRICH
University of Geneva

CAMBRIDGE
UNIVERSITY PRESS

CAMBRIDGE
UNIVERSITY PRESS

University Printing House, Cambridge CB2 8BS, United Kingdom

One Liberty Plaza, 20th Floor, New York, NY 10006, USA

477 Williamstown Road, Port Melbourne, VIC 3207, Australia

314-321, 3rd Floor, Plot 3, Splendor Forum, Jasola District Centre, New Delhi – 110025, India

79 Anson Road, #06–04/06, Singapore 079906

Cambridge University Press is part of the University of Cambridge.

It furthers the University's mission by disseminating knowledge in the pursuit of education, learning, and research at the highest international levels of excellence.

www.cambridge.org
Information on this title:www.cambridge.org/9781108477024
DOI: 10.1017/9781108655705

First published 2020

Printed in the United Kingdom by TJ International Ltd, Padstow Cornwall

A catalogue record for this publication is available from the British Library.

Library of Congress Cataloging-in-Publication Data
Names: Huggett, Nick, editor. | Matsubara, Keizo, 1977– editor. |
Wüthrich, Christian, editor.
Title: Beyond spacetime : the foundations of quantum gravity / edited by
Nick Huggett, University of Illinois at Chicago, Keizo Matsubara,
Uppsala University, Christian Wüthrich, University of Geneva.
Other titles: Beyond space time
Description: Cambridge, United Kingdom ; New York, NY, USA : Cambridge
University Press, 2020. | Includes bibliographical references and index.
Identifiers: LCCN 2019038856 (print) | LCCN 2019038857 (ebook) |
ISBN 9781108477024 (hardback) | ISBN 9781108655705 (epub)
Subjects: LCSH: Quantum gravity.
Classification: LCC QC178 .B49 2020 (print) | LCC QC178 (ebook) | DDC 530.14/3–dc23
LC record available at https://lccn.loc.gov/2019038856
LC ebook record available at https://lccn.loc.gov/2019038857

ISBN 978-1-108-47702-4 Hardback

Contents

Contributors

Suddhasattwa Brahma
Department of Physics, McGill University, Montreal, Canada

Robert H. Brandenberger
Department of Physics, McGill University, Montreal, Canada

Radin Dardashti
Interdisciplinary Centre for Science and Technology Studies, University of Wuppertal, Wuppertal, Germany

Richard Dawid
Department of Philosophy, Stockholm University, Sweden

Sebastian De Haro
Trinity College, University of Cambridge, Cambridge, UK

Fay Dowker
Department of Physics, Imperial College London, London, UK

Henrique Gomes
Department of History and Philosophy of Science, University of Cambridge, Cambridge, UK; Perimeter Institute for Theoretical Physics, Waterloo, Canada

Sean Gryb
Faculty of Philosophy, University of Groningen, Groningen, Netherlands

Daniel Harlow
Department of Physics, Massachusetts Institute of Technology, Cambridge, Massachusetts, USA

Nick Huggett
Department of Philosophy, University of Illinois at Chicago, Chicago, Illinois, USA

Keizo Matsubara
Department of Philosophy, Uppsala University, Uppsala, Sweden

Daniele Oriti
Arnold Sommerfeld Center for Theoretical Physics and Munich Center for Mathematical Philosophy, Ludwig Maximilian University of Munich, Munich, Germany

Carlo Rovelli
Centre de Physique Théorique de Luminy, Marseille, France

Ko Sanders
School of Mathematical Sciences and Centre for Astrophysics and Relativity, Dublin City University, Dublin, Ireland

Lee Smolin
Perimeter Institute for Theoretical Physics, Waterloo, Canada

Karim Thébault
Department of Philosophy, University of Bristol, Bristol, UK

Tiziana Vistarini
Department of Philosophy, Communication and Performing Arts, Roma Tre University, Rome, Italy

David Wallace
Department of History and Philosophy of Science and Department of Philosophy, University of Pittsburgh, Pittsburgh, Pennsylvania, USA

Christian Wüthrich
Department of Philosophy, University of Geneva, Geneva, Switzerland

1

Introduction

NICK HUGGETT, KEIZO MATSUBARA, AND
CHRISTIAN WÜTHRICH

This volume is one of the fruits of a three-year research project, *Space and Time after Quantum Gravity*, funded by the John Templeton Foundation.[1] Our goal was to explore the idea that attempts to quantize gravity either significantly modify the structures of classical spacetime or replace them— and spacetime itself— altogether. It is a premise of our work that philosophy and physics are intertwined, so that advances in physics entail revisions in philosophy but also require conceptual—that is, philosophical—advances and refinement. Hence our project activities were focused on bringing interested physicists and philosophers into conversation.

Thus, in addition to their research, project members organized numerous colloquia, workshops, and schools and ran three essay contests (our work is archived at www.beyondspacetime.net). From the researchers who participated in these events we selected a group that represents the cutting edge of a range of topics concerning the nature of spacetime in the new physics of quantum gravity and invited them to contribute to a pair of volumes. One—*Philosophy beyond Spacetime* (Wüthrich, Le Bihan, and Huggett, forthcoming)—deals more directly with the implications of quantum gravity for traditional philosophical concerns. This volume deals more with questions that require philosophical analysis, arising in the development of different approaches to quantum gravity. This distinction is a somewhat hazy one; several articles could have fitted equally well in either volume. But roughly speaking, the former volume should interest a wider range of philosophers, and the present volume a wider range of physicists (also being the more technical of the two); physicists and philosophers with interests in our foundational questions should find both volumes valuable.

---

[1] Grant number 56314 from the John Templeton Foundation, performed under a collaborative agreement between the University of Illinois at Chicago and the University of Geneva. The contents of the work produced under this grant are solely the responsibility of the authors and do not necessarily represent the official views of the John Templeton Foundation.

Even with two volumes, we could select only a small proportion of the researchers who were involved with the project, and not every topic, and far from every speaker, could be included here. So we have attempted to select a representative collection of papers that cover (1) research in the most active foundational areas in the field and (2) a range of approaches and questions within each topic. We hope, then, to provide a fairly comprehensive snapshot of the state of the field, to encourage further dialogue between physics and philosophy, and to promote further work.

The chapters in this volume are organized around three main themes: the possible 'emergence' of spacetime, the role of time in quantum gravity, and more specific interpretational issues raised by quantum gravity. The remainder of this introduction sketches these themes and the contributions. The following sketches focus on some (not all) important ideas in order to show how the papers develop common themes from different angles; they are not intended to replace reading the chapters, which contain much more than can be discussed here! Rather, we hope that the sketches will whet the reader's appetite for what follows.

1.1 Spacetime Emergence

The first part addresses the question of how the classical spacetime of general relativity (GR) and quantum field theory (QFT) might be derived or emergent in theories that attempt to quantize gravity: we shall say 'quantum theories of gravity' (QTG) in order to be clear that the category includes any approach that aims to unify gravity and the quantum (and not only those that attempt to apply quantization strategies to GR). One question is the different senses in which classical spacetime might be derived from, or emerge from, or reduced to a more fundamental theory, without the full structures of classical spacetime. Another question approaches the issue diachronically, asking whether classical spacetime could have been 'created' from something nonspatiotemporal at the big bang.

A traditional framework for thinking about the derivation of classical spacetime is given by the Bronstein cube (Bronstein 1933; see also Figure 2.1 in this volume), which can be thought of as picturing a system of physical theories as limits of one another. The dimensions are labeled with c, G, and $\hbar$, so that they represent nonrelativistic, nongravitational, and classical limits, respectively. The eight vertices are populated by various theories; for instance, Newtonian mechanics, special relativity and GR, and particle and field quantum mechanics; but, of course, the most significant vertex for our purposes is that occupied by a theory of everything (or at least 'more') incorporating a QTG. QFT (in flat spacetime) can be found in the $G \rightarrow 0$ limit of this theory, and GR in the $\hbar \rightarrow 0$ limit. Put this way, the picture seems to embody a fairly straightforward answer to the challenge of deriving spacetime;

classical spacetime is an effective description of a QTG, which holds in a formal limit and is a good approximation when the effects of the parameters in question can be experimentally ignored. But, of course, that is much too quick (even for known theories): What is the theory? Does the parameter actually appear in a way that lends itself to taking such a limit? And what is the physical significance of the parameter in the theory, such that we can argue that we live in a regime in which it can be neglected? These, especially the last one, are not purely formal questions but are the issues of interpretation that confront attempts to derive classical spacetime.

In the second chapter of this volume, Daniele Oriti argues that the cube in fact fails to capture an important formal and physical possibility; namely that the physical elements or atoms of a QTG may form spacetime only in special aggregate states, which have a spatiotemporal description in a large N, 'hydrodynamic' limit. In short, we need to add a fourth dimension, parameterized by the number of degrees of freedom, N, yielding a Bronstein–Oriti *hyper*cube of QTG. Traditional programs for QTG start with the ordinary cube in mind and so attempt either to quantize GR (as in the original loop quantum gravity [LQG] program), or to gravitize QFT (as in the first string revolution). But, as Oriti points out, the renormalization group revolution in statistical mechanics has yielded a formal and conceptual understanding of large N systems that was not available to Bronstein in 1933. Moreover, as these programs have developed, they have started to indicate that the fundamental degrees of freedom may not be obtained by the direct approach of quantizing or gravitizing; for instance, string dualities can be interpreted as indicating some structure that 'quotients' the apparent differences in spatiotemporal structure between duals. Oriti surveys similar clues from other programs.

His 'fourth dimension' gives substance to the idea of spacetime emergence. That is, if a set of physical quantities approximate those of a more fundamental theory in the limit in which a constant vanishes, there is a straightforward epistemic interpretation of the reduction; our observations are simply not fine grained enough to be sensitive to perturbations arising from the parameter, in the circumstances. That is a simple, really quantitative, sense in which one theory reduces to another. A large N limit might be of the same kind, but as Oriti explains, it highlights another possibility, suggested by various concrete proposals. That is, that the atoms of the theory might be intrinsically nonspatiotemporal and take on a spatiotemporal aspect only in suitable large N configurations. Note that for such a theory, the claim that the atoms are not spatiotemporal is not based on a direct interpretation of their degrees of freedom but rather on the fact that they simply do not constitute spatiotemporal structure in all states; if they were intrinsically spatiotemporal they would have to constitute something spatiotemporal however they were configured. (Of course, this argument depends *in*directly on the interpretation, specifically on claims about how the atoms can be physically combined to produce spacetime.)

This possibility carries a deeper, qualitative kind of reduction of spacetime from nonspacetime, and in the $N \to \infty$ limit, which underwrites a sense of 'synchronic emergence' often found in the literature. Though what is also generally expected is a formally and conceptually well-controlled map between the theories, constituting a 'reduction' in the classic sense, rather than the strong emergence found in other parts of the philosophy literature.

As Oriti explains and illustrates, a theory in which atoms may or may not combine to constitute spacetime offers a further, even stronger sense of emergence. Namely, there may be the possibility of a transition from a nonspatiotemporal to a spatiotemporal state, at the big bang perhaps: diachronic emergence, or 'geometrogenesis'. Indeed, the work of Oriti and his collaborators on group field theory strongly suggests just this. Of course, geometrogenesis is formally and conceptually very puzzling, for the very concept of a transition seems to imply time throughout the process, but by assumption there is no time before geometrogenesis!

The possibility that the history of the universe includes the emergence of the temporal from the atemporal also arises in the third chapter, by Suddhasattwa Brahma. (The more general question of time in QTG is discussed in Part II.) He presents results developed within the framework of loop quantum cosmology (LQC), which implements high-level principles drawn from LQG. As such, it is to a considerable extent neutral on the nature of the atoms of spacetime, their formal expression and conceptual significance; by assuming certain general features to be consequences of the underlying theory, LQC does not directly speak to the manner in which they are derived. The reasons for adopting such an approach are of course to obtain a framework in which concrete empirical consequences can be derived, without needing full knowledge of the fundamental theory or details of how to take appropriate limits; the results are assumed. While we cannot see a full story of emergence from studying such a theory, it is still a quantum theory, and as Brahma explains, LQC does entail a significant result about the derived nature of spacetime. (Moreover, because LQC is based on general principles, the lesson holds of any theory that realizes them. Brahma argues that the results do not depend on idealizing assumptions in the derivation—e.g., of sphericity—but follow from the physical principles of the theory alone.)

Specifically, the assumptions made appear to suffice for the resolution of classical singularities, at the big bang and (it seems likely) in black holes. Moreover, as Brahma explains, the resolution involves a transition from a Lorentzian metric signature in the classical region to a fully Euclidean metric signature in the region of the singularity; without a change in the number of dimensions, there is spacetime classically, but only space in the quantum region! (As the chapter discusses, a similar idea occurs in the distinct context of the Hartle–Hawking no boundary proposal [Hartle and Hawking, 1983].) We have in a sense the emergence of time from the nontemporal, but three points should be noted: First, we do not have full-blown geometrogenesis because the quantum regime is not strictly nongeometrical, as in the cases Oriti discusses. Second, as a result, it is in principle possible to consider one of the spatial dimensions as that in which the space–to–spacetime transition occurs, potentially providing the basis on which that issue can be resolved. However, third, one should not expect a well-defined Euclidean signature metric in the quantum regime resolving the singularity, but rather a fuzzy one; so the situation is not straightforward. To make progress on these questions, as Brahma discusses, one would need to open up the question of the atoms of the theory; and perhaps in that case one would see that geometrogenesis does after all underlie the process.

Classical singularities are, of course, one of the consequences of classical physics of greatest interest in QTG. Not exactly anomalies in the sense of a failure of the laws (if one is prepared to accept manifolds with singular points removed) but places at which one expects a more fundamental theory to diverge substantially from GR, yielding novel predictions. For instance, a QTG might predict specific traces in the cosmic microwave background (CMB) or in the Bekenstein–Hawking thermodynamics of black holes. The first possibility is the topic of Robert Brandenberger's chapter.

He explains the nature and content of the CMB and the conclusions about the origins of the universe that can be drawn from it using the theory of cosmological perturbations, in whose development he played a central role. In particular, isotropy implies that today's Hubble radius is smaller than the future horizon of early points, while causality requires that currently observed structures were within the Hubble radius at early times. Moreover, there are two further criteria, inferred from the observed power spectrum of the CMB: first, acoustic oscillations require that the universe has been isotropic above the Hubble scale for a long time, and second, any theory of the early universe must explain the scale invariance of the spectrum. Inflation is the conventional response to these constraints, but in the context of QTG it can only be an effective theory, to be understood in terms of some deeper quantum account of gravity. For instance, one might well expect to find a mechanism for inflation within string theory, but despite the efforts of theorists, no definitive mechanism has been found (e.g., Baumann and McAllister, 2015; see also Bojowald [2002] for a proposed account of inflation within LQC). (Moreover, inflation is not without problems.)

However, as Brandenberger explains, there are alternative accounts of the early universe to inflation, which also satisfy the CMB criteria; these solutions typically try to take into account proposals for more fundamental physics, QTG. For instance, the initial singularity could be smoothed out with a bounce solution, in which the universe extends through the big bang into an earlier classical spacetime. Unlike the

LQC bounce discussed by Brahma in Chapter 3, Brandenberger focuses on models based on string theoretic concepts, including the possibility of a T-dual universe on the other side of the big bang. While these proposals are speculative, they illuminate the way in which new physics in a QTG might resolve the puzzles of the CMB. Moreover, they again illustrate the idea that spacetime might be emergent in a diachronic sense, from a quantum state at the big bang; though again, whether they involve full geometrogenesis from nonspatiotemporal atoms depends on details of the scenarios that are not yet understood. (Some of the philosophical implications of this situation have been further explored in Huggett and Wüthrich [2018].)

In Chapter 5, Daniel Harlow returns to the question of synchronic emergence—the derivation of spacetime as an effective structure rather than its creation. Specifically, he addresses two important lessons for QTG that black holes may be teaching us. He first argues that one can best understand the enormous difficulty encountered in quantizing gravity by considering the tension between GR and QM caused by the possibility of black holes. Specifically, a rod capable of measuring Planckian lengths must have a sub-Planckian position uncertainty, hence a minimum momentum uncertainty according to QM. Assuming a low (with respect to the speed of light) velocity, a minimum mass follows, which is easily seen to exceed the Planck mass. But a Planck length-sized object of mass greater than the Planck mass is inside a black hole, according to GR, and incapable of measuring lengths. That is, black holes exemplify the difficulty in defining quantum observables for arbitrarily small regions in QTG.

Harlow's second lesson is how black holes help illuminate the nature of holographic duality (the latter is discussed further in Chapters 12 and 13) and plausibly show the existence in string theory of the synchronic emergence described by Oriti. Harlow makes the point that the duality is (if correct) an exact correspondence, holding between *fundamental* quantum theories on the boundary and the bulk of anti–de Sitter spacetime, known as AdS/CFT duality: a conformal field theory on the boundary and some form of string theory in the bulk. On the other hand, the bulk gravitational *field* arises as a derived, *effective* theory of the fundamental bulk quantum theory. The value of AdS/CFT duality is that the bulk quantum theory is not understood well enough to carry out such a derivation, but the boundary CFT is under enough control to allow the exploration of emergent gravitational—spacetime—features.

As Harlow points out, the philosophical literature has focused on the question of whether the duality between exact theories can be an asymmetric relation of emergence, generally concluding that instead it is some symmetric relation of physical equivalence. Harlow's central claim is that this focus ignores the fact that bulk spacetime physics is derived from boundary physics and hence indirectly from exact bulk physics, an asymmetric relation of derivation.

Using a simple black hole model, in the formal framework of quantum information theory, Harlow goes on to illustrate how this relation is one of spacetime emergence. Briefly, three qutrits (states in 3-dimensional Hilbert spaces) live on the boundary, comprising a $3^3 = 27$-dimensional Hilbert space of a boundary quantum theory. The effective bulk theory is represented by a single qutrit, living in a 3-dimensional subspace of the full theory, corresponding to the few degrees of freedom of a classical black hole. But the fundamental bulk theory is dual to that on the boundary and so also lives in a 27-dimensional Hilbert space; what has happened to the other 24 dimensions? It's not at all surprising that the effective theory has fewer degrees of freedom than the fundamental; that's more-or-less what it means to be effective, in a general sense. Rather the question is, since these extra degrees of freedom are not those of effective bulk gravity, what bulk physics do they describe? Harlow's work indicates that they represent microstates of a bulk black hole within the fundamental bulk theory. This toy model then represents the situation envisioned by Oriti; one has effective spacetime only to the extent that the quantum state of the system has a component in the appropriate subspace; other degrees of freedom belong to a fundamental, nonspacetime theory. Insofar as the model accurately represents nonperturbative bulk string theory, AdS/CFT duality shows that that too is a theory of emergent spacetime.

1.2 Time in Quantum Theories of Gravity

It has long been understood that a successful QTG could have significant implications for our understanding of the nature of time. Many of the difficulties—especially those related to the problem of time—in constructing such a theory seem to stem from the tension between needing a classical time parameter in the dynamics, yet quantizing time by quantizing the metric. In Part II, we have collected four chapters that focus on time in the construction of QTG: what the implications might be and how the conception might have to be changed in order to successfully quantize gravity.

The chapters draw on philosophical thought about the nature of time in this effort, showing nicely the interaction between the two disciplines. In particular, a central theme is the question of whether various QTG do or should realize a form of temporal becoming. Physics typically views time from the point of view of analysis, in which quantities 'flow' only in the sense of taking on different values at different times, with rates understood as limiting ratios $\Delta f(t)/\Delta t$. This picture seems adequate, and indeed natural, in a classical spacetime background, since it mirrors the mathematical treatment of physical quantities. Traditional temporal becoming is the view that there is more to the passage of time than this picture captures, that later states are in some further sense *produced* by earlier ones, or the later times are *created* after the earlier, or that successive presents *come to be*. The italics indicate that these terms don't merely redescribe the standard, analytic, account of physical time; what more they denote depends, of course, on the specific account offered (Savitt [2017] provides a survey). Becoming is also often combined with 'presentism', the view that in some substantive way the present is more real than other times: reality is becoming. Such a view is often contrasted with a 'block' conception of time, according to which the present is merely a matter of perspective, within a full space*time*.

These concepts are unpacked more fully in the following chapters, in relation to QTG. Three of them see becoming, in three different conceptions, as important to quantizing gravity. The final chapter in Part II takes an even more radical view: no becoming, but no block either, just (in some sense) a collection of frozen moments, fundamentally speaking, temporally unconnected.

In Chapter 6, Carlo Rovelli discusses how he thinks that spacetime—particularly time—should be understood in LQG (the chapter also includes a useful appendix summarizing the theory for nonspecialists). He argues that a number of confusions regarding space and time arise because people mean different things when using the expressions *space* and *time*; he describes the concepts as 'stratified, multi-layered'. To counter these confusions, he distinguishes five senses of time (and parallel senses of space).

Relational time involves only the relations between events; the temporal position of one event is specified by temporal adjacency with the occurrence of another. This conception of time is common to many theories, including LQG. In contrast, *Newtonian time* is a fixed metrical structure, independent of the unfolding of events and indeed of whether anything changes at all; it is exemplified by both Newtonian and special relativistic physics. Things are very different with the introduction of dynamical *general relativistic time*, which is understood in terms of clock time between events, which of course depends both on a dynamical metric and the path of the clock (as in the special theory as well). Rovelli explains how, with some subtleties due to quantum effects, this conception of time holds in LQG, thereby preserving what he takes to be an important lesson of GR.

In addition, he distinguishes *irreversible time*, connected with thermodynamics, statistical mechanics, and the entropy gradient, and *experiential time*, our experience or feeling that time flows. For Rovelli, these should be distinguished because they do *not* have any direct bearing on the nature of time from a distinctively LQG perspective but have to do with statistical and neurological effects, respectively. They are thus distinguished because bringing them into the current discussion can sow confusion.

With these distinctions drawn, the chapter unpacks the notion of time in LQG, focusing on the importance of temporal becoming. First, while accepting that the relativity of simultaneity undermines an absolute present and presentism—something challenged by Lee Smolin in his chapter—Rovelli argues that a block conception of time, devoid of becoming is not the inevitable consequence. Instead, he identifies the transition amplitudes of LQG with the coming to be of one state from another; moreover, since these are between spacetime states they can be further identified with regions of spacetime. Such a scheme does not require the global now of classic presentism, but it does rest on becoming at a local here and now and so is not a block universe picture either. Thus according to Rovelli, the choice between presentism and the block is a false one! In his view, time passes, things become, but locally rather than globally; this is the lesson for time from LQG. The remainder of the paper is devoted to showing in more detail how his interpretation of time (and the related understanding of space and spacetime) play out in LQG: the picture that emerges is one in which the universe is a 'network of quantum processes'.

One might ask whether his account of time leans more heavily toward a kind of local presentism or a block universe. The answer depends on how one fleshes it out. On the one hand, the system of transitions that make up a universe has something of the structure of block universe. However, Rovelli rejects questions of whether all regions or just the here-now is real, as a merely conventional one about the definition of 'real'. On the other, if describing quantum transitions as becoming is not merely verbal, but denotes some strong ontological status, then the view is more sympathetic to presentism. Here Rovelli's view of experiential and irreversible time (and his 2017 view that it may be perspectival) suggests that he does not subscribe to a 'thick' notion of becoming either.

We now turn to the chapter by Fay Dowker, which also addresses the question of temporal becoming and the block universe, but in the context of causal set theory (CST), which she argues realizes temporal becoming in a strong ontological sense, against defenders of a block universe. According to CST, the universe is constituted by a casual set, a discrete structure consisting of elements with causal or temporal relations between them; the manifold picture of spacetime used in GR is an approximation, applicable in some regimes. As the theory currently stands there is no full quantum version of the dynamics; instead what is given is a classical but stochastic description that gives rise to classical sequential growth models, according to which the causal set grows dynamically—'becomes'—by the addition of new elements. Dowker sees the births of new events as something that objectively happens, underwrites the irreversibility of time, and that moreover could be a physical underlying objective process that explains experiential time. In other words, her account of becoming not only has a different source from Rovelli's, she argues that quantum gravity has implications for conceptions of time that Rovelli thinks it does not.

Now, traditional conceptions of time flowing or becoming typically rely on some form a global present, either in presentist or growing block accounts. The problem is, of course, that a global spacetime present allows for an objective time parameter and so is not generally covariant; this seems to be an undesirable step backwards toward a prerelativistic understanding of time. Dowker argues that CST, however, provides an alternative model of flow without such a global now. Elements of the causal set are objectively created only before or after one another when they are in each other's causal pasts or futures, but there are no facts of the matter about the order in which elements that are not causally related were created. In turn, this is encoded in the equality of transition probabilities for paths that reorder the creation of such elements. The resulting temporal becoming is what Rafael Sorkin (2006) has dubbed 'asynchronous becoming', a localized form of becoming in a multiplicity of 'nows'. (An earlier version of Dowker's proposal has been critically discussed in Callender and Wüthrich [2017].)

Dowker argues that CST can thus accommodate both being (the baby) and becoming (its birth)—unlike the block universe, which fails to capture the latter aspect. In this manner, it reconciles two sides of a long-standing debate about which of these features ought to be given priority by embracing the essence of both. Arguably, one would expect *being* to refer to the objective structure of spacetime or of a causal set, while *becoming* would be rendered subjective by virtue of being relativized to a frame or a worldline. Surprisingly, Dowker defends the opposite view that being is subjective, whereas becoming is objective. According to her, the birth process of an atom of spacetime, and hence the becoming, is independent of any observers or frames as it constitutes an objective physical process. Conversely, there is no objective world of being; being is derivative in that it depends on a prior process of birthing, and what is objective is only each atom's past as that is what has become as of this atom. Thus being is relative to each atom and in this way subjective. Finally, Dowker asserts that this view of 'asynchronous becoming' is possible only in a discrete spacetime and hence not available in GR.

In the next chapter, Lee Smolin lays out the philosophical framework of his work over the past 20 or more years, a research program aimed both at providing a realist interpretation of quantum mechanics and at quantizing gravity. (While the chapter discusses how the two aims are intertwined, here we will focus on the latter.) At the foundation of this work is a commitment to an aspirational form of relationalism, a methodological imperative (rather than a priori truth) to seek to remove arbitrary—'absolute'—elements from physical theories. In part, he sees this principle in the history of science: eliminating absolute spacetime structure, including point identity in favor of equivalence-up-to-diffeomorphism, to give one example among the many he presents. In part, he sees it as guiding the search for a QTG.

Also central to his program is a form of temporal becoming that privileges time over space, a view developed in earlier publications (e.g., Smolin, 2013; Smolin and Unger, 2014). He does not use the expression temporal becoming explicitly, but his description of the view is clear: "the aspect of time I assert is irreducible is its activity as the generator of novel events from present events . . . The thick present is continually growing by the addition of novel events. At the same time other events in the thick present, having exhausted their potential to directly influence the future, slip from the present to join the always growing past."[2] Moreover, the present, but not the past or future, is real, so this is a form of presentism. Smolin's language expresses the difference between the traditional physical conception of time based on the real numbers and that appropriate to the dynamics of his approach; indeed, he explicitly rejects the possibility of the kind of fixed, deterministic dynamical laws typified by the familiar differential equations of physics. Instead, things do not just happen one after the other, but each (irreversibly) generates the next. Thus Smolin proposes a stronger conception of becoming than Rovelli's, one similar to that of causal set theory, explored by Dowker and colleagues.

Of course he is aware of the challenges to the 'present' presented by the relativity of simultaneity, but he points to the theory of shape dynamics, to demonstrate that they can be addressed. This classical theory (its quantization is discussed by Gomes in the following chapter) is based on preferred spatial slices but is locally indistinguishable from general relativity. Although shape dynamics is relational, as we shall see in Chapter 9, it is most naturally thought of as a theory in which space is fundamental and time derived, an example of timeless relationalism. As we noted, instead Smolin advocates the primacy of time over space: temporal relationalism. (Clearly this sense of time is stronger than Rovelli's relational time. Extending Smolin's terms, Rovelli would seem to advocate a form of 'spacetime relationalism', taking neither space nor time as more fundamental.)

In addition to time and causation (or generation), Smolin also proposes that energy and momentum are fundamental: the fundamental states live in momentum space rather than physical space. Even at the classical level, spacetime can be reconstructed, exemplifying the sense in which space can emerge from time and the nonspatial. (To editorialize: insofar as momentum space is nonspatial.) On quantizing, one discovers that locality—meaning point coincidence of trajectories in this case—is relative; as simultaneity is motion dependent, locality is energy dependent. For Smolin, this result shows that spacetime itself is as observer-dependent as simultaneity, an effective construct in limited regimes.

These ideas provide a cohesive conceptual framework for the program of temporal relationalism, and a number of the contributions that Smolin and others have

---

[2] The present is 'thick' in the sense that it contains events that are causally related.

made to quantum gravity. The final section of his chapter gives a comprehensive overview of the program and what has been accomplished, in the light of this framework. We will just emphasize the more recent developments from his study of energetic causal set models, which directly implement the tenets of temporal relationalism: from an underlying irreversible nonspatial dynamics, a reversible particle dynamics in a Lorentzian spacetime emerges as an effective structure. Moreover, the Einstein field equation can be derived from the thermodynamics of the model. Thus energetic causal models exemplify the way in which *space*time might emerge from a theory of temporal becoming.

In the final chapter of Part II, Henrique Gomes investigates the picture of time arising from his recent work on the foundations of QTG, drawing a radically different picture. The first half discusses a train of thought that points to the version of shape dynamics developed by Gomes, Gryb, and Koslowski (2011). Tracing the root of the problem to the difficulty in quantizing the causal structure of relativity, Gomes sketches the challenges facing QTG: the problem of time for Hamiltonian approaches and the problem of parameterizing the space of 4-dimensional Lorentzian spacetimes for covariant approaches. In both cases, Gomes notes that resolutions can be found in the space of classical solutions ('on shell'), but that this is inadequate to a quantum theory; he draws the lesson that causal structure should not be built in to a fundamental quantum theory but recovered in an effective theory.

Gomes argues that shape dynamics implements this idea classically by replacing space*time* symmetries with spatial symmetries: position relationalism and scale relationalism (which are the unique symmetries acting solely on configuration space, rather than on phase space). But, he asks, how are we to understand time— the *dynamics* of shape dynamics—in the theory? The original, classical theory proposed the simplest solution: introduce an independent time parameter. To quantize, one could use this parameter to define a Schrödinger equation. But in his chapter Gomes is dissatisfied with that approach; such an absolute time violates the relationalism that he sees as the cure for the problems of quantizing spacetime structure. In this then, he takes the opposite interpretational view from Smolin.

Instead, Gomes applies to time the idea that classical structures need only be recovered effectively. In the latter part of the chapter, he develops a formal framework to realize this idea, and develops an interpretation with profound consequences for time. The formalism involves a notion of quantum path integral that is independent of a time parameter and a measure over configuration space, which lead to a Born rule for the theory. This framework involves selecting a privileged 'in state', which Gomes notes is key to the notion of a record. Concretely, it allows him to define a formal notion of a record, the *apparent* trace in the present, of a past transition from the initial to the present state. Significantly, this definition requires a semiclassical approximation, and so such records are inherently semiclassical.

His interpretation of this framework is that fundamentally there is no duration, only instantaneous configurations, and all possible ones are on an equal ontological footing, akin to a many-worlds interpretation. Fundamentally, there is no time at all. Like Barbour (1999) but unlike Smolin, Gomes thinks that the past of any configuration is fully reducible to the records (or time capsules) contained in that configuration; however, unlike Barbour, Gomes's time can only be reconstructed effectively at the semiclassical level, since records are semiclassical. In sum, there is effective time, but only an instant. An extended past can be projected from the effective apparent records held by that instant, but it is merely a just-so story; even effectively, only the instantaneous records themselves are real. In this view, space is fundamental, and time barely real, and there is neither passage nor the block.

Of course, this image of time diverges greatly from our ordinary conception, and in particular clashes violently with our concepts of personal history, an issue that Gomes also takes up.

In this part of the book we have thus seen four different responses to how we should understand time in QTG. Some authors—in particular Dowker and even more strongly Smolin—award to time a more pronounced and special role than to space, in a departure from the orthodox understanding of relativity. In contrast, we take Rovelli to advocate a view in which space and time are on more equal footing—that is by both being emergent from the underlying quantum theory in basically the same way. Finally, Gomes takes space to be the more fundamental aspect of reality.[3]

1.3 Issues of Interpretation

The first two parts of this book investigated in various ways the implications of QTG for the nature and emergence of space and time, but such theories have raised other important questions for the interpretation, epistemology, and metaphysics of science, some of which have been hotly debated. The chapters in Part II address several such issues: the 'information loss' paradox(es), the meaning of string theory's dualities, and the implications of QTG for the logic and metaphysics of possible worlds.

First, what is the physical nature of black holes? Often thought to be a key to quantum gravity, black holes appear to admit a thermodynamic treatment, suggesting that, perhaps, a QTG ought to provide a description of its microstates. The issue of black hole thermodynamics is taken up by David Wallace in his contribution. He argues, contrary to recent philosophical criticisms (Maudlin, 2017), that black hole

---

[3] For a similar observation about the diverging views on the nature of time that can be found among researchers working on developing a QTG, see Matsubara (2017).

information loss (the failure of temporal reversibility, manifested as non-unitarity in QM) is indeed paradoxical, though not in the way often presented. Overall, the point is that one expects to understand the Bekenstein–Hawking entropy of a black hole in Boltzmannian terms as the logarithm of the fundamental microstates and indeed Hawking radiation (effectively) as a decay channel of the fundamental state. But this picture relies on treating a black hole both as possessing thermodynamical properties such as temperature and being quantum mechanical, and these features come under pressure in the information paradox.

Popular presentations present the 'paradox' as an issue for the *end point* of evaporation, at the Hawking time. In starkest terms, a black hole in a pure state undergoes unitary evolution (Hawking radiation) until it is entirely gone, and all that remains is thermal radiation, a mixed state. Yet it is a mathematical impossibility for a unitary process to turn a pure state into a mixed one. But as has been pointed out for over 20 years (and recently insisted on), because of the classical singularity the end point of the process is not a Cauchy surface, and so insisting on a deterministic, unitary evolution is at best to make a substantive, controversial claim. However, Wallace explains that the emission of thermal radiation from a black hole produces a paradox in the physical principles believed by many (though not all) to describe the process—*well before the Hawking time or even without complete evaporation* (and so is not resolved by considering the end point).

On the one hand are statistical mechanical principles. The discovery of Hawking radiation elevated the Bekenstein thermodynamical description from analogy to reality by showing that the description remains valid when black holes interact with other thermodynamical systems; in particular, they can exchange heat. Since thermodynamics in general is understood in terms of a microphysical description—with entropy as the logarithm of microstates—one concludes that the same is true of black holes, that black hole thermodynamics describes, in the large, the statistics of black hole microstates. (Hence the programs to derive, in string theory and LQG especially, the entropy and radiation spectrum of black holes from posited microstates.) In particular, as Wallace explains, the black hole is often modeled in the statistical mechanical 'membrane paradigm' as a surface located around the horizon, containing the microphysical degrees of freedom, which are transformed to thermal radiation and lost, decreasing their Boltzmann entropy. Whatever the underlying theory, if the microstates are quantum, this process must be unitary. (It is worth emphasizing that the membrane paradigm is most popular within string theory, which is the real target of Wallace's chapter.)

On the other hand, derivations of Hawking radiation rest on the principles of QFT in curved spacetime. These extend the well-tested principles of QFT in flat spacetime, but do not concern situations in which a full quantum theory of gravity is needed, of extreme curvature or energy density. They entail that Hawking radiation is in a thermal state, a mixture with unentangled modes. As noted, a unitary evolution cannot produce a mixed state from a pure one, so modes of the Hawking radiation are understood as entangled with degrees of freedom of the black hole, which are traced out in the usual way when one observes an entangled subsystem.

But there is a limit to this understanding; according to the statistical mechanical membrane paradigm that Wallace endorses, at a certain point—the 'Page time'—the decreasing Boltzmann entropy of the black hole means that there are no longer degrees of freedom with which the thermal modes can entangle. Yet the principles of QFT used in the derivation of Hawking radiation entail the continued production of a mixed state—in violation of unitarity, and so of a quantum description of the situation.

As Wallace reviews, the Page time is much shorter than the Hawking time for complete evaporation (roughly half), and the scenario can even be modeled in evaporating black holes without complete evaporation. So the qualms about the popular form of the information paradox do not arise; Wallace's explication of the Page paradox is thus much sharper, apparently calling for giving up either QFT in curved spacetime or the membrane paradigm. Neither option is attractive (at least within string theory). The former seems to imply that quantum gravitational effects are relevant whenever spacetime is curved, not only when the curvature is Planckian; the latter threatens to make black holes an exception to the statistical mechanical approach to microphysics. Wallace's essay reviews in detail how tight and difficult a bind this is and how attempts to resolve it with appeals to 'black hole complementarity' seem to lead to the 'firewall paradox'.

In Chapter 11, Richard Dawid discusses the current and future status of string theory. He makes an argument to the effect that the difficulty in giving a complete description of string theory—its 'chronic incompleteness' he calls it—can be explained if string theory is a final theory. (It should be noted that Dawid uses the expression *string theory* in a wide sense, including future developments of the theory and other associated ideas such as M-theory.) He presents a number of observations suggesting that string theory is a likely candidate for being a final theory of physics: First there is the universality of string theory; that is, it does not seem to be restricted to describing only a certain class of phenomena. Second, string theory does not have any fundamental dimensionless free parameters. Third, string theory has a minimal length scale.

Dawid sees these three features are in stark contrast to past physical theories. First, past physical theories have been expected to be applicable to only a restricted domain of phenomena. Second, typically our theories have had freely adjustable dimensionless parameters that could be chosen to fit what we observe in nature. The values of these parameters may be explainable by a more fundamental theory or by the way in which the theory is embedded in a specific physical background. The lack of such parameters is taken by Dawid to suggest that string theory is a final theory. Finally, our previous theories have been thought to be replaceable and seen as effective theories arising from more fundamental theories that are valid at smaller length scales. With the minimal length scale of string theory—suggested by the T-dualities that allow a small scale to be eliminated in favor of a description in terms of larger distances—Dawid argues that there is no reason to expect string theory to be similarly replaced by a more fundamental theory.

Just as he sees these features pointing to string theory as a final theory, he also argues that they make it hard to fully understand and articulate, leading to chronic incompleteness. Central to his argument is the way in which string theory is primarily understood in terms of perturbative calculations around near classical limits. Such backgrounds themselves are put in by hand and are not part of the dynamical description. The absence of freely adjustable parameters in the theory— even though we seem to have many allowed groundstates—means that we cannot tune the theory to a classical state of some unknown more fundamental physics. Thus for a deeper description, the backgrounds must be understood in terms of the nonperturbative dynamics of string theory itself. But while string dualities provide additional insights into the nonperturbative physics, they are not sufficient to this task. Broadly speaking, the challenge to developing such a deeper account of string theory is that it needs to handle situations that cannot be well approximated by any classical picture, thus making our commonsense understanding of the situation even less useful for guiding us in the right direction.

Turning an apparent failure—chronic incompleteness—into evidence in favor of string theory as a final theory is at least controversial! But as with Dawid's earlier work, it should contribute to an ongoing discussion of what might characterize a final theory and what research programs are worth pursuing. (And even whether it is worth pursuing such a theory, rather than taking smaller steps to greater but not final understanding.)

The next two chapters explore the interpretational and methodological significance of AdS/CFT (or gauge/gravity) duality mentioned earlier—the core of recent research in string theory—in two different ways. (Daniel Harlow addressed its significance for spacetime emergence Chapter 5.) First, Sebastian De Haro addresses the relation of duality to physical equivalence and the implications of such equivalence. This is a topic that has been addressed by philosophers, but De Haro provides a rigorous framework for the conceptual situation that permits clear and precise answers to the important questions, including whether spacetime is primitive or derived in string theory. Whereas Harlow raised the question of whether one side of the duality was *derived* from the other, here the issue is different: when are duals *equivalent*, and what follows if they are?

De Haro defines duality in general to be a formal relation between theories, a partial isomorphism of a certain kind: more specifically, an isomorphism between parts of two theories, with sufficient structure—a space of states, set of observables, and dynamics—to themselves be called theories. So for instance, in the case of a pair of simple harmonic oscillators (SHOs) with masses and spring constants related $(m, k) \rightarrow (1/k, 1/m)$, a duality maps $(x, p) \rightarrow (p, -x)$. A table of value pairs over time would be the same if it displayed (x, p) for one system and $(p, -x)$ for the other, and so the duals share this common structure. And in general a duality picks out a 'common core theory', a part that the two theories share up to isomorphism.

Now, of course, for a real bob on a spring, the SHOs are still different systems since they have inter alia observably different masses; it's just that if one were given only a table of value pairs over time, one could not tell if they described (x, p) for one system or $(p, -x)$ for the other. The point is that physical equivalence—intuitively, 'telling the same *story* about the world'—is not a purely formal matter but also depends on what the duals *mean* and whether they mean the same. At one time the answer would simply have been 'if they make the same observable predictions', but part of the value of De Haro's contribution is to offer an account of the interpretation of theories that does make this distinction precise, without appeal to such crude verificationism.

The reason real dual SHOs are distinct is that in our world, their common core can be embedded in—or 'extended', in De Haro's terms, to—a larger theory that gives external meaning to their terms: mass, spring constant, momentum, position. But one can envision a world in which the common core instead described *everything*; possible states are fully distinguished by the value pairs, with the same allowed histories and interpreted as the values of the only two canonical observables of the world. Then the duals are nothing but different tools for computing the dynamics, with their differences (in m and k, and x and p) as nothing but empty conventions needed to turn the mathematical handle. Then there would be no larger theory of the world (without uninterpreted surplus structure) in which the core could be embedded, so it could not receive an external interpretation but rather an internal one, like the interpretation we just gave, which simply maps the elements of the theory to their worldly referents. De Haro gives a precise account of the situation that will enable more focused discussion of its consequences.

Given this framework, (at least) two interpretational issues remain. First, could we ever reasonably believe that the common core of a pair of duals was not extendable? That it captured all the physical structure of the world (in its domain), so that it was not just part of a broader (perhaps more fundamental) theory? As Jeremy Butterfield emphasizes, in a paper for the companion volume to this one (Wüthrich, Le Bihan, and Huggett, forthcoming), the formal existence of a duality alone does not show that. (De Haro does not claim otherwise.) Here one may be tempted to use methodological principles such as ontological simplicity to move from duality to 'unextendability'. Second, suppose that the world were such that the common core of a pair of duals indeed has no external interpretation; does it follow that the duals are physically equivalent? Perhaps instead they could describe a pair of worlds in which different physical quantities are instantiated in isomorphic patterns. This is not a question of physics, but metaphysics: how properties are identified across possible worlds. De Haro argues that the existence of two such worlds would violate the identity of indiscernibles. If one answers yes to the previous questions for a pair of duals, then the duals are physically equivalent, with their content exhausted by an internal interpretation of their core.

In the final part of the chapter, De Haro applies his framework for duality to gauge/gravity duality, showing that while the common core contains some weak spatiotemporal structure, it does not contain most of the structure of a spacetime theory. Thus if one answers the two previous questions positively and adopts an internal interpretation of string theory, then AdS spacetime is not fundamental, but merely a conventional description, and in that sense emergent.

Next, in their jointly written chapter, Radin Dardashti, Richard Dawid, Sean Gryb, and Karim Thébault address a number of questions regarding how one should think about the possible empirical consequences of such AdS/CFT duality. While the original reason for studying the duality was to deal with quantum gravity and Planck scale physics, the formalism and mathematical results have more recently been used for dealing with other questions in physics than QTG; for instance, quark-gluon plasmas. The authors find it important to distinguish three different contexts in which an AdS/CFT duality could be applied and want to explain why different conclusions would be warranted in the different contexts if the dual theories were shown to be empirically adequate.

In the first context, in which AdS/CFT duality is used to describe fundamental physics (in a sense they explain), the authors argue that empirical success would not give us any reason to prefer one of the dual pictures over the other. Their argument is that each of the dual theories would be confirmed as much as the other in the Bayesian sense. Furthermore, in the context of fundamental theories the authors do not think there is any good principled reason for assigning one of the dual theories with different priors to the other. The authors consider this as a good reason for not prioritizing one picture over the other when it comes to ontology. Either the ontological picture to which one should be committed is to be articulated on the basis of a structure that is shared between the two pictures—a common core—or one should accept some form of dual ontology where in some sense the ontology of both theories should be equally acknowledged. In both cases, the upshot is that the duals are different descriptions of one single theory.

In the second context, the duality is supposed to relate effective theories for which there exists a single more fundamental theoretical description. In this context, the situation is somewhat different, and the authors open up the possibility that one of the dual pictures could justifiably be seen as a better description of the underlying physical reality: if the ontological picture suggested by one of the dual theories was closer to that of the more fundamental theory. In the effective context, arguments could also be introduced concerning whether one or the other picture could more plausibly be embedded in a larger, more encompassing description of reality. However, while the authors describe this possible way in which one of the duals could be given priority, they also point out that this way of reasoning relies on a rather strong form of scientific realism. Thus if weaker forms of scientific realism instead were assumed—where ontic commitments play a less central role—then the two dual theories would still be considered as on par and equally confirmed.

The third and final context is the instrumental one in which the duality is used for the purpose of making approximate predictions in another theory, where this other theory is not one of the two duals. The authors focus on the application of AdS/CFT duality to quark-gluon plasmas. These are governed by quantum chromodynamics (QCD), but the implications cannot be easily calculated because perturbation theory is not applicable in the relevant regime. However, it can be argued that a CFT would give results similar to those of QCD; then the CFT results can be calculated using the dual AdS description to make *approximate* predictions for the plasma. These results are approximate because the duality with QCD is not exact, so that we have an example of the third context. The authors argue that the empirical success of predictions of this kind do not confirm either of the dual descriptions; instead it confirms only the conjunction of QCD and the approximation scheme based on the duality that is used. Furthermore, there is no reason to take this kind of successful approximate prediction as constituting evidence supporting string theory as a QTG.

The final two chapters discuss the possible formalism of QTG, in particular relation to the metaphysics of 'possible worlds'. These of course play an important role in the thinking of such figures as Leibniz but, as Tiziana Vistarini explains in Chapter 14, especially in the modern modal logic of possibility and necessity. In the work of Lewis, for instance, one can interpret the claim that 'P is possible' in terms of an 'accessibility relation': as saying that P is true in some possible world accessible to the actual world. (For instance, if one defines a world to be physically accessible iff the laws of physics are true in it, then P is physically possible if there is some physically accessible world in which P is the case.) Moreover, from accessibility one can develop a graded relation of degree of similarity, so that we can talk of possible worlds being more or less similar. Lewis (1973) introduced this notion for an account of the logic of 'counterfactual conditionals': 'if P had been the case, then Q would have been the case' is true iff in the most similar world(s) in which P is true, so is Q.

The manifold of possible worlds thus invoked, and the metaphysics of modality it represents, is thus derived from the logic of modal sentences and so, as Vistarini explains, arguably on the logic of ordinary language, with inevitable imprecisions (for instance in the cardinality of worlds). The crux of her chapter is to argue that the moduli space of string theory—the space of possible models or theories—provides a physically grounded and metaphysically substantive extension of the space of possible worlds. She explains how this space has a topological structure, relative to a given model, induced by the space of deformations of that model, within moduli space. Crucially, she shows how this topology is strong enough to define a partial ordering on the points of moduli space, which she proposes interpreting as a similarity relation and hence points of moduli space as possible worlds of the theory. This proposal, of course, raises a host of philosophical questions, which the chapter starts to address. For one thing, the space of worlds is precisely defined, allowing precise answers to questions of its structure; for instance, the similarity relation has a countable spectrum. For another, the worlds described by ordinary language—the 'manifest image'—should, in some way, be reducible to those of fundamental physics, including their structure of possibility; Vistarini sketches how such a reduction of modality might go in string theory, through a revised form of Humean supervenience.

Of course, many of the essays of this volume have proposed significant modifications to the classical spacetime picture of GR in a QTG. In the last chapter, Ko Sanders proposes another, using the tools of mathematical category theory to reformulate a classical spacetime theory, with an eye to a different route to quantization. Again, as in other approaches it is important that the classical picture can be recovered in appropriate regimes where we know that this picture is accurate.

As a starting point for the analysis, Sanders uses the framework of locally covariant quantum field theory (LCQFT). This framework is similar to algebraic quantum field theory (AQFT) but uses the tools of category theory for the purpose of encoding the features of locality and general covariance. In contrast to AQFT—where only one quantum system is given a description in terms of a C^*-algebra—LCQFT associates to each object in the category Loc of globally hyperbolic Lorentzian manifolds a corresponding object in the category Alg of C^*-algebras. In this axiomatic framework QFTs can be formulated to take gravitation into account without actually quantizing gravity, a way of formulating QFT in curved spacetimes.

In general terms, Sanders suggests that to go beyond LCQFT and to formulate a bona fide QTG one could try to preserve much of the structure used in LCQFT but replace the category Loc with another category. Using this other category the classical manifold description would be a good approximate description only in certain regimes. He does not propose such a category in his chapter but proposes searching for it as a research program. Overall then, as with the papers in the first part of this volume, Sanders presents yet another example of a picture where the traditional picture of spacetime is an emergent and not fundamental feature of reality, this time using the framework of category theory.

The chapter also proposes in some detail that categories could serve as models of modal logic. More specifically he claims that the category Phys—whose objects are mathematical descriptions of physical systems—ought to be such a model. Here the possibilities that are modeled are not full possible worlds but rather physical systems; these could be extended to whole worlds but we do not have to deal with the whole worlds when articulating possibilities. Sanders argues that this aligns better with the actual practice of physicists since it is not typically the case that one needs to describe a full possible world when describing a physical system.

Acknowledgments

We offer our thanks first to everyone who spoke to the project, at any of the events we hosted; we learned enormously from all of you, and you made our project a success (all these talks can be viewed on our website). We offer a special thanks to those who accepted our invitation to write essays for these volumes. The interdisciplinary nature of the work means that many were writing for different audiences and approaching questions in different ways from their usual work, and we are very grateful for their intellectually open and sincere efforts; we think that the results are meaningful contributions to an important dialogue. We want to especially acknowledge the winners of our essays contests: in this volume the outstanding chapters by Suddhasattwa Brahma, Sebastian De Haro, Henrique Gomes, Ko Sanders, and David Wallace. The contest could not have happened without the help of our assistants and the efforts of our (anonymous) judges, who gave careful recommendations and thoughtful feedback, that helped us select the winning essays from a strong field and provided material assistance to the authors—thank you. The preparation of this volume owes a lot to Nick Gibbons, Sarah Lambert, and the team at Cambridge University Press, to whom we are very grateful for their help (and patience). It also owes a huge debt to our Editorial Assistant at UIC, Niranjana Warrier, whose combination of facility with the material and modern manuscript production was invaluable; we truly don't know how this would have got done without you. Finally, there are the administrators at UIC and UNIGE who worked hard to facilitate our efforts; and of course the John Templeton Foundation, whose financial support of the sciences made the project possible.

References

Barbour, J. (1999). *The End of Time: The Next Revolution in Physics*. New York: Oxford University Press.

Baumann, D. and McAllister, L. (2015). *Inflation and String Theory*. Cambridge: Cambridge University Press.

Bojowald, M. (2002). Inflation from quantum geometry. *Physical Review Letters*, 89(26): 261301.

Bronstein, M. (1933). K voprosu o vozmozhnoy teorii mira kak tselogo (On the question of a possible theory of the world as a whole). *Uspekhi Astronomicheskikh Nauk*, Sbornik 3: 3–30.

Callender, C. and Wüthrich, C. (2017). What becomes of a causal set? *British Journal for the Philosophy of Science*, 68: 907–925.

Gomes, H., Gryb, S., and Koslowski, T. (2011). Einstein gravity as a 3D conformally invariant theory. *Classical and Quantum Gravity*, 28(4): 1–24.

Hartle, J. B. and Hawking, S. W. (1983). Wave function of the universe. *Physical Review D*, 28(12): 2960–2975.

Huggett, N. and Wüthrich, C. (2018). The (a)temporal emergence of spacetime. *Philosophy of Science*, 85: 1190–1203.

Lewis, D. (1973). *Counterfactuals*. Oxford: Blackwell.

Matsubara, K. (2017). Quantum gravity and the nature of space and time. *Philosophy Compass*, 12(3): e12405.

Maudlin, T. (2017). (Information) paradox lost. arXiv:1705.03541.

Rovelli, C. (2017). Is time's arrow perspectival? Pages 285–296 of: Chamcham, K., Silk, J., Barrow, J. D., and Saunders, S. (eds.). *The Philosophy of Cosmology*. Cambridge: Cambridge University Press.

Savitt, S. (2017). Being and becoming in modern physics. In: Zalta, E. N. (ed.). *The Stanford Encyclopedia of Philosophy*, fall ed. plato.stanford.edu/archives/fall2017/entries/spacetime-bebecome.

Smolin, L. (2013). *Time Reborn: From the Crisis of Physics to the Future of the Universe*. London: Allen Lane.

Smolin, L. and Unger, R. (2014). *The Singular Universe and the Reality of Time*. Cambridge: Cambridge University Press.

Sorkin, R. D. (2006). Geometry from order: Causal sets. *Einstein Online*, 2: 1007.

Part I

Spacetime Emergence

2

The Bronstein Hypercube of Quantum Gravity

DANIELE ORITI

2.1 Introduction

The quest for quantum gravity has undergone a dramatic shift in focus and direction in recent years. This shift followed, and at the same time inspired and directly produced many important results, further supporting the new perspective. The purpose of this chapter is to outline this new perspective and to clarify the conceptual framework in which quantum gravity should then be understood. I will emphasize how it differs from the traditional view and new issues that it gives rise to, and I will frame within it some recent research lines in quantum gravity.

Both the traditional and new perspectives on quantum gravity are nicely captured in terms of a 'diagram in the space of theoretical frameworks'. The traditional view can be outlined in correspondence with the Bronstein cube of physical theories (Bronstein, 1933; Stachel, 2003). The more modern perspective, I argue, is both a deepening of this traditional view and a broader framework, which I will outline using (somewhat light-heartedly) a Bronstein hypercube of physical theories.

2.2 The Bronstein Cube of Quantum Gravity

The Bronstein cube of quantum gravity (Bronstein, 1933) is shown in Figure 2.1. It lives in the cGh space, identified by the three axes labeled by Newton's gravitational constant G, the (constant) velocity of light c, or, better, its inverse $1/c$, and Planck's constant h. Its exact dimensions do not matter, the axes all run from 0 to infinity, but its corners can be identified with the finite values that the same constants take in modern physical theories.

The figure does not represent specific physical theories or models (despite some of the labels), but more general theoretical *frameworks*. Its conceptual meaning can be understood by moving along its corners, starting from the simplest theoretical framework, i.e., classical mechanics, located at the origin $(0,0,0)$ (understood as hosting all theories and models formalized within this framework, be they

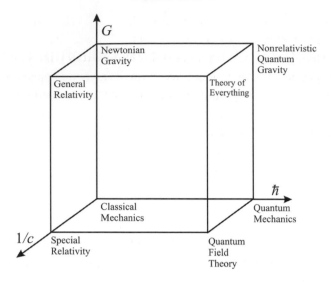

Figure 2.1 The Bronstein cube. Reproduced, with many thanks to the author, from S. Hossenfelder, http://backreaction.blogspot.com/2011/05/cube-of-physical-the-ories.html.

about fields, particles, forces). Moving from the origin along the G-axis, we start including in our theoretical framework gravitational physics, i.e., the effects of the gravitational interactions on the same entities dealt with in classical mechanical models. The very moment these become non-zero, we are in the realm of classical Newtonian gravity. If we move instead from the origin along the $1/c$ axis, we start taking into account relativistic effects, i.e., due to the finite propagation speed of physical signals (information), bodies, and interactions, bounded by the velocity of light. If both relativistic effects and gravitational ones are taken into account, we reach the corner presided by general relativity, a classical, relativistic mechanics including also the gravitational interactions of all mechanical systems. Historically, this is the corner reached with the first revolution of twentieth-century physics. The other revolution came with the realization that an altogether different 'direction' exists, in the physical world: quantum phenomena, those (roughly) due to the existence of a finite lower bound for the 'action' of a system, for the area it can occupy in phase space, corresponding to the Planck constant h. Thus, moving from the origin along the h-direction, we find quantum mechanics, the modern framework for all physical systems, with its associated amount of weirdness and marvels, which we have not yet grown fully accustomed to. Actually, it becomes the modern framework for all physical systems once we also take into account relativistic features, by moving away from the nonrelativistic side of the cube, entering the domain of quantum field theory. This is indeed the modern framework of physics. Or is it? Not really, of course, since we know that all systems are quantum and relativistic, but we also know that gravity exists and that, in the more modern understanding coming from general relativity (GR), the gravitational field, the spacetime geometry that is identified with it, and thus spacetime itself, is a physical, dynamical entity. Modern physics is somehow framed either close to the GR corner, or around the quantum field theory (OFT) corner, but it cannot be said to correspond to any single domain within the Bronstein cube, lacking a quantum theory of gravity. We would then like to be able to move along both the h-direction and the G-direction, incorporating both gravitational effects (including very strong ones) and quantum effects into a single coherent description of the world. The corner we would reach by constructing a quantum gravity theory would be that of a theory of everything, not in the sense of any ontological unification of all physical systems into a single physical entity (although that is a possibility and a legitimate aspiration for many theoretical physicists), but simply in the sense that in such framework we could in principle describe in a formally unified way all known types of phenomena: quantum, relativistic, gravitational.

Obviously, this is but an extremely rough sketch of theoretical physics. It does not account even remotely for the complexity of phenomena that are actually described by the mentioned frameworks. And it does not say anything about the very many subtleties involved in actually moving from one framework to the other, and back from there. One example should suffice to illustrate these limitations: the classical limit, which naively should allow us to reduce a quantum (description of a) system to its classical counterpart. Taking this limit and understanding how the classical world emerges from the quantum one is a notoriously thorny topic, involving mathematical complications and conceptual ones, including the issue of measurement, decoherence, etc. Other limitations have to do with the fact that all the quantities appearing in the picture are dimensionful, and thus they do not correspond to directly observable/measurable quantities. Plus, no mention is made of the actual nature of the systems considered, which however modifies greatly what can actually be described (and how) within each framework. For example, we know that a relativistic description of quantum interacting particles is problematic and requires moving to a field-theoretic framework, being then understood only as an excitation of a quantum field. Thus not every entity can live in every corner of the Bronstein cube.

2.3 The Problem of Quantum Gravity from the Bronstein Cube

Before thinking about such an extension, let us dig more into the perspective on quantum gravity as encoded in the Bronstein cube. This is the straightforward view that sees quantum gravity as obtained from quantizing the geometry (metric/gravitational field) of spacetime and its dynamics, by whatever quantization method; i.e., as quantized GR. The quantum gravity corner can be reached by incorporating quantum effects starting from the GR corner or by adding the gravitational aspects of the world (equivalently, nontrivial spacetime geometries) in a quantum description of it. This corresponds to the strategy of all the traditional approaches to quantum gravity (Kiefer, 2005): canonical quantum gravity, including, at least in its original form, the connection-based version of this program corresponding to loop quantum gravity (Ashtekar and Lewandowski, 2004; Bodendorfer, 2016); covariant path integral formalisms, including discretized versions of the same quantum Regge calculus (Hamber, 2009); and (causal) dynamical triangulations (Ambjorn et al., 2012), to the extent in which the lattice structures are understood only as regularization tools. It also includes the asymptotic safety program (Niedermaier and Reuter, 2006), based on the nonperturbative completion of the formulation of (perturbatively quantized) gravity as an effective field theory. It could include also (depending on the interpretation) the noncommutative geometry program (Majid, 2006; Lizzi, 2008), where one quantizes the geometric structures of spacetime directly, without relying on their role as encoding the gravitational interaction. The situation with string theory (Polchinski, 1998; Blau and Theisen, 2009) is more ambiguous, due to the huge variety of formalisms and research directions now under such umbrella label; still, the understanding of string theory as quantized GR may at least apply to its very early perturbative versions, since the interpretation as quantum gravity theories was due to the existence, in their spectrum, of graviton excitations, quanta of the gravitational field from an effective quantum field theory point of view. If string theory is used in this perturbative, semiclassical form to study the gravitational side of the anti–de Sitter spacetime conformal field theory (AdS/CFT) correspondence (which in itself may not require string theory at all), then also the latter would not present immediately a challenge to the usual picture of quantum gravity. Still, as I will discuss in the following, the AdS/CFT correspondence itself can be seen as a reason to believe that string theory itself requires a more radical departure from the conventional understanding of spacetime and geometry.

This perspective makes perfect sense and exhausts the range of possibilities if the step from classical to quantum gravity does not entail a change of fundamental degrees of freedom; i.e., if the spacetime geometry, the metric field, and gravity are primary entities and the task is to understand their quantum properties. Even in this case, of course, understanding their quantum properties may reveal a number of surprising and very exotic aspects of the world. The step from classical to quantum, when dealing with such a fundamental entity like spacetime geometry, is by all means a challenging one, both mathematically and conceptually. The issue of time, the debate between relationalism versus substantivalism in spacetime theories, the problems with diffeomorphism invariance, on top of the purely technical issues faced by quantum gravity theorists are there to testify the magnitude of the challenge. These issues are already challenging in a classical GR context, where the theory is complete and the physics is well understood. In a quantized GR context, what the theory is *expected* to involve, even leaving aside its incomplete status, raises a host of new and even more severe difficulties (Kiefer, 2014). One example will suffice: what is left of usual physics, of the customary understanding of the world, in a theory with indefinite and fluctuating causal structures, an immediate consequence of superposing quantized geometries, even assuming that each of them maintains a continuum and close-to-classical character? This explains the difficulties in constructing a theory of quantum gravity on such basis, despite the many results obtained over a span of decades.

2.4 Beyond the Bronstein Cube: The Idea of Emergent Spacetime

The point is, however, that the perspective discussed here does not capture the range of problems faced by modern approaches to quantum gravity, neither at the technical nor at the conceptual level. It does not capture where quantum gravity stands with respect to the rest of fundamental physics either. Let me explain why by first reviewing briefly a number of hints challenging the view discussed earlier. They were produced by research *within* the Bronstein cube, but at the same time pushing against its walls, so to speak, noticing their fictitious nature and thus strongly suggesting that there is more to be done and discovered outside it.

These recent results are of two different types. First, they come from research directions not directly aiming at constructing a full theory of quantum gravity but focusing on semiclassical gravitational physics and sometimes on systems that are not gravitational at all but that give surprising insights on the possible nature of geometry and gravity. Second, they come directly from quantum gravity approaches, often of the 'conservative' tradition living inside the Bronstein cube, which nevertheless end up producing challenges to the very perspective that inspired them.

The first group of results can be taken as suggesting that the continuum geometric structure of spacetime, on which general relativity and quantum field theory are based, is not fundamental and that some sort of discrete quantum counterpart should replace it in a full theory of quantum gravity. If this is the case, spacetime as we know it would be an approximate, emergent notion from something else, which would be then not be spatiotemporal in the usual sense (although it may retain some features of the spacetime we are accustomed to). The key points here are the discrete nature of the more fundamental degrees of freedom and the need to see spacetime (and its geometry) as emergent.

The second group of results offers a number of proposals for what the more fundamental degrees of freedom could look like and for which features of continuum spacetime and geometry could be dropped in the more fundamental description. They also explore a range of more or less radical departures from accepted behavior that we may need to be accustomed to if we want to understand the more fundamental nature of space and time (and how to understand the world in their absence).

Among the first group, the oldest results can be taken to be those establishing the existence of singularities in gravitational physics. They may be taken to imply only that quantum corrections to gravitational dynamics have to be taken into account, but they may also be taken as a suggestion that something more radical happens: a breakdown of the continuum spacetime description itself. The divergences of quantum field theory too admit a conservative as well as a more radical interpretation. The correct coupling of quantum field theories for matter (and other interactions) with a properly quantized version of GR may be all that is needed to cure them, introducing a natural cut-off scale. Or they may be an indication that some more radical form of discreteness replaces the continuum nature of quantum fields, including the gravitational field. Indeed, several scenarios incorporating a minimal length (or a maximal energy scale) have been proposed (Hossenfelder, 2013), as effective descriptions of quantum gravity, and they end up challenging many more aspects of standard spacetime physics, including for example locality, which is at the root of quantum field theory and of the whole of continuum spacetime physics. This challenge to locality is not surprising, since the hypothesis of a minimal length was proposed from the very beginning as a consequence of the impossibility of exact localization when the gravitational effects of quantum measurements are taken into account. Among these scenarios, many rely on noncommutative geometry tools (Majid, 2006; Lizzi, 2008). Thus they also offer a first example of a quantum gravity approach that can be understood at first in a conservative way, and that turns out to be more radical than imagined in its implications. To this group belong also the very many results dealing with black hole thermodynamics and in particular with black hole entropy (Carlip, 2014). These are far too many (and interesting) to review them here. However, the main message, for our concerns, is simple. A black hole, in the end, is a region of spacetime. If it has entropy and it is a standard statistical mechanical system, then this entropy accounts for its microstates. These microstates could be (a subset of) modes of the quantized metric field or of some relevant matter field (inside or close to the horizon). But they could also be microstates associated with a different set of dynamical entities that only at macroscopic scales look like any of the two. Moreover, if this entropy is finite, this microstructure should have some built-in fundamental discreteness and thus be of a very different nature than ordinary spacetime (and geometry). Similarly radical are the results that support a holographic nature for the degrees of freedom constituting a black hole and that stem from a combination of classical and semiclassical arguments. In turn, semiclassical black hole physics has inspired a number of research directions investigating the more general thermodynamical properties of spacetime and the possibility that spacetime/gravitational dynamics (including the whole of GR) is itself to be understood as the thermodynamics, or hydrodynamics in some of the approaches, of unknown microscopic degrees of freedom (Jacobson, 1995; Chirco, Eling, and Liberati, 2010; Padmanabhan, 2015). In this view, the spacetime metric, thus the gravitational field, would be a coarse-grained variable accounting for such microscopic degrees of freedom, and spacetime should be understood as a sort of a fluid-like collective entity. Another independent research area implicitly makes the same suggestion: analogue gravity models in condensed matter systems (Barcelo, Liberati, and Visser, 2005), in particular in the context of quantum fluids. These systems reproduce, at the hydrodynamic level, several phenomena with an equivalent description in terms of semiclassical physics on a curved geometry, including semiclassical black hole–like physics. Thus they support the suggestion that the gravitational physics is an effective, emergent description of a different type of physics, and spacetime is indeed a fluid-like system, the result of the collective behavior of nonspatiotemporal entities.

Quantum many-body systems have also brought us a different type of surprise, showing an intriguing connection between entanglement and geometry, suggesting that the latter can be reduced to the former instead of being treated as fundamental (Van Raamsdonk, 2010). For example, the entanglement entropy associated with a region A on the flat boundary of an AdS space, computed within a simple CFT, is proportional to the area of the minimal surface inside the bulk AdS space with the same boundary as A. The mutual information between two spatial regions on the same flat boundary scales inversely with the geodesic distance between the two regions, measured again in the bulk AdS; the very connectivity between two regions of spacetime has been conjectured to be due to the entanglement between (the quantum degrees of freedom of) the two regions.

Many of these results have been obtained in the context of the AdS/CFT correspondence (Ramallo, 2015). This can be seen as an approach to quantum gravity (at least the sector of it corresponding to AdS boundary conditions), which, despite relying so far mostly on standard field theory methods, suggests a more radical view of spacetime and gravity (understood as curved geometry), in which the latter is again emergent from a system that, while defined on a continuum flat spacetime, is not gravitational. AdS/CFT is then also a first example of the second group of results, pointing to a view of the quantum gravity problem beyond the Bronstein cube (for a discussion on the conceptual challenges raised by the AdS/CFT correspondence, see De Haro, Mayerson, and Butterfield, 2016; De Haro, 2017).

String theory (Polchinski, 1998; Blau and Theisen, 2009) is often used to describe the gravitational side of the AdS/CFT correspondence, and it has been another independent source of radical challenges to the conventional view of spacetime and of quantum gravity. These range from the implications of T-duality for the notion of spatial distance itself to the equivalence between different spacetime topologies encoded in mirror symmetry to the generalized geometries that seem to be needed to describe various effective configurations of string theories (Hohm, Luest, and Zwiebach, 2013). The upshot is that, while we do not know what sort of fundamental degrees of freedom underlie string theories, we know that they will not be spatiotemporal or geometric in any standard sense and that spacetime and geometry as we know them are both collective and emergent notions, in such context (Seiberg, 2006; Huggett and Vistarini, 2015).

Generalized geometries, in particular fractal geometries (Calcagni, 2012), have also been studied extensively, because the running of spacetime dimensions found in several quantum gravity approaches suggests a role for them in the full theory, maybe as an intermediate regime between the fundamental, nonspatiotemporal one and the emergent spacetime of standard field theory.

In fact, quantum gravity approaches, even when starting as conservative quantizations of GR, ended up proposing concrete candidates for the fundamental degrees of freedom underlying spacetime, which are, on their own, not spatiotemporal in the standard sense. Loop quantum gravity (Ashtekar and Lewandowski, 2004) has fundamental quantum states encoded in spin networks—graphs labeled by group representations, with histories corresponding to cellular complexes labeled by the same algebraic data (spin foams) (Baratin and Oriti, 2012; Perez, 2013). The same type of quantum states (and discrete histories) are shared with group field theories (Krajewski, 2011; Oriti, 2012, 2017a). These states and histories, in appropriate regimes, can be put in correspondence with piecewise-flat (thus discrete and singular) geometries, but in the most general cases they will not admit even such protogeometric interpretation. The latter are, in turn, the building blocks of simplicial quantum gravity approaches like quantum Regge calculus (Hamber, 2009) and (causal) dynamical triangulations (Ambjorn et al., 2012), which indeed can be seen as strictly related to group field theories and their purely combinatorial counterparts, random tensor models (Gurau and Ryan, 2012; Rivasseau, 2016). In all these quantum gravity formalisms, therefore, continuum spacetime has to emerge from structures that are fundamentally discrete and rather singular (from the continuum geometric perspective), and in some cases, purely combinatorial and algebraic. A different type of fundamental discreteness, not less radical, is the starting point of the causal set approach (Dowker, 2013).

The main lesson seems to be that continuum spacetime and geometry have to be replaced, at the fundamental level, by some sort of discrete, quantum, nonspatiotemporal structures and have to emerge from their collective dynamics in some approximation (Esfeld and Lam, 2013; Huggett and Wüthrich, 2013; Crowther, 2014; Oriti, 2014; Wüthrich, 2014).

One could say that, since these 'atoms of space' are assumed to be quantum entities, spacetime is understood in all these formalisms as a peculiar quantum many-body system (Oriti, 2017c), that only at macroscopic scales will look like the smooth (indeed, fluid-like) object we are accustomed to. There is, therefore, an obvious general coherence between this picture of spacetime painted by quantum gravity approaches and the more indirect (but also more closely related to established physics) indications obtained by the semiclassical considerations, e.g., in black hole physics.

2.5 The Bronstein Hypercube of Quantum Gravity

The suggested existence of new types of fundamental entities, different from continuum fields, has one general consequence for our understanding of the quantum gravity problem. Given such fundamental (nongeometric, nonspatiotemporal) degrees of freedom, there is one new direction to explore: from small to large numbers of such fundamental entities. We know (from quantum many-body systems and condensed matter theory) that the physics of few degrees of freedom is very different from that of many of them. When taking into account more and more of the fundamental entities and their interactions, we should expect new collective phenomena, new collective variables more appropriate to capture those phenomena, new symmetries and symmetry-breaking patterns, etc. And it is in the regime corresponding to many fundamental building blocks that we expect a continuum geometric picture of spacetime to emerge, so that the usual continuum field theory framework for gravity and other fields will be a good approximation of the underlying nonspatiotemporal physics.

Notice that this is to a large extent independent of whether the discrete structures are understood as physical entities or simply as regularization tools. This will affect, of course, whether one assigns a physical interpretation to all the results of their collective behavior or not and whether or not one tries to eliminate any signature of the discrete structures leading to them. But the existence of the mentioned new direction remains a fact, as it remains true that one has to learn to move along this new direction if one wants to recover a continuum picture for spacetime and geometry (and with them, a gravitational field with relativistic dynamics).

To have a better pictorial representation of what quantum gravity is about, then, the Bronstein cube should be extended to an object with four (a priori) independent directions, to a Bronstein hypercube, as shown in Figure 2.2.

Daniele Oriti

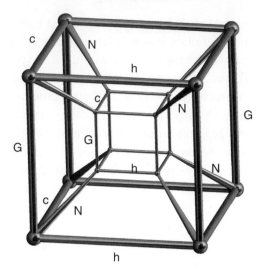

Figure 2.2 The Bronstein hypercube. Figure © Robert Webb/Stella software, www.software3d.com/Stella.php /CC-BY-SA-3.0.

The fourth direction is labeled N, to indicate the number of quantum gravity degrees of freedom that need to be controlled to progressively pass from an entirely nongeometric and nonspatiotemporal description of the theory to one in which spacetime can be used as the basis of our physics. A complete theory of quantum gravity will sit at the same corner in which it was sitting in the Bronstein cube (which is obviously a subspace of this hypercube), but the same theory admits a partial, approximate formulation at any point along the N-direction ending at that corner. Only, the more one moves away from it, the less the notions of continuum spacetime and geometry will fit the corresponding physics. One could say that the definition of a theory of quantum gravity will be provided in the opposite corner (looking only at the two ends of the N-direction, while keeping both G, h, and $1/c$ finite), because it is at this point that the definition of the fundamental degrees of freedom of the theory and of their basic quantum dynamics will be put on the table. This is sensible, but it is also true that providing a complete definition of the same theory amounts to making sure it is well defined up to the opposite end, even though the same theory will always be used in some approximation or truncation.

As we had anticipated, proposing the Bronstein hypercube as the proper arena for quantum gravity means stating that one needs to bring in the lessons and tools of statistical mechanics and condensed matter theory, i.e., the third revolution of last century's physics. It is in that context that we have learned to control the rich physics of many quantum interacting degrees of freedom. We could label then the new direction of the Bronstein hypercube by the Boltzmann's constant, also in order to emphasize the above point (Cohen-Tannoudji, 2009). It could also be a way to make a link with information theory (another crucial area of developments in modern physics), with the implicit link between number of degrees of freedom of a system and its (Boltzmann) entropy, in turn hinting at the physical nature of such information content. This relabeling would have the advantage of characterizing the hypercubic extension of the Bronstein cube by the addition of a fourth fundamental constant, in many ways on equal footing as the other three. It is indeed useful to think in these terms. We do not use this relabeling explicitly simply because we want to maintain the focus on the number of (quantum gravity) degrees of freedom to be controlled in different regimes of the theory, rather than with any specific context in which the new degrees of freedom manifest their physical nature. Another reason for not adopting a terminology directly reminiscent of statistical concepts is the following. We should not confuse the task of 'moving along the N-direction', that is, of understanding the continuum limit of the fundamental degrees of freedom of quantum gravity, and the emergence of continuum spacetime in the process, with the distinct issue, albeit related and definitely important, of defining a general relativistic (quantum) statistical mechanics, including the gravitational field and its thermal fluctuations (Montesinos and Rovelli, 2001; Rovelli, 2013; Chirco and Josset, 2016). Understanding the continuum physics of quantum gravity degrees of freedom may involve formulating a proper statistical framework for them, and this framework would have to be covariant in the obvious sense of not depending on any preferred spatiotemporal frame, local or global, for the simple reason that they would not be defined in any spatiotemporal context to start with. The key distinction has to do with the nature of the entities whose statistical framework one is considering. These fundamental quantum gravity degrees of freedom are not smooth spacetimes (or geometries) or their straightforward quantized version. Thus it is the difference between the (quantum) statistical treatment of the gravitational field (with coupled matter fields) and the (quantum) statistical treatment of 'non-spatiotemporal building blocks of spacetime' (together with 'nonspatiotemporal building blocks of matter') (Kotecha and Oriti, 2018).

One obvious limitation of the description in terms of the number of fundamental degrees of freedom N is that this notion is ambiguous. Not only do what count as fundamental quantum gravity degrees of freedom depend inevitably on the quantum gravity formalism under consideration, but the very notion of relevant degrees of freedom of a system is intrinsically ambiguous. For any quantum system, in fact, it depends on the vacuum state chosen (on the chosen irreducible representation of the fundamental algebra of its quantum observables), on the adopted scale of description, and on the observables chosen as relevant for capturing the physics one is interested in. The first aspect shows that the starting point adopted for describing the system is not God-given, and one should keep this ambiguity under check; but one starting point needs to be chosen and once this is done, as in all quantum gravity approaches we know, our arguments about the need to move along the N-direction are valid. These ambiguities imply that there is no single hypercube of physical theories, because what one finds in its corners and along its edges depends not only on the specific quantum gravity framework being chosen but also on specific criteria used to choose a set of degrees of freedom, a certain vacuum state, relevant symmetries, and so on *within* each specific quantum gravity framework. The choice of all these elements is complex and leads, in general, to inequivalent theories. Once more, the Bronstein hypercube, just like the Bronstein cube, should be understood as only a representation of the *conceptual environment* within which specific, candidate physical theories of quantum gravity are to be studied and interpreted. The second aspect includes a good part of the difficulties that we have to solve when moving along the N-direction. The number of fundamental degrees of freedom is a good proxy, in quantum gravity, for the usual notion of scale in usual spacetime physics, and the need for a change in description at different scales is specifically an important part of the notion of emergence, including the problem of the emergence of continuum spacetime.

2.6 Understanding the Bronstein Hypercube

The first point that using the Bronstein hypercube allows us to emphasize is the crucial distinction between classical and continuum limits. This is where the novelty of the new perspective on quantum gravity is most apparent. In a theory of quantum gravity in which the fundamental entity remains the gravitational field or the geometry of continuum spacetime, the problem of recovering the usual physics of GR (coupled to quantized fields) is the problem of controlling the classical approximation of the more fundamental quantum description of the same. If the fundamental discrete and quantum entities are not directly spatiotemporal, the usual spacetime physics should emerge after taking some sort of continuum approximation, which may or may not be taken in conjunction with a semiclassical approximation. In fact, we have many examples in physics in which such a continuum limit must be taken while maintaining the quantum properties of the fundamental constituents and in which doing otherwise entirely misses important macroscopic physics.

Let us give some examples. Consider some nonrelativistic many-body quantum system of interacting atoms in flat space and with gravitational interaction switched off, i.e., the side of the Bronstein hypercube corresponding to $c \gg 1$, $G \sim 0$. This definition is, of course, so generic that it includes an infinity of systems; in practice, all of condensed matter and solid state physics and more. In the corner corresponding to small number of quantum atoms, we have a bunch of discrete quantum entities, and we can take two directions out of it. If we neglect their quantum properties (going toward the $h \sim 0$ area of the Bronstein hypercube), we obtain the classical mechanics of a few (point) particles. If we now take a continuum approximation by increasing the number of particles to infinity, we obtain a continuum classical system that, e.g., in the case of a fluid, could be described by classical hydrodynamics. Starting from the same corner but taking instead a continuum/hydrodynamic limit first, we could end up with a very peculiar continuum system like a quantum fluid, for example a Bose condensate (if we were dealing with spinless atoms at low temperatures), characterized by peculiar but very much physical features like superfluidity (or superconductivity), even at the macroscopic level. Or we could end up with even more exotic macroscopic phenomena, as in new phases of matter with topological order. All these macroscopic physical properties would be invisible if we were to take the continuum approximation after taking the classical approximation, or by taking the two simultaneously. Moreover, the two directions may be, in general, noncommuting, in the sense that taking the same two approximations in a different order may give different results. The distinction between the two is therefore crucial. A similar story can be told, starting with the same system in the same corner of the Bronstein hypercube, in terms of its statistical treatment, noticing the differences between classical and quantum statistical mechanics. And a similar picture can be drawn in the relativistic case, thus looking at the opposite side of the Bronstein hypercube, corresponding to $c \sim 1$.

The distinction between the two independent directions of the Bronstein hypercube corresponding to h and to N is crucial also because they are traveled using very different types of mathematical techniques and conceptual tools and because we encounter very different types of new physics along the two paths. It is even more crucial in quantum gravity, for two main reasons. First, in the context of a theory so much under construction and so incomplete, and with so little guidance from empirical observations, it is very dangerous to not pay enough attention to the path we are taking or to have the wrong expectations about what we are supposed to do to make progress or about what phenomena we should look for at each stage. Second, in the case of quantum gravity we have really no obvious reason to expect that the two directions commute.

The specifics depend of course on the quantum gravity formalism being considered. As an example, one can consider the relativistic and gravitational counterpart of the earlier atomic case, thus with $c \sim 1$ and $G \sim 1$, within the context of loop quantum gravity and/or group field theories. The corner with few quantum gravity degrees of freedom in the full quantum regime corresponds to a description of the world in terms of simple (superpositions of) spin networks associated with graphs with a smallish number of nodes and with a quantum dynamics captured by amplitudes (spin foam models or lattice gravity path integrals) associated with (superpositions of) cellular complexes with limited combinatorial complexity (this could be increased, of course, but this description implies that it remains limited enough that we do not need to use different, collective, or coarse-grained entities and variables). The straightforward classical approximation of the same structures results in a description in terms of classical piecewise-flat geometries (character-ized, in the case of spin networks, by first-order classical variables: edge lengths or triangle areas and discrete connection variables), and with a dynamics encoded in solutions of discrete geometric equations, e.g., Regge geometries, possibly coming from the discretization of some continuum gravity action. Better, this is what we expect given the partial results we have so far (Freidel and Geiller, 2013; Han and Zhang, 2013), but it is not completely established, and it remains an interesting challenge for the community. A further continuum limit is then needed to obtain an effective description in terms of (some possibly modified version of) GR and matter field theories. When the correspondence with Regge geometries (or similar) is solid, one can rely on the results obtained in that context for studying such continuum limit. When this is done, we are back in the sector of the Bronstein hypercube corresponding to the Bronstein cube, and we could consider quantizing our continuum gravitational theory. This already shows the difference between clas-sical and continuum approximations in this quantum gravity context. Seen from the perspective encoded in the Bronstein hypercube, and given the same starting point, i.e. the same fundamental structures, the route toward a complete quantum theory of gravity would instead require taking into account more and more spin network degrees of freedom, at the quantum dynamical level, reaching (at least formally) the opposite corner of the hypercube along the N-direction. This would be the regime of very large (possibly infinite) superpositions of quantum spin network states, including very refined graphs, with correspondingly complex interaction processes, which would then be effectively described in terms of collective variables like continuum fields (including the gravitational field) and field theoretic dynamics, e.g., some modified version of GR, up to any additional quantum corrections. As I emphasized earlier, we should not expect that the result will be the same as that obtained by quantizing GR, and we should actually expect that this is not the case, if not under additional assumptions or approximations. Why this is the case should be clearer in the following, once we take a deeper look at what it is involved in moving along the N-direction.

2.7 How to Move along the N-Direction

The N-direction is the path along which emergence of spacetime and geometry should take place. The notion of emergence itself is a thorny topic in philosophy of science (Batterman, 2006, 2011; Bedau and Humphreys, 2008; Butterfield, 2011a, 2011b; Butterfield and Bouatta, 2012). The possibility that spacetime is not a fun-damental but an emergent notion, and that the emergence process should then be understood in a nonspatiotemporal manner, raises a host of conceptual puzzles at both ontological and epistemological level. Some of them have been discussed in earlier work (Oriti, 2014). Here, I want to focus more on the physical aspect of this idea. That is, I want to discuss how we move along the N-direction, technically, and what we could expect to find, when we do so.

So, first, how do we move along the N-direction, from less to more degrees of freedom? One main technical tool is the renormalization group, which is usually phrased as mapping a theory seen at a given scale to its counterpart at a different scale (this requires us to go beyond the naive view of renormalization as a way to 'cure' or 'hide' the theory's divergences). It is accompanied by several approximation schemes, by which we can extract suitable descriptions of the theory, capturing the key observables we are interested in at the given scale. As mentioned earlier, the notion of scale in the usual spacetime physics, e.g., in ordinary quantum field theory, intertwines the number of degrees of freedom of the system with some geometric or spatiotemporal quantity, like energy or distance, simply because such theories deal with degrees of freedom that are localized in spacetime and are associated to a well-defined notion of energy. In quantum gravity, while the specifics of any renormalization scheme will depend on the approach being considered, we should expect only a more abstract notion of scale, more or less reduced to a counting of degrees of freedom, to be available. This does not mean that such notion of scale cannot be tentatively interpreted in some protospatiotemporal manner, but it means that such interpretation will be fully justified only in the regime of the theory where continuum spacetime and geometry are shown to emerge.

Computing the full renormalization group flow of a given theory, in fact, amounts (formally) to defining the full (quantum) dynamics of the same (Kopietz, Bartosch, and Schütz, 2010; Rivasseau, 2014a); this means removing (again, formally) any truncation that may initially be applied to it to have mathematically well-defined quantities, in particular any truncation to a finite (small) number of its fundamental degrees of freedom. In this sense, it contains both a continuum limit (usually associated with the limit of large momenta and small distances) and a thermodynamic limit (usually corresponding to large volumes and infinite number of atoms or field values), when the two notions both make sense but differ, as in usual spacetime physics. We do not know if this is the case in quantum gravity, in general, and the question can be addressed only by considering specific quantum gravity formalisms and providing a definition of both limits. I do not distinguish the two, in what follows, referring simply to the 'continuum limit'.

The renormalization group flow is always computed within some approximation scheme. This is a technical necessity, but it also contains an important physical insight: what matters at each step is to control the (approximate) behavior of key observables, and only the aspects of the theory that are most relevant to them, neglecting the rest. The important insight is that a lot of what the theory contains, in principle, does not affect the relevant physics, at least not significantly. The relevant collective variables and observables may differ from those in the initial definition of the same theory (and that enter the computation of the same collective quantities). Such approximate, coarse-grained description, alongside other forms of truncation, is thus not just a technical tool that physicists have to adopt for lack of computational power or skills but is where they show or test their physical understanding of the system. These approximations, moreover, are both a prerequisite for the understanding of the renormalization group flow of the theory and directly suggested by it, since the renormalization group flow itself gives indications on which dynamics and which observables (e.g., order parameters) are relevant in different regimes (scales).

Next, we ask the second question: what should we expect to find, once we move along the N-direction via renormalization group tools and as we approach the full quantum gravity corner of the Bronstein hypercube? A continuum spacetime and geometry is the goal, of course. And it is what we should find if our quantum gravity formalism is to be physically viable. But we should also expect to find much more than that. Physical quantum systems, when they are interacting (thus nontrivial) and possess an infinite (or very large) number of degrees of freedom, do not have a unique continuum limit. How they organize themselves when large numbers of their constituents are taken into account depends on the value of their fundamental coupling constants (or other external parameters). Their collective behavior leads to different macroscopic phases, separated by phase transitions. The effective, emergent physics of different continuum phases can be very different, and they are also stable (by definition) under moderate changes of their defining parameters and, of course, under the dynamics of the system. They are in many ways different possible worlds—inequivalent collective realizations of what we were initially considering a single physical system when focusing on small numbers of its fundamental constituents. The questions in any quantum gravity approach, once the fundamental entities and their quantum dynamics are identified and the renormalization group flow can be, in principle, set rolling, are as follows: What are the macroscopic phases? Are any of them effectively described in terms of smooth geometry (with matter fields) and spacetime? With these new questions come many others, for example, concerning the physical meaning of the phase transition(s) separating a nongeometric from a geometric phase, that of the different geometric phases themselves, if more than one such phase appears, and the possible observational signatures of this potentially new physics.

Before I briefly survey recent progress on these issues, let me draw two general consequences of this line of reasoning. First, moving along the N-direction, i.e., toward the full definition of the theory, brings potentially new physics and requires, possibly, a change in description of the system at each step. Second, the result of such journey is not unique but it is given by different possible continuum limits, different continuum theories with different effective physics. In this sense, the Bronstein hypercube should not be expected 'to close' to form a hypercube at all, but a multiplicity of possible hypercubes at best. And this is assuming that moving along the h-direction starting from the GR corner (or the one corresponding to some other continuum classical gravitational theory) gives any consistent result at all, and that, in addition, the h- and N-direction commute. This is also the point, however, where our attitude toward the nonspatiotemporal structures that our quantum gravity formalism is built on, whether we regard them as physical or mere mathematical (e.g., regularization) tools, becomes crucial. In the latter case, in fact, only the continuum phase(s) with a geometric interpretation will be deemed physical, and the others ignored, and the result of quantizing the classical theory will have to be the same by definition, since the extension from the Bronstein cube to the hypercube would be a mere technical expedient, with the true physics remaining captured by the perspective associated with the Bronstein cube.

One more interesting and challenging aspect of the hypercube perspective on quantum gravity concerns symmetries, which are crucial tools in theory building (Brading and Castellani, 2003). Investigating the role of symmetries in quantum gravity from the point of view of the Bronstein hypercube means investigating two (related) issues: first, the fate of diffeomorphism invariance, the defining symmetry of general relativity, in going from the Bronstein cube to the Bronstein hypercube, and then within the latter; second, the nature of new quantum symmetries and their use in moving along the N-direction and in recovering continuum spacetime and geometry in the end. Diffeomorphism invariance stands and falls, in many ways, alongside continuum spacetime and geometry, whether one sees it as a mere mathematical convenience or as a fundamental requirement of physical theories (Norton, 2003; Pooley, 2010). This means that, in the Bronstein cube, it will follow geometry and the gravitational field in being 'quantized', to reach the quantum gravity corner. This is nowhere more clear than in the canonical approach, where defining and imposing diffeomorphism invariance at the quantum level is equivalent to defining and imposing the quantum gravitational dynamics. It also means that, when changing the nature of the fundamental degrees of freedom of the theory, thus moving to candidate fundamental entities different from spacetime fields defined on some differentiable manifold, diffeomorphisms will cease to be defined altogether. Formally, at least, they will play no role in the fundamental theory. This does not mean, however, that diffeomorphism symmetry will not still be important, at the conceptual level. To start with, the fundamental dynamics will have to be some form of constrained or relational dynamics, like in continuum, diffeomoprhism invariant GR. This is simply due to the fact that, like continuum diffeomorphism invariant GR, the fundamental quantum gravity theory will not admit any notion of time that is not given, at best, in terms of some internal (and approximate) degree of freedom of the theory, used as a relational clock, and thus will not take the form of a standard time evolution. This implies that the experience and mindset developed in the context of diffeomorphism invariant, continuum quantum theories will be very much relevant. Further, depending on the nature of the candidate fundamental degrees of freedom of the theory, the same may be characterized by new quantum symmetries that are some sort of pregeometric counterpart of diffeomorphisms. They could be some discretized version of them, like in approaches based on simplicial piecewise-flat structures (Dittrich, 2009, 2011; Baratin, Girelli, and Oriti, 2011), or of more exotic type, e.g., purely combinatorial (Ambjorn et al., 2012; Dowker, 2013). In any case, the requirement that they reduce to diffeomorphisms in the appropriate continuum (and classical) limit, or that, at least, they can be traded for diffeomorphisms in the same limit, will play a crucial role in identifying such new symmetries. In turn, the identification of new fundamental symmetries, whatever their relation to diffeomorphisms, will certainly play an important role in the construction of the fundamental theory and in the reconstruction of continuum spacetime and geometry from it. They could be the key instrument for defining the fundamental theory space within which the dynamics of the pregeometric atoms of space should be studied; they could be the tool for characterizing the different universality classes or continuum phases in which they organize themselves when moving along the N-direction; they could single out conserved quantities or other collective observables around which the emergent continuum geometric dynamics could be built. Work on all these aspects can be found in the literature, for example in group field theory (GFT) and random tensor models (Rivasseau, 2014b; Bahr, Dittrich, and Geiller, 2015; Dittrich and Geiller, 2015, 2017; Kegeles and Oriti, 2017; Kegeles, Oriti, and Tomlin, 2017).

2.8 A Brief Survey of Recent Results along the N-Direction

The renormalization of quantum gravity models is a very active area of research, in many of the quantum gravity formalisms based on nonspatiotemporal, discrete building blocks.

It has been a central research topic in simplicial quantum gravity approaches since their very inception, also because in these approaches the discrete structures are usually seen as unphysical regularization tools so the continuum limit is mandatory before thinking of any physics. In quantum Regge calculus (Hamber, 2009) one fixes the triangulation at the onset and has its edge lengths as dynamical variables, weighted by the (exponential of the) Regge action (a discretization of the Einstein–Hilbert action for GR). The strategy for taking the continuum limit, usually limited to Euclidean geometries, is then analogous to the lattice gauge theory one, adapted to a varying lattice geometry, in which lattices are progressively refined by increasing their complexity while keeping some macroscopic quantity fixed, e.g., the total 'spacetime volume'. There is no consensus on whether a continuum phase with a smooth geometry is found, and some of the phases identified so far are consistently interpreted as rather degenerate or singular from the continuum spacetime perspective, on the basis of simple indicators like dimension estimators (e.g., spectral or Hausdorff dimension). In dynamical triangulations one takes a complementary approach, and with the same starting point, it restricts attention to equilateral triangulations, which are summed over with the same Regge weight. The continuum limit amounts to computing the full sum over triangulations concentrating it on the finer ones (sending the fixed edge length to zero while keeping the total volume fixed). The results are consistent (and inconclusive) in the Euclidean setting but become much more interesting when Lorentzian discrete geometries are considered and additional 'causality' conditions are imposed, basically providing the triangulations summed over with a fixed foliation structure. Then, one finds strong indications that (at least) one smooth geometric phase is produced in the continuum limit, as it can be detected by dimension estimators and rather coarse geometric quantities, like the total spatial volume.

Random tensor models (Gurau and Ryan, 2012; Rivasseau, 2016) produce in their perturbative expansion (around the fully degenerate configuration, corresponding to no spacetime at all) the same type of dynamical configurations and equilateral triangulations, and with the same weight, as the Euclidean dynamical triangulations approach. They provide a generating functional for them and, accordingly, offer a new set of statistical and field-theoretical tools to study their continuum limit. The results are so far broadly consistent with what has been found from the pure simplicial gravity perspective, in the simple large N limit of the tensors. The new set of tools, however, promises more, and it has already hinted at a deeper level of analysis, with many preliminary results on double scaling limits, phases beyond what has been found in the large N, etc.

In canonical loop quantum gravity, maybe due to its traditional understanding as a straightforward quantization of continuum GR, the issue of renormalization and continuum limit has received less attention. A canonical renormalization group scheme at the full dynamical level has been proposed (Lang, Liegener, and Thiemann, 2017), but most of the results have been so far limited to the kinematical setting, neglecting the quantum dynamics of the theory. These results include the continuum limit that is used to define the kinematical Hilbert space of the theory and that singles out the so-called Ashtekar–Lewandowski vacuum as the basis for constructing spin network excitations (Ashtekar and Lewandowski, 2004).

Its construction is a nontrivial mathematical achievement, but its physical nature is dubious. From the continuum perspective, it is a state corresponding to a totally degenerate geometry and a highly fluctuating connection, thus far away from anything resembling our spacetime. This prompts one to look for the sector of the theory corresponding to highly excited states over such vacuum and encoding many of the fundamental spin network degrees of freedom. It also suggests that, when this is done, the relevant description of the same theory will be very different, and possibly will involve a phase transition to a new, more geometric phase. Lacking a full renormalization group analysis, the issue of possible new vacua/phases could be studied only at the kinematical level, but has already produced very interesting results. New kinematical vacua with a nondegenerate (constant) geometry, but still with highly fluctuating connection, were constructed and analyzed in some detail (Koslowski, 2007; Koslowski and Sahlmann, 2012), shown to define inequivalent representations (thus genuinely new phases) and already to offer a more sensible physical interpretation in terms of continuum spacetime geometries. They can be understood as a sort of condensate of spin networks excitations, thus resonating with earlier (Oriti, 2007) and more recent (Oriti, 2017b) work in the group field theory (re)formulation of spin network dynamics. Even more recently, a different, complementary type of new vacua have been constructed (Bahr et al., 2015; Dittrich and Geiller, 2015, 2017; Dittrich, 2018), corresponding instead to a fixed curvature and fluctuating geometry (triad/flux variables) and with excitations corresponding to curvature defects. The simplest such vacuum can be associated with a simple BF topological field theory and zero curvature, while vacua corresponding to non-zero constant curvature (a cosmological constant?) seem to be given by condensates of curvature defects.

Spin foam models (Baratin and Oriti, 2012; Perez, 2013) can be understood as a covariant dynamics for canonical loop quantum gravity states, i.e., spin networks, thus they are an alternative setting for defining the renormalization group flow and the continuum limit of the theory, encoding also the quantum dynamics. They can also be understood as lattice gravity path integrals in first-order (tetrad+connection) variables, thus in direct relation with simplicial quantum gravity approaches. Renormalization of spin foam models has been tackled in two (complementary) frameworks. The first (Dittrich, 2012, 2017; Bahr et al., 2013; Dittrich, Mayerson, and Butterfield, 2016; Bahr, 2017; Bahr and Steinhaus, 2017; Delcamp and Dittrich, 2017) treats them as analogous to lattice gauge theories on a fixed lattice, thus in line with the way the continuum limit of quantum Regge calculus is studied. However, it brings on board mathematical methods and insights of canonical loop quantum gravity and also takes advantage of the direct resemblance with gauge theories. More recently, key tools from quantum information theory and many-body systems, like tensor networks, have also started to play an important role. Coarse-graining steps and renormalization are encoded in maps between spin foam amplitudes associated with different scales, where the notion of scale here is tied to the combinatorial complexity of the underlying lattice, and they therefore provide a dynamical counterpart of the kinematical continuum limit that defines the Hilbert space of the canonical theory. The results are so far mostly confined to simplified models, rather than with the full-fledged amplitudes proposed for 4-dimensional quantum gravity, but they are already very interesting and possibly indicative of more general lessons. For example, one finds hints of a nontrivial phase diagram with a degenerate geometric phase and a nontrivial phase of topological nature, thus tentatively supporting the kinematical results on possible vacua in the canonical theory. The other way to tackle spin foam renormalization is to see them as Feynman amplitudes of group field theory models and to focus on the renormalization of the latter.

Group field theory renormalization has been in fact a very active and rapidly growing research direction for almost 10 years, now, with many results (Freidel, Gurau, and Oriti, 2009; Carrozza, Oriti, and Rivasseau, 2014; Benedetti, Ben Geloun, and Oriti, 2015; Ben Geloun and Rivasseau, 2016; Ben Geloun, Martini, and Oriti, 2016; Carrozza, 2016; Carrozza and Lahoche, 2017; Carrozza, Lahoche, and Oriti, 2017; Lahoche and Oriti, 2017). The strategy is to rely on the close-to-standard field theoretic formulations of these models as field theories on Lie group manifolds (not interpreted, of course, as spacetimes) and use their intrinsic notion of scale as a 'distance on the group manifold' or, conversely, the conjugate momentum/resolution scale. Indeed, on such premises one can apply standard renormalization group techniques, suitably adapted to the peculiar combinatorially nonlocal nature of the GFT interactions. The trivial Fock vacuum of the theory, around which one sees spin network excitations and develops the perturbative expansion of a given group field theory model, is again a fully degenerate one, with no topological or protogeometrical excitations; thus it is in the nonperturbative sector of the theory that one looks for continuum spacetime, geometry, and physics. Perturbative renormalization is, however, where one can find a consistency check of the quantum theory, a way to constrain model-building ambiguities and deal more directly with spin foam amplitudes. This activity relied heavily on the parallel results on random tensor models, in particular the large-N expansion, and has been also focused on simplified models. It has produced rigorous proofs of renormalizability of a wide range of tensorial GFTs via multiscale methods: abelian and nonabelian, with local gauge invariance, thus having Feynman amplitudes corresponding to lattice gauge theories (and spin foam models) and without it, in low (e.g., 3-dimensional) as well as higher (e.g., 6-dimensional) topological dimensions. More recently, work on nonperturbative renormalization has gained traction, with the development of the functional renormalization group (FRG) formalism for GFTs. The same range of models studied perturbatively has been studied by the FRG method, and there is by now a large body of results establishing renormalizability as well as nontrivial phase diagrams for many tensorial GFTs, their asymptotic freedom or safety in the UV, but also solid hints of Wilson–Fisher-type fixed points in the IR, suggesting the existence of a condensate phase in the continuum limit. This phase would be directly relevant for the extraction of effective spacetime physics, especially in the cosmological setting, as I discuss later in this chapter. For full-blown GFT models for 4-dimensional spacetime, we have mainly partial results on radiative corrections (Bonzom and Dittrich, 2013; Riello, 2013; Doná, 2018), but the mathematical technology at our disposal is improving fast, and we can also rely on related analyses of GFT models on homogeneous spaces and on existence proofs of phase transitions in the GFT formulation of topological BF theories in any dimension (the basis of a lot of model building for 4-dimensional quantum gravity models) (Baratin et al., 2014).

While the renormalization analysis of quantum gravity models has grown in attention and results, comparatively little work has been done so far on the extraction of continuum physics from them. By this we intend the extraction of effective dynamics for collective observables with a spatiotemporal and geometric interpretation, in the regime in which large numbers (possibly infinite) of the fundamental constituents are accounted for, and thus employing a set of explicit approximations and coarse graining operated at the level of the fundamental theory itself. I do not refer here to the many works in which tentative physics is extracted from either simple models of continuum spacetime and gravitational dynamics that are only inspired but not derived from the fundamental theory (e.g., loop quantum cosmology and several models of quantum black holes) or truncations of the fundamental theory dealing with very small numbers of the fundamental entities (e.g., restricted to simple spin network graphs or simple spin foam or simplicial gravity lattices).

One example of this type of work is the extraction of an effective minisuperspace dynamics and of a de Sitter-like (spatial) volume profile from the causal dynamical triangulations approach, obtained by explicit numerical evaluation of this (very coarse grained) observable (Ambjorn et al., 2012).

Another example of recent work in this direction is GFT condensate cosmology (Gielen, Oriti, and Sindoni, 2013, 2014; De Cesare, Pithis, and Sakellariadou, 2016; Gielen and Sindoni, 2016; Oriti, Sindoni, and Wilson–Ewing, 2016b; Pithis and Sakellariadou, 2017; De Cesare et al., 2018; Gielen and Oriti, 2018), based on two main assumptions—one of perspective, one more technical. The first is that cosmology has to be looked for in the hydrodynamics of the fundamental theory, as the most suitable approximation for close-to-equilibrium and most coarse-grained dynamics, in the continuum limit. The second is that the relevant class of continuum states (implicitly, the relevant continuum phase of the theory) for the extraction of gravitational physics is that captured by condensates of the microscopic building blocks (GFT quanta, i.e., spin network vertices or basic simplices). The first assumption suggests concepts and technical tools to be used. The second, supported also by the hints coming from the renormalization analysis of GFT models, makes it possible to go directly from the microscopic definition of the quantum dynamics of any given model, including the more promising 4-dimensional gravity ones, to an effective dynamics for cosmological observables in the continuum limit. For quantum condensates, in fact, the continuum hydrodynamics corresponds, in the simplest (mean field, Gross–Pitaevskii) approximation, to the classical equations of motion of the underlying field theory for the atoms, and the same happens in GFT models. This strategy led to many results, over the last five years. They include the extraction of a modified Friedmann dynamics for homogeneous and isotropic geometries (and scalar matter), whose physics is captured by relational observables,[1] with the correct classical limit at late times; a quantum bounce replacing the classical big bang singularity, as long as one remains within the hydrodynamics approximation of the full theory; the possibility of a long-lasting accelerated phase of expansion after such bounce (a sort of purely quantum gravity–induced inflation) without the need for introducing any inflation-like field; some preliminary study of the dynamics of anisotropies, showing their natural suppression as the universe grows; the first extensions of the formalism to cosmological perturbations, which seems to indicate how a scale invariant spectrum is the natural outcome of the dynamics, as long as one remains close to homogenous condensate states. To

---

[1] Let us add some clarifications about the use of relational (dynamical) variables in a pregeometric, nonspatiotemporal quantum gravity formalism. The use of relational clocks and rods is a convenient strategy for defining a diffeomorphism invariant and physical notion of time and space, often adopted in general relativity and in its quantized counterparts. I find it a perfectly acceptable solution to such issue (even if such clocks and rods will not be idealized or perfect, since they remain physical and quantum entities, and thus will offer only an approximate substitute for temporal or spatial coordinates). It is a solution, that is, to the issue of 'disappearance of space and time', in the sense proper to (quantum) GR of disappearance of any absolute, nondynamical notion of space and time (as codified by a preferred frame), due to diffeomorphism invariance and background independence (Rovelli, 2018). In the relational strategy, instead of any such absolute space or time, we have dynamical fields, which may include the metric field itself, playing the role of clocks and rods. This strategy requires, for its adoption, the usual continuum setting of the Bronstein cube, in which dynamical fields are basic entities. The adoption of the same strategy in a pregeometric, nonspatiotemporal setting as pictured in the Bronstein hypercube should be understood in the following generalized sense. One looks at the fundamental degrees of freedom of the quantum gravity formalism at hand, which are nonspatiotemporal by definition and do not correspond directly to continuum fields. Among those, one identifies the ones that, in a continuum limit and in an approximate sense (that is, after moving along the N-direction toward the standard continuum, spatiotemporal setting), will correspond to emergent dynamical fields and that can then be used as relational clock and rods, i.e., in terms of which one will then assign spatiotemporal localization properties to the *other* degrees of freedom of the quantum gravity system. Such use as relational notions of space and time will take place, in other words, only after a spatiotemporal GR-like description has emerged (or in order to test its emergence). It is possible that the application of this strategy requires the introduction of additional degrees of freedom in a given model of pregeometric model, but it does not affect the need to move along the N-direction, for the whole set of degrees of freedom, to show the emergence of spacetime along the way (thus it does not mean that one is introducing space or time 'by hand' in the fundamental definition of the theory, just as the fact that eventually spacetime and geometry emerge approximately in a pregeometric quantum gravity model does not imply that they were there from the very beginning).

these, one could add the first generalization of the scheme to spherically symmetric geometries and black hole horizons (Oriti, Pranzetti, and Sindoni, 2016a; Oriti et al., 2018), with more interesting results. We are just at the beginning of the exploration of the emergent continuum physics of GFT models (and of their spin foam counterpart), clearly, but the path seems promising.

2.9 Conclusions

I have argued that the proper setting for thinking about quantum gravity and for exploring the many issues it raises (mathematical, physical, conceptual) is broader than the traditional one of quantizing GR, well captured by the Bronstein cube. It is best pictured as a Bronstein hypercube, in which the nonspatiotemporal nature of the fundamental building blocks suggested by most quantum gravity formalisms (and even by semiclassical physics) and the need to control their collective dynamics are manifest. This allows the proper focus on the problem of the emergence of continuum spacetime and geometry from such nonspatiotemporal entities. I have also argued that modern quantum gravity approaches are well embedded into this conceptual scheme and have already started producing many results on the issues that are put to the forefront by it. The quantum gravity world is therefore even richer and more complex but also more exciting than traditionally thought, and we are already actively exploring it. More surprises should be expected.

References

Ambjorn, J., Goerlich, A., Jurkiewicz, J. and Loll, R. (2012). Nonperturbative quantum gravity. *Phys. Rept.*, 519, 127–210.

Ashtekar, A. and Lewandowski, J. (2004). Background independent quantum gravity: A status report. *Class. Quant. Grav.*, 21, R53.

Bahr, B. (2017). On background-independent renormalization of spin foam models. *Class. Quant. Grav.*, 34, no. 7, 075001.

Bahr, B., Dittrich, B. and Geiller, M. (2015). A new realization of quantum geometry. arXiv:1506.08571 [gr-qc].

Bahr, B., Dittrich, B., Hellmann, F. and Kaminski, W. (2013). Holonomy spin foam models: Definition and coarse graining. *Phys. Rev.*, D87, no. 4, 044048.

Bahr, B. and Steinhaus, S. (2017). Hypercuboidal renormalization in spin foam quantum gravity. *Phys. Rev.*, D95, no. 12, 126006.

Baratin, A., Carrozza, S., Oriti, D., Ryan, J. and Smerlak, M. (2014). Melonic phase transition in group field theory. *Lett. Math. Phys.*, 104, 1003–1017.

Baratin, A., Girelli, F. and Oriti, D. (2011). Diffeomorphisms in group field theories. *Phys. Rev.*, D83, 104051.

Baratin, A. and Oriti, D. (2012). Group field theory and simplicial gravity path integrals: A model for Holst-Plebanski gravity. *Phys. Rev.*, D85, 044003.

Barcelo, C., Liberati, S. and Visser, M. (2005). Analog gravity. *Living Rev. Rel.*, 8, 12.

Batterman, R. (2006). Reduction and renormalization. In: *The Robert and Sarah Boote Conference in Reductionism and Anti-Reductionism in Physics*, Pittsburgh, April 22–23, 2006.

Batterman, R. (2011). Emergence, singularities and symmetry breaking. *Found. Phys.*, 41, no. 6, 1031–1050.

Bedau, M. and Humphreys, P. (eds.). (2008). *Emergence*. Cambridge, MA: MIT Press.

Benedetti, D., Ben Geloun, J. and Oriti, D. (2015). Functional renormalisation group approach for tensorial group field theory: A rank-3 model. *JHEP*, 1503, 084.

Ben Geloun, J., Martini, R. and Oriti, D. (2016). Functional renormalisation group analysis of tensorial group field theories on $\mathbb{R}^d$. *Phys. Rev.*, D94, no. 2, 24017.

Ben Geloun, J. and Rivasseau, V. (2013). A renormalizable 4-dimensional tensor field theory. *Commun. Math. Phys.*, 318, 69.

Blau, M. and Theisen, S. (2009). String theory as a theory of quantum gravity: A status report. *Gen. Rel. Grav.*, 41, 743–755.

Bodendorfer, N. (2016). An elementary introduction to loop quantum gravity. arXiv:1607.05129 [gr-qc].

Bonzom, V. and Dittrich, B. (2013). Bubble divergences and gauge symmetries in spin foams. *Phys. Rev.*, D88, 124021.

Brading, K. and Castellani, E. (eds.). (2003). *Symmetries in Physics: Philosophical Reflections*. Cambridge: Cambridge University Press.

Bronstein, M. (1933). K voprosu o vozmozhnoy teorii mira kak tselogo (On the question of a possible theory of the world as a whole). *Uspekhi Astronomicheskikh Nauk*, Sbornik 3, 3–30.

Butterfield, J. (2011a). Emergence, reduction and supervenience: A varied landscape. *Found. Phys.*, 41, 920–959.

Butterfield, J. (2011b). Less is different: Emergence and reduction reconciled. *Found. Phys.*, 41, 1065–1135.

Butterfield, J. and Bouatta, N. (2012). Emergence and reduction combined in phase transitions. *AIP Conf. Proc.*, 1446, 383.

Calcagni, G. (2012). Geometry of fractional spaces. *Adv. Theor. Math. Phys.*, 16, 549.

Carlip, S. (2014). Black hole thermodynamics. *Int. J. Mod. Phys.*, D23, 1430023.

Carrozza, S. (2016). Flowing in group field theory space: A review. *SIGMA*, 12, 070.

Carrozza, S. and Lahoche, V. (2017). Asymptotic safety in three-dimensional SU(2) group field theory: Evidence in the local potential approximation. *Class. Quant. Grav.*, 34, no. 11, 115004.

Carrozza, S., Lahoche, V. and Oriti, D. (2017). Renormalizable group field theory beyond melonic diagrams: An example in rank four. *Phys. Rev.*, D96, no. 6, 66007.

Carrozza, S., Oriti, D. and Rivasseau, V. (2014). Renormalization of an SU(2) tensorial group field theory in three dimensions. *Commun. Math. Phys.*, 330, 581–637.

Chirco, G., Eling, C. and Liberati, S. (2010). Non-equilibrium thermodynamics of space-time: The role of gravitational dissipation. *Phys. Rev.*, D81, 024016.

Chirco, G. and Josset, T. (2016). Statistical mechanics of covariant systems with multi-fingered time. arXiv:1606.04444 [gr-qc].

Cohen-Tannoudji, G. (2009). Universal constants, standard models and fundamental metrology. *Eur. Phys. J. ST.*, 172.

Crowther, K. (2014). Appearing out of nowhere: The emergence of spacetime in quantum gravity. arXiv:1410.0345 [physics.hist-ph].

De Cesare, M., Oriti, D., Pithis, A. and Sakellariadou, M. (2018). Dynamics of anisotropies close to a cosmological bounce in quantum gravity. *Class. Quant. Grav.*, 35, no. 1, 015014.

De Cesare, M., Pithis, A. and Sakellariadou, M. (2016). Cosmological implications of interacting group field theory models: Cyclic universe and accelerated expansion. *Phys. Rev.*, D94, no. 6, 064051.

De Haro, S. (2017). Dualities and emergent gravity: Gauge/gravity duality. *Stud. Hist. Philos. Mod. Phys.*, 59, 109–125.

De Haro, S., Mayerson, D. and Butterfield, J. (2016). Conceptual aspects of gauge/gravity duality. *Found. Phys.*, 46, no. 11, 1381–1425.

Delcamp, C. and Dittrich, B. (2017). Towards a phase diagram for spin foams. *Class. Quant. Grav.*, 34, no. 22, 225006.

Dittrich, B. (2009). Diffeomorphism symmetry in quantum gravity models. *Adv. Sci. Lett.*, 2, 151.

Dittrich, B. (2011). How to construct diffeomorphism symmetry on the lattice. *PoS QGQGS*, 2011, 012.

Dittrich, B. (2012). From the discrete to the continuous—Towards a cylindrically consistent dynamics. *New J. Phys.*, 14, 123004.

Dittrich, B. (2017). The continuum limit of loop quantum gravity—A framework for solving the theory. Pages 153–179 of: Ashtekar, A. and Pullin, J. (eds.). *Loop Quantum Gravity*. Singapore: World Scientific.

Dittrich, B. (2018). Cosmological constant from condensation of defect excitations. *Universe*, 4, no. 7, 81.

Dittrich, B. and Geiller, M. (2015). A new vacuum for loop quantum gravity. *Class. Quant. Grav.*, 32, no. 11, 112001.

Dittrich, B. and Geiller, M. (2017). Quantum gravity kinematics from extended TQFTs. *New J. Phys.*, 19, no. 1, 013003.

Dittrich, B., Mizera, S. and Steinhaus, S. (2016). Decorated tensor network renormalization for lattice gauge theories and spin foam models. *New J. Phys.*, 18, no. 5, 053009.

Doná, P. (2018). Infrared divergences in the EPRL-FK spin foam model. *Class. Quant. Grav.*, 35, no. 17, 175019.

Dowker, F. (2013). Introduction to causal sets and their phenomenology. *Gen. Rel. Grav.*, 45, no. 9, 1651–1667.

Esfeld, M. and Lam, V. (2013). A dilemma for the emergence of spacetime in canonical quantum gravity. *Stud. Hist. Philos. Mod. Phys.*, 44, no. 3, 286–293.

Freidel, L. and Geiller, M. (2013). Continuous formulation of the loop quantum gravity phase space. *Class. Quant. Grav.*, 30, 085013.

Freidel, L., Gurau, R. and Oriti, D. (2009). Group field theory renormalization—The 3d case: Power counting of divergences. *Phys. Rev. D*, 80, 044007.

Gielen, S. and Oriti, D. (2018). Cosmological perturbations from full quantum gravity. *Phys. Rev. D* 98, no. 10, 106019.

Gielen, S., Oriti, D. and Sindoni, L. (2013). Cosmology from group field theory formalism for quantum gravity, *Phys. Rev. Lett.*, 111, 031301.

Gielen, S., Oriti, D. and Sindoni, L. (2014). Homogeneous cosmologies as group field theory condensates, *JHEP*, 1406, 013.

Gielen, S. and Sindoni, L. (2016). Quantum cosmology from group field theory condensates: A review. *SIGMA*, 12, 082.

Gurau, R. and Ryan, J. (2012). Colored tensor models—A review. *SIGMA*, 8, 20.

Hamber, H. (2009). Quantum gravity on the lattice, *Gen. Rel. Grav.*, 41, 817–876.

Han, M. and Zhang, M. (2013). Asymptotics of spinfoam amplitude on simplicial manifold: Lorentzian theory. *Class. Quant. Grav.*, 30, 165012.

Hohm, O., Luest, D. and Zwiebach, B. (2013). The spacetime of double field theory: Review, remarks, and outlook. *Fortsch. Phys.*, 61, 926.

Hossenfelder, S. (2013). Minimal length scale scenarios for quantum gravity. *Living Rev. Rel.*, 16, 2.

Huggett, N. and Vistarini, T. (2015). Deriving general relativity from string theory. *Philos. Sci.*, 82, no. 5, 1163–1174.

Huggett, N. and Wütrich, C. (2013). Emergent spacetime and empirical (in)coherence. *Stud. Hist. Philos. Mod. Phys.*, 44, 276–285.

Jacobson, T. (1995). Thermodynamics of spacetime: The Einstein equation of state. *Phys. Rev. Lett.*, 75, 1260–1263.

Kegeles, A. and Oriti, D. (2017). Continuous point symmetries in group field theories. *J. Phys.*, A50, no. 12, 125402.

Kegeles, A., Oriti, D. and Tomlin, C. (2017). Inequivalent coherent state representations in group field theory. *Class. Quant. Grav.*, 35, no. 12, 125011.

Kiefer, C. (2005). Quantum gravity: General introduction and recent developments. *Annalen Phys.*, 15, 129–148.

Kiefer, C. (2014). Conceptual problems in quantum gravity and quantum cosmology. *ISRN Math. Phys.*, 2013, 509316.

Kopietz, P., Bartosch, L. and Schütz, F. (2010). *Introduction to the Functional Renormalization Group*. Berlin: Springer.

Koslowski, T. (2007). Dynamical quantum geometry (DQG programme). arXiv:0709.3465 [gr-qc].

Koslowski, T. and Sahlmann, H. (2012). Loop quantum gravity vacuum with nondegenerate geometry. *SIGMA*, 8, 026.

Kotecha, I. and Oriti, D. (2018). Statistical equilibrium in quantum gravity: Gibbs states in group field theory. *New J. Phys.*, 20, no. 7, 073009.

Krajewski, T. (2011). Group field theories. *PoS QGQGS*, 2011, 005.

Lahoche, V. and Oriti, D. (2017). Renormalization of a tensorial field theory on the homogeneous space SU(2)/U(1). *J. Phys.*, A50, no. 2, 025201.

Lang, T., Liegener, K. and Thiemann, T. (2017). Hamiltonian renormalisation I: Derivation from Osterwalder-Schrader reconstruction. *Class. Quant. Grav.*, 35, no. 24, 245011.

Lizzi, F. (2008). The structure of spacetime and noncommutative geometry. arXiv:0811.0268 [hep-th].

Majid, S. (2006). Algebraic approach to quantum gravity II: Noncommutative spacetime. Pages 466–492 of: Oriti, D. (ed.). *Approaches to Quantum Gravity*. Cambridge: Cambridge University Press.

Montesinos, M. and Rovelli, C. (2001). Statistical mechanics of generally covariant quantum theories: A Boltzmann-like approach. *Class. Quant. Grav.*, 18, no. 3.

Niedermaier, M. and Reuter, M. (2006). The asymptotic safety scenario in quantum gravity. *Living Rev. Rel.*, 9, 5.

Norton, J. (2003). General covariance, gauge theories and the Kretschmann objection. Pages 110–123 of: Brading, K. and Castellani, E. (eds.). *Symmetries in Physics: Philosophical Reflections*. Cambridge: Cambridge University Press.

Oriti, D. (2007). Group field theory as the microscopic description of the quantum spacetime fluid: A new perspective on the continuum in quantum gravity. *PoS QG-PH*, 030.

Oriti, D. (2012). The microscopic dynamics of quantum space as a group field theory. In: Ellis, G., Murugan, J. and Weltman A. (eds.). *Foundations of Space and Time*. Cambridge: Cambridge University Press.

Oriti, D. (2014). Disappearance and emergence of space and time in quantum gravity. *Stud. Hist. Philos. Mod. Phys.*, 46, 186–199.

Oriti, D. (2017a). Group field theory and loop quantum gravity. Pages 125–151 of: Ashtekar, A. and Pullin, J. (eds.). *Loop Quantum Gravity*. Singapore: World Scientific.

Oriti, D. (2017b). The universe as a quantum gravity condensate. *C. R. Phys.*, 18, 235–245.

Oriti, D. (2017c). Spacetime as a quantum many-body system. arXiv:1710.02807 [gr-qc]

Oriti, D., Pranzetti, D. and Sindoni, L. (2016a). Horizon entropy from quantum gravity condensates. *Phys. Rev. Lett.*, 116, no. 21, 211301.

Oriti, D., Pranzetti, D. and Sindoni, L. (2018). Black holes as quantum gravity condensates. *Phys. Rev. D*, 97, no. 6, 066017.

Oriti, D., Sindoni, L. and Wilson-Ewing, E. (2016b). Emergent Friedmann dynamics with a quantum bounce from quantum gravity condensates. *Class. Quant. Grav.*, 33, no. 22, 224001.

Padmanabhan, T. (2015). Gravity and/is thermodynamics. *Curr. Sci.*, 109, 2236–2242.

Perez, A. (2013). The spin foam approach to quantum gravity. *Liv. Rev. Rel.*, 16, 3.

Pithis, A. and Sakellariadou, M. (2017). Relational evolution of effectively interacting GFT quantum gravity condensates. *Phys. Rev.*, D95, no. 6, 064004.

Polchinski, J. (1998). *String Theory*. Cambridge: Cambridge University Press.

Pooley, O. (2010). Substantive general covariance: Another decade of dispute. Pages 197–209 of: Suarez, M., Dorato, M. and Redei M. (eds.). *EPSA Philosophical Issues in the Sciences*. Berlin: Springer.

Ramallo, A. (2015). Introduction to the AdS/CFT correspondence. *Springer Proc. Phys.*, 161, 411.

Riello, A. (2013). Self-energy of the Lorentzian EPRL-FK spin foam model of quantum gravity. *Phys. Rev.*, D88, 2, 024011.

Rivasseau, V. (2014a). *From Perturbative to Constructive Renormalization*. Princeton, NJ: Princeton University Press.

Rivasseau, V. (2014b). The tensor theory space. *Fortsch. Phys.*, 62, 835–840.

Rivasseau, V. (2016). Random tensors and quantum gravity. *SIGMA*, 12, 69.

Rovelli, C. (2013). General relativistic statistical mechanics. *Phys. Rev. D*, 87, 084055.

Rovelli, C. (2018). Space and time in loop quantum gravity. arXiv:1802.02382 [gr-qc].

Seiberg, N. (2006). Emergent spacetime. arXiv:hep-th/0601234.

Stachel, J. (2003). A brief history of space-time. Pages 15–34 of: Ciufolini, I., Dominici, D. and Lusanna, L. (eds.). *2001: A Relativistic Spacetime Odyssey*. Singapore: World Scientific.

Van Raamsdonk, M. (2010). Building up spacetime with quantum entanglement. *Gen. Rel. Grav.*, 42, 2323.

Wüthrich, C. (2014). Raiders of the lost spacetime. *Einstein Stud.*, 13, 297.

3

Emergence of Time in Loop Quantum Gravity

SUDDHASATTWA BRAHMA

3.1 Introduction

It is not difficult to imagine a mind to which the sequence of things happens not in space but only in time like the sequence of notes in music. For such a mind such conception of reality is akin to the musical reality in which Pythagorean geometry can have no meaning.

—Tagore to Einstein, 1920

We are yet to come up with a formal theory of quantum gravity that is mathematically consistent and allows us to draw phenomenological predictions from it. Yet, there are widespread beliefs among physicists working in fundamental theory regarding some aspects of such a theory, once realized. These premonitions about the final form of a quantum gravitational theory comes from two somewhat mutually exclusive ideas. First, experience in dealing with other fundamental forces of nature and their quantization, which have resulted in the Standard Model of particle physics, has led to certain expectations regarding the outcome of quantizing gravity. This is, of course, quite natural and what one expects to happen. The other major source of prejudice, however, comes from the rather maverick nature of gravity as described through its classical theory—general relativity (GR). The foundational idea of relativity that *gravity is geometry* requires a rather careful handling in the way one approaches to 'quantize' such a theory. It has often been said that GR is one of the most beautiful theories in physics; and like any beautiful object, its significance lies in the eyes of the beholder. Let us give an example to illustrate this point.

Particle physicists considered the primary difficulty of quantizing gravity to be the ultraviolet (UV) divergences that cannot be canceled by a finite number of counterterms, as is done for every other fundamental force. This is what is commonly known as the *non-renormalizability* of gravity (Collins, 1986). Thus they went looking for a more suitable way to describe the dynamics of a massless, spin-2 particle—namely, the graviton—just as for other such gauge bosons in the Standard Model. Broadly speaking, this search has culminated in the foundation of string theory (Polchinski, 2007a, 2007b; Green, Schwarz, and Witten 2012), which describes not only gravity but also all the other fundamental forces through the spectra of string oscillations, as this fundamental object (the string) travels through a d-dimensional spacetime.[1] Implementing the excitations as extended objects (strings) with nonlocal interactions, one gets rid of the infinities, the unwieldy mathematical objects standard field theory is beset with. Without going through the extremely rich history of the development of string theory in detail, it is safe to say that one of the most ambitious goals for a string theorist is to *unify* all the fundamental forces of nature via one common framework. In the more recent past, string theory has also led to rather amazing mathematical dualities whereby one is allowed to study the quantum gravity theory in a bulk spacetime by simply examining a quantum field theory, without gravity, on its boundary (Becker, Becker, and Schwarz, 2006). This remarkable feature, termed the 'gauge/gravity duality', or more generally the 'holographic principle', can be looked upon as one of the most beautiful artifacts of quantum gravity, arising from underlying mathematical symmetries.

On the other hand, relativists[2] take the idea of *background independence* as a guiding principle for quantizing gravity rather seriously (Thiemann, 2008). The idea here is that one cannot quantize the graviton as a particle moving on a specified spacetime, as is done for the other fundamental gauge fields in nature. Rather gravity supplies us with *both* the stage as well as the actors on it: the background spacetime as well as the particle mediating gravity has to emerge from the fundamental theory (Ashtekar and Lewandowski, 2004). This beautiful idea led to several theories that do not assume a fixed form of a background spacetime, most notably in loop quantum gravity (LQG). Under the rather minimal requirement of a differential manifold, a suitable reformulation of GR is proposed such that the fields describing it do not, a priori, require a background metric. This does not imply that there is no background independence in string theory; it is, however, implemented in a much more indirect manner and is in a rather nascent stage at the moment (for instance in string field theory [Rastelli, 2005]). It is important to emphasize that background independence, as a motivation, is not a monopoly of LQG, but there are other approaches, such as asymptotic safety or causal dynamical triangulations, which implement this feature in different ways. I mention this example to cite what was at the heart of looking for an alternative to string theory. Similarly, although

---

[1] Historically, string theory did come about to describe strong interactions before one found a more suitable candidate for the latter in quantum chromodynamics.

[2] Obviously, both these classifications of particle physicists and relativists are made rather loosely in order to make a larger point.

direct manifestations of the 'gauge/gravity duality' is yet to be revealed in LQG, there is nothing to suggest that such a principle is lacking in the theory[3] (see for instance, Han and Hung [2017]).

So we find that two of the major approaches to quantum gravity not only do not share their technical and conceptual foundations but even diverge in their ambitions for the resulting theory. And, of course, the diversity in research for a quantum theory of gravity is also rather rich with many of the other approaches (not mentioned here) bringing distinct aspects of gravity into play. How can such ideas converge in their expectations from the final version of their individual quantum gravity theories? In order to understand this, let us first recall what quantizing the other fundamental forces has taught us. Quantum theories, typically, help us in resolving singularities which appear in their classical counterparts. This general expectation, as can be understood from its origin, is shared by all theories of quantum gravity. Since some well-defined physical quantity, such as energy density, diverges at the singular point, one concludes that some of the laws of physics have broken down at the singularity and we require a new set of rules to describe the dynamics in those regimes. Let us recourse to a (perhaps over-)simplified example to make our case. Hooke's law, describing the force exerted by a spring, can be written as $k = F/x$. This implies if we are to calculate the tension in the spring between two infinitesimally close points, then it would indeed be infinite! But this is not a true infinity but rather only a mathematical manifestation of the fact that we have applied the classical Hooke's law beyond its realm of validity. This can be seen if one recalls that the spring is built out of atoms, and once we consider subatomic lengths, laws of atomic physics has to come in to replace the classical Hooke's law. Thus the fundamental minimum distance scale of atomic spacings (which is something that can be derived and is not postulated, from the Schrödinger equation governing the quantum theory) comes in to save the day. A similar thing is expected to happen for the (curvature) singularities found in GR (Hawking and Ellis, 2011), most famously the initial singularity in cosmology termed the big bang and the one found inside the core of black holes. Any quantum gravitational theory attempts to bring in a new set of laws that would cure the theory of such (classical) pathologies.

Next we have to deal with the more subtle point of the expectation of a non-classical, fuzzy nature of quantum spacetimes, which is also expected by and large from almost all theories of quantum gravity.[4] This is a more surprising common factor between quantum gravity theories since it is realized through

---

[3] However, a crucial difference between the two theories remain in that LQG is less ambitious than string theory in not attempting an unification of all the known forces in nature.

[4] One notable exception is the asymptotic safety program (Reuter and Saueressig, 2012).

distinct mathematical procedures having origin in the different theories. Almost all *background independent* quantizations of gravity talk of a discrete, granular description of spacetime at some fundamental level. The guiding principle for these theories is that spacetime, at the most primordial level, is an irregular substratum made up of the fundamental building blocks, or the *atoms of spacetime*. Naturally, the explicit form of these elementary degrees of freedom depend on the specific framework one deals with,[5] but, on the whole, one loses the smooth, continuous description of classical backgrounds geometries at sub-Planckian scales. On the other hand, although string theory is at least formulated over smooth manifolds, there is a growing consensus that spacetime in this case (and may indeed be for any consistent theory of gravity) is built out of quantum entanglement (for instance, see Lin et al. [2015]). Even speculative proposals that string theory is a coarse-grained approximation of a more statistical phenomenon of microscopic degrees of freedom, with gravity being an emergent interaction, have recently surfaced (Verlinde, 2011). The bottom line is that all of this points toward a description of gravity at fundamental scales that is quite exotic compared to its classical counterpart as a smooth spacetime continuum.

In this essay, I aim to use the input of singularity resolution from LQG and show that demanding mathematical consistency leads to the emergence of non-Lorentzian geometry in the theory. Specifically, my goal is to show that time' is an emergent concept in LQG, in the sense that one effectively transitions from a 4-dimensional fuzzy' Euclidean space to the usual $(3 + 1)$-dimensional Lorentzian spacetime in high curvature regimes. Although this shall be demonstrated only for symmetry-reduced systems in LQG, care needs to be taken that they are not oversimplified toy models, bland enough to miss key subtleties of the full theory. I emphasize that my goal is not to comment on the robustness of singularity resolution in LQG, which has been established to a great extent in the existing literature (for instance, see Ashtekar, 2009; Ashtekar, Corichi, and Singh, 2003; Bojowald, 2001; Gambini and Pullin, 2013) but rather to extract only a crucial ingredient commonly required for tackling singularities in the theory. The other input to be assumed is that the theory avoids quantum gravitational anomalies and satisfies a well-defined notion of covariance (adapted to a canonical setup), which is a strict requirement, not just of LQG, but any consistent quantum gravity theory.[6] These two minimal requirements, when combined, would be shown to give rise to a rather remarkable change in our understanding of the underlying spacetime, leading to emergence of time in LQG.

---

[5] In LQG, for instance, one has discrete spin-network states giving rise to a discrete area and volume spectra (Thiemann, 2008).

[6] What I mean precisely by this shall be spelled out more explicitly in a later section but it is important to note that this is different from the idea of Lorentz covariance for a flat, Minkowski spacetime.

3.2 Singularity Resolution in LQG

Setting aside the mathematical beauty of GR, it has also been an extremely successful theory and has passed all proposed experimental tests with flying colors thus far. Yet, a more fundamental theory is required that incorporates not only the dynamical nature of geometry but also the features of quantum physics. Indeed, a brief glance at Einstein's equations, $G_{\mu\nu} = (8\pi G/3)T_{\mu\nu}$, is enough to convince one of that. Since the right-hand side of the equation consists of quantum matter described by the Standard Model, the left-hand side must be replaced by some suitable quantum version of gravity. The more obvious reason for invoking a quantum gravity theory obviously arises when dealing with curvature singularities contained in solutions to Einstein's equations, where the equations themselves fail. This is a more serious problem than the divergences that arise in other gauge theories describing, for instance, the electromagnetic field. In Minkowskian physics, if such a field becomes singular at a given point, it does not affect the underlying spacetime and, consequently, has no effect on the evolution of the other fields in the theory. However, since *gravity is geometry* according to GR, when the gravitational field becomes singular, the picture of the continuum spacetime itself breaks down and all of physics gets disrupted. Now, we no longer have a stage on which to describe the dynamics of the other matter or gauge fields. This points toward the fact that our picture of gravity as a smooth continuum of spacetime must itself not be applicable to points arbitrarily close to such a singularity and what we require is a quantum theory of geometry (see, for instance, Thiemann [2008] and Bojowald [2010] for details).

Let us focus on the simple example of the $k = 0$, Friedmann–Lemaitre–Robertson–Walker (FLRW) geometry to understand how singularity-resolution takes place in LQG (Ashtekar and Singh, 2011; Bojowald, 2013). (In this context, we would refer to this theory as loop quantum cosmology [LQC] rather than a coarse-graining of the cosmological degrees of freedom from the full quantum theory.) The requirements of consistency from LQC are twofold: on one hand, the quantum theory needs to have a regular behavior for the quantum state in the higher curvature regime and resolve the big bang singularity. On the other hand, these quantum corrections, which are strong enough to overcome the gravitational collapse due to Planck scale energy densities, have to die off extremely fast beyond the Planck scale so that we recover the well-known behavior of GR. Both these conditions are met by LQC and thus we have a well-defined quantum cosmological model. Inside the deep quantum regime, the density and curvature reach a maximum value, as opposed to increasing indefinitely as in the classical theory due to novel *quantum geometry effects*. This might be interpreted as an 'effective' repulsive force that overcomes the classical gravitational attraction and thus resolves the singularity by preventing the quantum evolution from going to the singular point. In the full quantum theory, one can show that singularity-resolution might be seen as the zero volume state getting decoupled from the dynamics of the quantum state corresponding to the universe. Although the main achievement of LQC is to incorporate novel quantum geometry corrections, heuristically adapted from LQG in the form of a minimum non-zero value of area (or the area-gap) in the left-hand side of Einstein's equations, we can shift the correction to the right-hand side to facilitate comparison with the standard Friedmann equation, which now reads:

$$H^2 = \frac{8\pi G}{3}\rho \left(1 - \frac{\rho}{\rho_c}\right), \tag{3.1}$$

where H is the Hubble parameter and ρ is the energy density, which gets an upper bound (the critical density, ρ_c, which is proportional to M_{Pl}^2) from these quantum effects.

Although we describe the broad picture of singularity-resolution in the specific case of LQC described earlier, the mechanism remains the same for other models where it has been possible to do the same. Specifically, singularity resolution has been achieved for a variety of cosmological models (Ashtekar, Corichi, and Singh, 2008; Ashtekar and Singh, 2011), such as the Bianchi models, FLRW geometries with cosmological constants, Gowdy systems with additional symmetries (de Blas, Olmedo, and Pawlowski, 2015) and for open or closed universes, in addition to the flat case mentioned earlier. Recently, similar techniques have been applied to black hole solutions, such as Schwarzschild (Gambini and Pullin, 2013) or Callan–Giddings–Harvey–Strominger (CGHS) models (Corichi, Olmedo, and Rastgoo, 2016), with the remarkable success of resolving the singularity inside the cores of such black holes much as in the manner described. However, for my purposes, I do not need to go into the full quantum theory of any of these cases in detail. Instead, I focus on the most crucial ingredient in LQG, which leads to such singularity resolution across a variety of different models.

In LQG, we have well-defined operators for holonomies (or parallel transports of connections) on the kinematical Hilbert space but not for the connections themselves.[7] Thus one requires a different way of expressing connections on the Hilbert space via their holonomies. Curvature of the connection, which shows up in the Hamiltonian of the theory, has to be regularized in terms of these holonomies. Indeed, a direct consequence of this can be seen in the minisuperspace quantization of LQC. The Wheeler–DeWitt (WDW) theory in this case, which arises from the Schrödinger quantization of the quantum mechanical phase space, is known to suffer from the same pathologies as the classical solution as far as the big

---

[7] Mathematically, this implies that the holonomy operators are not *weakly continuous*.

bang is concerned. On the other hand, the Stone—von Neumann uniqueness theorem guarantees that the quantum kinematics for a finite dimensional phase space isomorphic to $\mathbb{R}^n$ (such as the quantum cosmological model) is unique. The way LQC can go past this obstruction is precisely via violating the assumptions of weak-continuity from the von Neumann theorem and thus, essentially, giving rise to a new form of quantum mechanics. In effect, all of this mathematical rigor of working with weakly continuous holonomies can be pinned down to give one effect: the inclusion of *holonomy corrections* in the theory.

In an effective theory of a loop-quantized model, the idea is to replace the connection (or, in practice, extrinsic curvature components) by a polymerization function in the Hamiltonian of the theory (Bojowald, 2012). This polymerization function is obviously not chosen ad hoc, but rather through a rigorous regularization of the curvature components in a specified representation of the internal gauge group (SU(2)), along with inputs of the area-gap from the full theory. Typically, the holonomies are calculated in the fundamental spin-(1/2) representation of SU(2) and the polymerization function, for some extrinsic curvature component K, takes the form:

$$K \rightarrow \frac{\sin(\delta K)}{\delta}, \tag{3.2}$$

where δ is related to the Planck length in a specified manner. This kind of nonperturbative correction, coming from LQG, is commonly known as holonomy correction (or equivalently, modification) function. Although the sine function has been obtained explicitly in the case of LQC, for the spin-(1/2) representation, we can allow for more general quantization ambiguities and replace connection components by any local and *bounded* function of it. This type of correction plays the most crucial role in singularity resolution and the bounded nature of the function comes from this requirement. It is important to emphasize that LQG does *not* resolve singularities by incorporating some arbitrary bounded functions of the connection, but rather *derives* them in some symmetry-reduced models. For the purposes of this essay, I simply choose to not work with a specific form of this function but rather keep it arbitrary, thus demonstrating that my conclusions are not tied to some specific fine-tunings arising from a particular model in LQG but are rather general in their ambit. Therefore, the first input for us is going to be incorporating such holonomy correction functions which lead to singularity resolution (and is thus bounded) $K \rightarrow f(K)$.

Before concluding this section, we need to make a comment about why this function has to be chosen to be local. It is tied to the same reason that we are yet to have a consistent formulation of any canonical quantum gravity theory in $(3 + 1)$-dimensions. It is only known how one can loop-quantize simpler, symmetry-reduced models and, only in such cases, can singularity resolution be demonstrated manifestly. In LQG, one is yet to understand how to explicitly evaluate holonomy operators in inhomogeneous directions, along edges of spin-network states. Although the full technical details of this problem is far beyond the scope of this essay, I shall revisit this issue in the simplified arena of Schwarzschild spacetimes.

3.3 Covariance in Canonical Quantum Gravity

It is well known that coordinate freedom is one of the stepping-stones of GR (and, indeed, of all general covariant theories going beyond GR), i.e., the theory remains invariant under local diffeomorphisms $x^\mu \rightarrow x^\mu + \xi^\mu(x)$. LQG is based on a Hamiltonian formulation of gravity, where spacetime is foliated into spatial hypersurfaces evolving along a time parameter. However, in canonical theories, coordinate transformations of spacetime tensors are replaced by gauge transformations that generate deformations of the spatial hypersurfaces in full spacetime. The Hamiltonian (or scalar) constraint is defined as the one that generates deformations along the direction normal to the hypersurface while the (spatial) diffeomorphism constraint generates those along tangential directions. In generally covariant theories, these four smeared constraints (per point) satisfy the (Dirac) hypersurface deformation algebra.[8] These constraints satisfy a first-class system forming a closed algebra, which implies that the Poisson brackets of the constraints vanish on the constraint surface (see Bojowald [2010], for details).

The phase space of canonical gravity is formed by the metric on the spatial hypersurface q_{ab} and its conjugate momenta proportional to the extrinsic curvature components K_{ab}. The full spacetime metric is typically parametrized by a lapse function M, the shift vector fields N^a and the spatial metric q_{ab}. The arbitrary function M labels the Hamiltonian constraint while N^a smears out the diffeomorphism constraint. The crucial observation is that the symmetry deforming the spatial hypersurfaces, tangentially along N^a and along the vector Mn^μ (n^μ being the unit normal to the hypersurface), are equivalent to Lie derivatives along spacetime vector fields therefore representing coordinate freedom. Gauge-covariance under hypersurface deformations (equivalent to the underlying spacetime diffeomorphisms) is ensured by the following algebra:

$$\{D[N_1^a], D[N_2^a]\} = D[\mathcal{L}_{N_1} N_2^a] \tag{3.3}$$

[8] Technically, it forms a Lie algebroid and not an algebra, a fact I use in the next section. However, in the meantime, I shall keep referring to it as an algebra as is common in the physics literature.

$$\{H\left[M\right], D\left[N^a\right]\} = -H\left[\mathcal{L}_N M\right] \tag{3.4}$$

$$\{H\left[M_1\right], H\left[M_2\right]\} = D\left[q^{ab}\left(M_1 \nabla_b M_2 - M_2 \nabla_b M_1\right)\right]. \tag{3.5}$$

Equations (3.3) and (3.4) demonstrate the action of infinitesimal spatial diffeomorphisms with the right-hand side given by the Lie derivatives $\mathcal{L}_{N_1} N_2^a = [N_1, N_2]^a$ and $\mathcal{L}_N M = N^a \partial_a M$. Equation (3.5) complicates the intuitive geometric picturization due to the appearance of structure functions, in the form of the inverse of the spatial metric. The Hamiltonian constraint does indeed provide the time reparametrizations required to supplement the spatial diffeomorphisms generated by $D[N^a]$ to form the full spacetime diffeomorphism symmetry of gravity, but only on the constraint surface.

It is important to emphasize that covariance, in the canonical context, is an off-shell property. Fields that do not satisfy the constraint equations $D[N^a] = 0$ and $H[M] = 0$, as derived from GR, still must be well behaved so as to not violate this algebra. This is in keeping with our understanding of 4-dimensional symmetries from Lagrangian theories, where to write down a covariant action one only needs to define a Lorentz-invariant measure and contract indices properly with *any metric*, without paying attention to whether they are solutions to Einstein's equations. The specific nature of the algebra of the constraints has to be satisfied by the constraint functions on the whole phase space and not just the constraint surface. Its form dictates the kinds of gauge transformations the constraints generate and how they are related to spacetime properties. All of this is, however, well understood in the classical formulation of GR as a canonical theory. An open question, recently addressed, is regarding the fate of such symmetries once quantum modifications from a particular theory are taken into account.

We first postulate the requirements for a canonical quantum theory of gravity to be covariant (Bojowald and Brahma, 2015; Bojowald, Brahma, and Reyes, 2015):

1. The classical Hamiltonian and diffeomorphism constraint, on including quantum corrections, must still act as generators that form a closed algebra free of quantum anomalies.
2. The quantum algebra of the new generators must have a classical limit that is identical to the classical hypersurface deformation one as defined in Equations (3.3)–(3.5).

The first condition simply ensures that we do not violate the gauge symmetries of gravity in the quantum theory. If the quantum constraint operators (or equivalently, effective versions of them including quantum corrections) still form a first-class system then we have the same number of gauge conditions required to eliminate spurious degrees of freedom as in the classical case. Thus the gauge generators, in the quantum theory, would also lead to the same dimension of the solution space and thereby avoid gauge anomalies. Since the absence of anomalies is the requirement for any consistent quantum (gauge) theory, condition (1) is the corresponding one for a quantum version of gravity viewed as a gauge theory. The second condition ensures that what we end up with, the quantum gravity theory, is a consistent quantum theory of spacetime. This is an even stricter requirement since this ensures that we end up with a theory that gives the correct spacetime structure in the classical regime. While the first condition deals alone with the important issue of anomaly freedom in quantum gravity theories, both of them together ensure that one ends up with a covariant quantization of gravity. It is conventional to assume that once one can get the quantum corrected brackets to have a closed algebra, then the quantization is covariant. What has been recently demonstrated (Bojowald, Brahma, and Reyes, 2015) is that a *background-independent* quantization requires not only anomaly freedom but also a well-defined classical limit so that the quantum theory is indeed one of spacetime. This is indeed the difference of full quantum gravity theory from a theory of matter on a quantum-modified spacetime. The latter, by itself, has severe requirements that the gravitational and the matter form of the constraints close to form first-class algebras, whereas the former requires not only this, but, in addition, also that they have a *matching version* of covariance (i.e., the two first-class algebras of the matter and gravitational constraints have the same form). Naturally, this has severe ramifications for the way one can covariantly couple matter to such quantum theories, which is beyond the scope of this essay. As a final reminder, I stress that our key requirement is *not* that the theory remains Lorentz invariant, which turns out to be a deformed symmetry in this context (Brahma et al., 2017), but simply that the gravitational gauge conditions are not violated.

I end this section with an important caveat: one cannot see the effects of covariance in a minisuperspace model. One deals with homogeneous degrees of freedom alone in such systems and thus the diffeomorphism constraint is trivially satisfied for such models. Covariance cannot be addressed in such models because, owing to lack of inhomogeneities, the relationship between both temporal and spatial variations of fields are absent in them. Mathematically, since the diffeomorphism constraint is trivially zero, one cannot examine any behavior of the constraint algebra in such settings. As a consequence, no restrictions on the dynamical equations of wave-functions can be imposed by demanding covariance in such models and any putative quantum effect can be included at will. However, such arbitrary quantum modifications would result in a minisuperspace model that cannot be embedded within a larger covariant full theory. Thus we shall have to deal with models that have at least one inhomogeneous component. In fact, *all known models* of LQG that resolve classical singularities (and are not completely homogeneous) are of this form. Thus we shall hereafter restrict ourselves to such midisuperspace models.

3.4 Emergence of Time in LQG

We have all our ingredients and are ready to state the main result. From Section 3.2, we infer that singularity resolution in LQG typically introduces a bounded function for extrinsic curvature components. We keep our assumptions to a minimum and do not even fix a specific form for the holonomy correction function. This kind of a general ansatz, requiring only that the correction function is bounded, then suggests that we are gathering the least information necessary from the particular theory (LQG) and keeping ample room for improvements to the quantization scheme within it. Our second requirement, following Section 3.3, is the necessary condition that the quantization procedure is spacetime covariant, in a well-defined sense. Based on these two rather general assumptions, we shall show that there is a remarkable deformation to the underlying spacetime structure for such quantum modifications.

3.4.1 Explicit Example: Spherically Symmetric Gravity

It is easiest to first concentrate on a concrete example to illustrate my point and then state the general result for other such systems. For spherically symmetric gravity adapted to Ashtekar–Barbero variables (Bojowald and Swiderski, 2006), one has a 2-dimensional phase space spanned by the triad variables E^x and E^ϕ in the radial and angular directions respectively, along with their canonically conjugate extrinsic curvature components K_x and K_ϕ. There is only one nontrivial component of the diffeomorphism constraint, given by:

$$D[N^x] = \frac{1}{G} \int dx \, N^x(x) \left(-\frac{1}{2}(E^x)' K_x + K_\phi' E^\phi\right), \qquad (3.6)$$

while the gravitational spherically symmetric Hamiltonian constraint takes the form:

$$H[M] = -\frac{1}{2G} \int dx \, M(x) \left(|E^x|^{-\frac{1}{2}} E^\phi K_\phi^2 + 2|E^x|^{\frac{1}{2}} K_\phi K_x\right.$$
$$\left. + |E^x|^{-\frac{1}{2}}(1 - \Gamma_\phi^2) E^\phi + 2\Gamma_\phi' |E^x|^{\frac{1}{2}}\right), \qquad (3.7)$$

with the spin-connection given by $\Gamma_\phi = -(E^x)'/2E^\phi$. The prime denotes a derivative with respect to the radial coordinate. We have suppressed the dependence of the field variables on the radial coordinate in the above expressions (due to spherical symmetry, we can safely integrate out their dependence on the angular coordinates). Thus we have a system of 2-dimensional phase space with 2 first-class constraints, thus resulting in no local physical degrees of freedom (the only global degree of freedom is the ADM mass of the system). The usual spatial metric of the system can be written in terms of the Ashtekar–Barbero variables as:

$$\mathrm{d}q^2 = \frac{\left(E^\phi\right)^2}{|E^x|}\, \mathrm{d}x^2 + |E^x|\left(\mathrm{d}\theta^2 + \sin^2\theta \mathrm{d}\varphi^2\right). \tag{3.8}$$

The classical constraint algebra can be easily evaluated using the Poisson structure of the phase space:

$$\{K_x(x),\, E^x(y)\} = 2G\delta(x,y), \qquad \{K_\phi(x),\, E^\phi(y)\} = G\delta(x,y). \tag{3.9}$$

The only quantity of the inverse spatial metric that shows up as a structure function in the algebra is $q^{xx} = |E^x|/(E^\phi)^2$ since the only derivative that gives a non-zero result is along the radial coordinate.

We now proceed with our program of replacing the (angular) extrinsic curvature components with local functions of itself, i.e., $K_\phi \rightarrow f\left(K_\phi\right)$ in the Hamiltonian constraint. As mentioned before, I do not consider holonomies corresponding to the K_x variables since they are calculated along the edges of a (1-dimensional) spin-network and are, therefore, nonlocal in nature and difficult to implement (Bojowald et al., 2014). However, it has been shown that one can suitably reformulate the constraints to eliminate the K_x variable from the Hamiltonian constraint altogether and have only the K_ϕ one (Gambini and Pullin, 2013). Thus we can consider the point-wise holonomy operators corresponding to the K_ϕ component alone, which act locally at the nodes of the spin networks, without loss of generality. With these modifications, the effective Hamiltonian constraint takes the form (Bojowald et al., 2015; Brahma, 2015):

$$H[N] = -\frac{1}{2G} \int \mathrm{d}x\, N(x) \left(|E^x|^{-\frac{1}{2}} E^\varphi f_1\left(K_\varphi\right) + 2|E^x|^{\frac{1}{2}} f_2\left(K_\varphi\right) K_x \right.$$
$$\left. + |E^x|^{-\frac{1}{2}}(1 - \Gamma_\varphi^2)E^\varphi + 2\Gamma_\varphi'|E^x|^{\frac{1}{2}}\right). \tag{3.10}$$

Note that we do not replace both the instances where K_ϕ shows up in (Equation 3.7) with the same function but allow for even more generalities by plugging in different correction functions f_1 and f_2 (the classical expressions are recovered for $f_1(K_\phi) = K_\phi^2$ and $f_2(K_\phi) = K_\phi$. However, we keep our diffeomorphism constraint unmodified from the classical case. The reason for this is that, in LQG, one does not have an infinitesimal quantum operator generating spatial diffeomorphisms represented on spin network states but rather spatial diffeomorphism invariance is implemented in the full quantum theory through finite unitary transformations. (See Laddha and Varadarajan [2011], and references therein for an attempt to define an infinitesimal diffeomorphism constraint in LQG.) The Hamiltonian constraint, on the other hand, acts infinitesimally only on diff-invariant spin network states in

LQG. We are interested in questions regarding the covariance of the theory and thus are interested in the off-shell structure of the constraint algebra, which should be a largely representation-independent question and not depend specifically on the spin-network states, the latter being after all only a choice of basis. In other words, we assume that the flow of the diffeomorphism constraint is not crucially different from the classical one as is done in all models of LQG, which achieve singularity-resolution.[9]

We are now ready to calculate the brackets between the quantum-corrected constraints to find the resulting form of the constraint algebra. The algebra of basic variables is now modified due to the inclusion of holonomy corrections for the angular connection component:

$$\{f(K_\phi)(x), E^\phi(y)\} = G\left(\frac{\mathrm{d}f}{\mathrm{d}K_\phi}(x)\right)\delta(x, y). \tag{3.11}$$

There is a priori no reason to assume that the quantum-corrected constraints would even form a closed algebra and thereby satisfy the requirement of anomaly freedom. However, it turns out that the calculation of the brackets reveals two things. First, we find a condition relating the two arbitrary holonomy modification functions and hence they are not both independent. This condition is a requirement for anomaly freedom, which gives (Bojowald et al., 2015; Brahma, 2015):

$$f_2(K_\phi) = \frac{1}{2}\frac{\mathrm{d}f_1}{\mathrm{d}K_\phi}. \tag{3.12}$$

More importantly, we find that the full algebra of the quantum-corrected constraints stays the same except for the bracket between two Hamiltonian constraints, which takes the form (Bojowald and Paily, 2013; Bojowald et al., 2015; Brahma, 2015):

$$\{H[M_1], H[M_2]\} = D\left[\beta\, q^{xx}\left(M_1\partial_x M_2 - M_2\partial_x M_1\right)\right], \tag{3.13}$$

where the classical structure function, q^{xx}, is modified by the factor:

$$\beta = \frac{1}{2}\frac{\mathrm{d}^2 f_1}{\mathrm{d}K_\phi^2}. \tag{3.14}$$

We make the following observations regarding this deformation of the quantum constraint algebra.

1. Since we kept both the holonomy modification functions arbitrary, there was no guarantee that the algebra would close. However, we find that it is not only anomaly free but closure also requires that one of the function be expressed in terms of the other.

---

[9] This presupposes that 3-dimensional space, but not necessarily spacetime, retains some features of the classical structure.

2. When we take the classical limit of the holonomy correction function f_1, we see that β goes to 1, reproducing the familiar hypersurface deformation algebra. This remarkable result satisfies our condition for a covariant quantization.

3. Classically, the curvature component can increase infinitely, which in turn implies that the energy density gets infinite at the classical singularity. However, due to the inclusion of the holonomy modification from LQG, the function f_1 reaches a maximum, and we avoid the classical singularity. The deformation function β *necessarily* turns negative when the holonomy modification function *precisely* reaches its maximum. This is because β is the second-order derivative of a function that reaches its maxima. This simple, yet striking, feature of the deformation function has long-ranging ramifications.

Once the deformation function changes sign, the bracket between two normal deformations[10] in the modified constraint algebra has the same sign as one gets from Euclidean gravity. This is referred to as 'signature change' in the literature since, although we start from the usual assumptions of a Lorentzian spacetime, due to quantum corrections from LQG, we end up in an Euclidean four-space in the deep quantum regime. In order to gain some physical intuition about it, in models of quantum cosmology, it has been shown that signature change occurs when the energy density is half the maximum of what it can achieve ($\rho_c/2$ of Equation 3.1). As one comes to this point, tracing backward from the well-understood realm of classical cosmology, one enters this realm, thereby avoiding the classical singularity but ending up in a fuzzy Euclidean regime. Mathematically, the equations of motion turn elliptic for the perturbation modes for 'times'[11] less than the critical time mentioned earlier.

This implies radical new ideas regarding the nature of underlying quantum spacetime and a new perspective of its geometrical structure above Planck scales. At this point, it is sufficient to point out that we have successfully derived a *deformed* notion of covariance for the quantized system, whereby one modifies gauge transformations by quantum corrections but does not allow them to be violated. As promised, I introduced only the crucial feature of bounded holonomy operators from LQG to achieve this goal. As a consequence of this, time appears as an emergent parameter below some energy scale during our effective transition from an Euclidean to the Lorentzian phase. Although all the calculations shown are effective constraints for simplicity, i.e., we do not take explicit quantum operators

---

[10] Throughout this essay, I have been using the word *deformation* to mean two different things, which should be clear from the context in which it is used.

[11] In the Euclidean regime, one no longer has a timelike Killing vector but can still formally associate one of the spatial dimensions as Euclidean time.

corresponding to the constraints into consideration, the same result holds when operator effects are taken into consideration as shown in Brahma (2015). Thus this result is certainly not the manifestation of some semiclassical approximation introduced within this scheme. (It has also been shown that fluctuations and higher moments of the quantum state cannot introduce perturbative loop corrections that can deform the structure functions of the constraint algebra [Bojowald and Brahma, 2016a].)

3.4.2 Ubiquity of Signature Change in LQG

Since we have shown that time emerges at a some particular scale for spherically symmetric gravity when effects required for singularity resolution are taken into account, how can one be sure that this is not due to some additional quantization choices introduced by us? We can answer this question in two ways. First, the only assumption used, as shown in the previous subsection, was that of bounded curvatures, which is at the heart of singularity resolution in LQG. However, although the framework was kept extremely general on purpose, we can do even better to exhibit the robustness of our claims. The same spherically symmetric system, or equivalently the Schwarzschild black hole model, has been quantized based on a completely different, but classically equivalent, set of first-class constraints where the most complicated part of the constraint algebra had been Abelianized (Gambini and Pullin, 2013). In other words instead of working with the familiar $H[M]$ and $D[N^x]$ constraints as introduced in Equations (3.7) and (3.6), the quantization was performed over a newly defined $C[L]$ constraint and the usual diffeomorphism $D[N^x]$ constraint. For this set, first-class algebra has the same form for Equations (3.3) and (3.4), (with H replaced by C), while the last relation is replaced by $\{C[L_1], C[L_2]\} = 0$. Thus we no longer have the structure functions appearing, and we have a true Lie algebra. (This is possible by choosing some judicious linear combination of the old constraints to define the new constraint $C[L]$.) It has been further shown that the Schwarzschild singularity has been resolved for a quantum theory based on these constraints, after polymerizing them according to LQG. One would then imagine that signature change would be impossible for such a system due to the disappearance of the structure functions. However, as explained in Bojowald et al. (2015), requiring that the resulting quantum theory be covariant (as defined in Section 3.3), one can show that not only does the structure functions reappear and get deformed but also we have a signature change for this model. Additionally, it was shown that if the holonomy correction functions were kept arbitrary for this partially Abelianized system, then the restriction (3.12) is the same one that is obtained even for the new system. Since the technical details of the two systems were largely different, except for the choice of bounded functions for curvature components, we conclude that signature change (and consequently, the appearance of time) does not depend on any additional regularization choices made by us in the previous subsection but is a rather general result for the Schwarzschild black hole model in LQG.

However, it might still be that there is some magical coincidence that leads to the fortuitous notion of deformed covariance in the spherically symmetric spacetime and therefore signature change is nothing but a manifestation of the symmetry of the system. To verify that this is not the case, we can go back to the original cosmological setup where singularity was first resolved within LQG. However, as explained before, one cannot ask questions regarding covariance in homogeneous models, the traditional setup for LQC. Fortunately, nature also demands that our universe is not exactly homogeneous but rather has small inhomogeneities that lead to formation of galaxies. We should then loop quantize a system that has perturbative inhomogeneities on top of a homogeneous FLRW background in cosmology to get a more realistic picture. Indeed such a quantum scheme leads to resolution of singularities once again through a nonsingular signature change (see, for instance, Bojowald and Mielczarek [2015] and Cailleteau et al. [2012]). The fundamental setup of these models is quite different from the one considered earlier in that one only introduces holonomy correction functions for the background connection component but leaves classical expression for the perturbative inhomogeneous extrinsic curvatures. Yet, one finds that a bounded version of background curvature component to be a sufficient condition for signature change in these models.

Finally, there is a vast arena of models in LQG that have recently been examined to show that singularity resolution cannot be divorced from signature change for real valued connections (Bojowald and Brahma, 2018, 2017; Brahma et al., 2017). In particular, a class of models shown to manifest signature change would be all 2-dimensional dilaton models, which include the Callan–Giddings–Harvey–Strominger (CGHS) black hole solution. Another example would be the (polarized) Gowdy system with local rotational symmetry. I mention these models in particular since the classical singularity in these models has been shown to be satisfactorily resolved through loop quantization (de Blas et al., 2015; Corichi et al., 2016). It has been demonstrated that for all these models the classical singularity is, in fact, replaced by a nonsingular signature change and thus time appears in all these models below some energy scale. Indeed, this work has been generalized even further to a general midisuperspace model, with one direction of inhomogeneity and no local physical degrees of freedom, to show that even for such a general model (which includes all the given examples and even more class of models with no well-defined classical analogs), holonomy corrections from LQG results in deformation of the structure function in a way that leads to signature change. The robustness of these findings has also been checked against different choices of canonical variables and for different choices of arbitrary dilatonic potential terms. It appears that time is indeed an emergent parameter, at least within the class of models for which one can have singularity resolution in LQG.

3.5 Discussion

Having shown that the same modification that resolves singularity in LQG is also responsible for signature change, at least in all models in which classical singularities can be resolved in LQG, let us discuss some of the aspects of the findings.

3.5.1 Mathematical Basis for Emergence of Time

This essay shows novel features in the background spacetime structure in quantum regimes due to the modification of the classical structure functions appearing in the quantum-corrected constraint algebra. However, it is natural to assume at first glance that one can absorb the deformation factor β in the inverse of the spatial metric to get an 'effective' spatial metric $\tilde{q}^{ab} := \beta q^{ab}$. However, such an effective metric from the relation:

$$\{H[M_1], H[M_2]\} = D\left[\tilde{q}^{ab}(M_1\nabla_b M_2 - M_2\nabla_b M_1)\right], \qquad (3.15)$$

cannot be part of a spacetime metric to form a classical line element of the form $ds^2 = -N^2 dt^2 + \tilde{q}_{ab}(dx^a + N^a dt)(dx^b + N^b dt)$. It is because the modified gauge transformations generated by the quantum corrected constraints for $\tilde{q}_{ab}$ do not match the coordinate transformations of the infinitesimal dx^a. This complication notwithstanding, one might venture to find field redefinitions of the lapse, shift, the original spatial metric and the extrinsic curvature, or a combination thereof, to absorb β and recover the classical hypersurface deformation algebra. In other words, one can try to define new field configurations as combinations of the old ones that would then give rise to the same constraint algebra structure as the classical case. Even if one is successful in finding such transformations, it does not imply that the deformation we find is spurious: It would still affect the equations of motion of particles on these deformed spacetimes. However, it would imply that the background can be treated effectively as a Lorentzian geometry. This *does not* mean that the background spacetime does not pick up quantum corrections but rather that, in spite of such modifications, they retain some notion of classical spacetime structures in that they can still be expressed as Lorentzian manifolds. However, it is notoriously difficult to come up with different canonical transformations to check whether β can be absorbed by one of them and therefore still retain a

Lorentzian geometry. (Even if one cannot find one such transformation for a given system, it does not rule out the possibility of one existing since there are infinite number of transformations that can be applied to define new variables from the old ones.)

Recently, this issue was addressed using a different approach (Bojowald et al., 2016): using the well formulated theory of Lie algebroids since the hypersurface deformation brackets provide an example of one. Indeed, irrespective of the specific choice of the quantum gravity theory, it was possible to classify different inequivalent spacetime structures that cannot be related by algebroid-morphisms. Since I do not want to obfuscate the central findings with more mathematical structures than necessary, I do not reproduce the details of that proof in this essay. However, I shall nevertheless state my main result, as follows. Although it would seem that an arbitrary phase-space deformation function β would imply virtually unrestricted quantum corrections, only $\mathrm{sgn}(\beta)$ remains the unique choice after an equivalence class of algebroids are taken into account. This has two main implications. First, as long as one has a deformation function that does not change sign, it is possible, in principle, to absorb this factor using some transformation and thus has an effective Lorentzian structure. However, once β changes sign, then one cannot absorb it globally, and a new version of quantum spacetime is obtained. In such a case, one has distinct Lorentzian and Euclidean patches, which form nonisomorphic Lie algebroids. This is the mathematical reason behind the emergence of time in such theories. Keep in mind, that this entire mathematical analysis has been done without reference to any specific quantum gravity theory. Our only input from LQG is to provide a quantum modification function, which results in a sign changing β, which triggers signature change.

Let us finally make some speculative comments regarding the emergence of time in these models of LQG. As has been emphasized throughout, for every symmetric model of LQG where one can resolve the classical singularity, one can explicitly prove signature change. These are precisely systems that model real-world cosmology (such as FLRW geometry with small inhomogeneities) or black holes (such as Schwarzschild or CGHS). Let us focus on the cosmological example first. Since the spacetime now changes sign, a natural question to ask is how do the cosmological perturbations propagate on top of such a quantum spacetime. First, in a strict sense, perturbations cannot propagate in the Euclidean phase due to a lack of a sense of time in this phase. However, one might identify one of the spatial directions as the one that undergoes this 'physical' Wick rotation to emerge as a time parameter. In that case, the change (I purposefully avoid using the word *evolution*) of the perturbations may still be calculated with respect to this parameter. However, a better method to think about the cosmological setting may be the following. For actual physical predictions, it might be sufficient to specify some initial value for cosmological perturbations infinitesimally close to the signature-changing hypersurface on the Lorentzian side and then evolve them on to the beginning of the inflationary phase. This way the initial state for inflation can be specified accurately from the underlying quantum gravity theory since there is no conceptual problem in evolving them throughout in the Lorentzian phase. Furthermore, to obtain some intuitive understanding of the deep quantum regime, one may even evolve them backward to evaluate their value on the signature-changing hypersurface. This would then be matched onto the values of these perturbation modes on the boundary of the fuzzy Euclidean phase. Recently, in the context of cosmological perturbations, it has been suggested (Barrau and Grain, 2016) that signature change implies dynamical instabilities in the form of super-Planckian excursions. However, this result assumes the existence of Lorentzian spacetimes with a well-posed initial-value problem, along with a corresponding standard line element, which, of course, is not correct when holonomy corrections become strong and the structure function changes sign. In other words, one should keep in mind that it is a four-dimensional boundary value problem once signature change takes place and use an effective line element, which is consistent with such a signature (Bojowald, Brahma, and Yeom, 2018).

The overall picture with black holes is somewhat less clear at present. However, if the core of black holes display a Euclidean region, it implies there is no deterministic evolution through this high curvature region. While singularity in the form of diverging curvature may be avoided, a challenging question persists in the sense of a spacetime incompletely determined by initial data. In order to extend spacetime across the Euclidean region, one requires additional data on the part of its boundary that borders on the future Lorentzian spacetime. This additional requirement is reminiscent of other proposals of black-hole models, for instance stretched horizons, introduced in the context of the black hole complementarity principle. In any case, the detailed analysis of anomaly-free black holes within LQG points toward a much subtler nonsingular description of quantum spacetime than usually postulated in simplified bounce models. Just as an outside observer finds the stretched horizon as a membrane, first storing and later releasing information in the form of microphysical degrees of freedom, additional information is encountered once an observer moves into the future of a Euclidean region embedded in spacetime. However, it should be pointed out that in the case of black hole models of LQG, there is as yet no microscopic theory that would restrict or determine possible data around Euclidean regions.

3.5.2 Necessity of Deformed Covariance

In their seminal work, Hojman, Kukař, and Teitelboim (1976; see also Kuchar [1974]) had shown that starting from the classical hypersurface deformation algebra, one can uniquely get the Einstein–Hilbert action (up to the cosmological constant), if one restricts to second-order derivatives of the field variables. This remarkable result shows the uniqueness of GR as a covariant gravitational theory, provided one does not consider higher derivative terms.[12] Thus to get a quantum (or at least, quantum-corrected effective) theory that obeys some notion of covariance and is yet different from GR one needs to have a some deformation in the hypersurface deformation algebra. Since we assumed that the diffeomorphism constraint remains unmodified, our only deformation could appear in the brackets between the Hamiltonian constraint; consequently, we *do* end up with a nonclassical version of spacetime, although our classical hypersurfaces retain their classical form. Hence this is an additional argument, which is consistent with the findings, that a deformed notion of covariance is *required* also from the point of the uniqueness theorem due to Hojman and co-workers.

It must be appreciated that modifying the constraints and yet attaining a closed algebra can turn out to be an extremely ambitious task to accomplish. This has been demonstrated recently in the works of Gomes and Shyam (2016), where the authors wished to generalize the constraints by including noncanonical kinetic terms. Such higher derivative kinetic terms were shown to be severely restrictive since the resulting algebra was shown to suffer from anomalies. The underlying reason for this goes back to the idea that modifying gravity is rather more difficult than what it seems like at first. Let us look at the problem from another angle. In the Lagrangian formulation one might say that the quantum-corrected effective (gravitational) action takes the form:

$$S[g] = \frac{1}{16\pi G} \int \mathrm{d}^4x \sqrt{-g}\, [R + \cdots], \qquad (3.16)$$

where the dots signify corrections to the Einstein–Hilbert action coming from loop corrections. Any such term must itself be covariant in the sense that it is built out of curvature invariants and has its Lorentz indices contracted in a proper way. But these corrections, local or nonlocal, still lead to the same Dirac algebra since the theory remains covariant after including them. However, there can be one further correction possible in this context: one might find a redefinition of the measure $\mathrm{d}^4x \sqrt{-g}$. This would mean that the notion of covariance in the theory is now changed. However, such changes must also follow certain rules, which in the canonical theory are implemented by requiring that there are no anomalies in the theory. This is precisely what one finds in LQG: a new deformed notion of covariance that renders the invariant line-element nonexistent in the deep quantum regime. However, if one finds that quantum corrections lead to the Dirac algebra

---

[12] Indeed, it is known that higher curvature theories also share the same classical hypersurface deformation algebra (Deruelle et al., 2010).

giving rise to anomalies, as is the case for Gomes and Shyam (2016), then such corrections must be discarded, at least in their present form. In this context, one must remember that our holonomy corrections are well defined local functions for extrinsic curvature components and not like higher time-derivative kinetic terms as used in Gomes and Shyam (2016). Hence we are easily able to avoid their no-go theorems.

3.5.3 What Does Quantum Spacetime Consist Of?

The most provoking question raised by my essay is regarding the explicit nature of the fundamental spacetime. I have shown that it *cannot* be described by conventional classical Lorentzian structures once holonomy modifications from LQG are taken into account. Although we start from a metric, and then define triads in the classical theory, it turns out that once these corrections are included, one cannot reverse the process to go back to the metric picture. In lower curvature regimes one has access to effective Lorentzian spacetimes, and one gets back the usual constructions. However, in the deep quantum regime, one can locally identify an Euclidean or a Lorentzian patch, but there is no global metric structure that can represent the full spacetime.

We know what quantum spacetime is *not*; it is not the usual Lorentzian manifold we have grown accustomed to. However, what remains to be investigated is the explicit nature of the atoms of spacetime. Can there perhaps be a noncommutative, or even fractal, geometry, replacing the Lorentzian one, at the elemental quantum level? One way to answer this question might be to assume that the spacetime has a, say, noncommutative character. Then one can evaluate the same Dirac algebra for such a mathematical construction. If the resulting deformation has a similar nature to that of LQG, it would imply that LQG hints toward such a more fundamental geometry. In that case, we would have to come up with a suitable coarse graining for such a quantum theory, which would be consistent with our understanding of smooth, continuous Lorentzian manifolds of classical gravity. There are reasons for suspecting such a relationship between LQG and exotic geometries coming from other considerations. If the hypersurface deformation algebra is deformed due to holonomy corrections as described here, it can be shown that the flat Minkowski limit of it gives rise to a noncommutative κ-Poincaré spacetime (Brahma et al., 2017; Mielczarek and Trześniewski, 2017). To make this relationship more robust, one needs to establish this correspondence beyond the flat limit.

It is currently a matter of debate as to how should one think about signature change arising from LQG. First, it needs to be emphasized that the main mathematical concept arising out of holonomy corrections is the idea of 'deformed covariance', with signature change being the main physical effect of that. It seems that LQG provides a physical Wick-rotation to be realized, as was envisioned in theories such as the Hartle–Hawking (HH) proposal (Hartle and Hawking, 1983).[13] However, there are also crucial differences in the signature change emerging in LQG and the HH proposal. The quantum wave-function typically obeys a difference equation in LQG, as opposed to a more-familiar differential equation in HH, and is closer spirit to the proposal by Vilenkin (1984). There is another way to approach the problem that relies on noticing the fractal-dimensional nature of the theory. One can estimate the effective (spectral) dimension of spacetime in quantum gravity theories using methods involving diffusion of particles on a given spacetime. These have led to the evaluation of the UV dimension of theories such as causal dynamical triangulations and asymptotic safety. Including holonomy modification to the constraints in LQG, one can calculate such UV dimension for the theory. But there is something unique in this case due to signature change. One gets a physical cutoff for momentum when evaluating integrals due to the fact that the deformation function changes sign. When this function goes to zero, on the signature-changing hypersurface, it gives an equation for an upper (UV) limit of the physical momentum that can be achieved in the Lorentzian phase. What we obtain after that is a fuzzy Euclidean space. Preliminary calculations have shown that the spectral dimension for LQG, including holonomy corrections, goes toward the magic value of 2. However, it does not flow smoothly from 2 to 4 in a linear manner, but rather exhibits a rich multifractional character. Simply put, it means that, at different energies, the dimension of spacetime is different in the theory.

But why do we keep referring to the Euclidean phase as fuzzy? Well, for one it is not known what is on the other side of this phase. There might have been a Lorentzian phase before it, as is indicated by the cyclic, bouncing models of minisuperspace quantizations of the background in LQG. However, it might just be that there was simply a quantum Euclidean phase till our usual Lorentzian phase popped up. Remember, in the Euclidean patch, one cannot ask temporal questions and hence it makes no sense to ask how long it lasted. There is another reason to suspect that this phase is not just a simple classical, Euclidean geometry. Although it can be seen from the deformed Dirac algebra that the signature of the background is the same as that for Euclidean geometry, the constraints themselves are not the ones arising from the Euclidean version of GR, but rather a from a modified version of it. More concretely, ongoing investigations show that there is an additional potential term that appears in the Euclidean Hamiltonian constraint after signature

---

[13] Let us illustrate this point with another well-understood analogy. Many interesting physical consequences are known for coupling between long- and short-wavelength cosmological modes in the early universe. In a similar vein, it has been shown that LQG can lead to such couplings spelling out a physical mechanism for it.

change. For instance, if one starts with usual spherically symmetric gravity (with holonomy corrections) in the Lorentzian phase, after signature change, we end up with twice[14] the spherically symmetric potential in the Euclidean phase. More generally, since the usual Lorentzian structure is replaced by a quantum spacetime lacking classical analogs, the physical meaning of such additional terms is under active research. We can, effectively, still understand some of its features through the deformed algebra structure and the modified constraints. However, if spacetime is truly composed of discrete packets that obey some exotic, noncommutative or fractal behavior, it is only natural that the resulting system cannot be fully understood by classical notions. This forms another reason for the epithet *fuzzy* to be assigned to this region—in the anticipation that the quantum geometry may turn out to be noncommutative and perhaps some version of the fuzzy sphere. However, what is now well understood for certain is that there is a fuzzy quantum region if we go to very high energies, which is nonsingular and is certainly *not* described by well-understood Lorentzian structures.

Finally, once there is no more an invariant line element, it is not possible to define an action without the usual invariant measure for the integral. This makes working in the Lagrangian picture impossible. However, the Hamiltonian picture remains well defined with rigorous definitions of gauge-invariant quantities, which might be evaluated to calculate observables. It seems that the Hamiltonian formulation, which is on the same footing as the Lagrangian one in the classical setting, is now somehow more preferred in the quantum theory. This is a surprising twist of turns in LQG which would be interesting to explore further. This is, however, what was once prophesied by Dirac[15] and it seems to be arising from a quantum theory of gravity.

3.6 Summary

We have shown that nonclassical structures can arise in quantum gravitational theories with explicit examples of symmetry-reduced models in LQG. As outlined in this essay, preliminary results suggest that emergent time is a concept central to LQG and therefore, this has serious consequences for physical systems such as the big bang cosmology or black hole singularities. Nonsingular signature change replaces classical singularity in a much more intricate way than what was previously predicted in strict minisuperspace quantizations, through a quantum bounce. A fuzzy Euclidean region inside the core of black holes has new possibilities for the

---

[14] The precise factor of 2 depends on tuning a fundamental parameter to 1 in the theory to simplify calculations. However, the extra potential term in the Euclidean part appears irrespective of this choice.

[15] "It would be permissible to look upon the Hamiltonian form as the fundamental one, and there would then be no fundamental four-dimensional symmetry in the theory" (Dirac, 1958).

resolution of the information loss paradox, especially in its partial similarity to the black hole complementarity paradigm (Bojowald and Brahma, 2018), whereas replacing the big bang singularity with an 'asymptotically silent' phase is reminiscent of the Hartle–Hawking wave-function and similar results from causal dynamical triangulations (Mielczarek, 2012). Thus nonsingular bounce from LQG seems to find common ground with other quantum gravity theories, which has been a very tough proposition for these different approaches traditionally. The real question, however, remains whether such exotic constructs are realized in nature. To further probe this question, one first needs to find a full theory of canonical quantum gravity that retains this feature of emergent time, going beyond simplified toy models.

References

Ashtekar, A. 2009. Singularity resolution in loop quantum cosmology: A brief overview. *J. Phys. Conf. Ser.*, **189**, 012003.

Ashtekar, A., Bojowald, M., and Lewandowski, J. 2003. Mathematical structure of loop quantum cosmology. *Adv. Theor. Math. Phys.*, **7**(2), 233–268.

Ashtekar, A., Corichi, A., and Singh, P. 2008. Robustness of key features of loop quantum cosmology. *Phys. Rev.*, **D77**, 024046.

Ashtekar, A., and Lewandowski, J. 2004. Background independent quantum gravity: A status report. *Class. Quant. Grav.*, **21**, R53.

Ashtekar, A., and Singh, P. 2011. Loop quantum cosmology: A status report. *Class. Quant. Grav.*, **28**, 213001.

Barrau, A., and Grain, J. 2016. Cosmology without time: What to do with a possible signature change from quantum gravitational origin? ArXiv e-prints.

Becker, K., Becker, M., and Schwarz, J. H. 2006. *String Theory and M-Theory: A Modern Introduction*. Cambridge: Cambridge University Press.

Bojowald, M. 2001. Absence of singularity in loop quantum cosmology. *Phys. Rev. Lett.*, **86**, 5227–5230.

Bojowald, M. 2010. *Canonical Gravity and Applications: Cosmology, Black Holes, and Quantum Gravity*. Cambridge: Cambridge University Press.

Bojowald, M. 2012. Quantum cosmology: Effective theory. *Class. Quant. Grav.*, **29**, 213001.

Bojowald, M. 2013. Mathematical structure of loop quantum cosmology: Homogeneous models. *SIGMA*, **9**, 082.

Bojowald, M., and Brahma, S. 2015. Covariance in models of loop quantum gravity: Gowdy systems. *Phys. Rev.*, **D92**(6), 065002.

Bojowald, M., and Brahma, S. 2016a. Effective constraint algebras with structure functions. *J. Phys.*, **A49**(12), 125301.

Bojowald, M., and Brahma, S. 2017. Signature change in loop quantum gravity: General midisuperspace models and dilaton gravity. *Phys. Rev.*, **D95**(12), 124014.

Bojowald, M., and Brahma, S. 2018. Signature change in 2-dimensional black-hole models of loop quantum gravity. *Phys. Rev.*, **D98**, 026012.

Bojowald, M., Brahma, S., Buyukcam, U., and D'Ambrosio, F. 2016. Hypersurface-deformation algebroids and effective space-time models. *Phys. Rev.*, **D94**(10), 104032.

Bojowald, M., Brahma, S., and Reyes, J. D. 2015. Covariance in models of loop quantum gravity: Spherical symmetry. *Phys. Rev.*, **D92**(4), 045043.

Bojowald, M., Brahma, S., and Yeom, D. 2018. Effective line elements and black-hole models in canonical (loop) quantum gravity. *Phys. Rev.*, **D98**(4), 046015.

Bojowald, M., and Mielczarek, J. 2015. Some implications of signature-change in cosmological models of loop quantum gravity. *JCAP*, **1508**(08), 052.

Bojowald, M., and Paily, G. M. 2013. Deformed general relativity. *Phys. Rev.*, **D87**(4), 044044.

Bojowald, M., Paily, G. M., and Reyes, J. D. 2014. Discreteness corrections and higher spatial derivatives in effective canonical quantum gravity. *Phys. Rev.*, **D90**(2), 025025.

Bojowald, M., and Swiderski, R. 2006. Spherically symmetric quantum geometry: Hamiltonian constraint. *Class. Quant. Grav.*, **23**, 2129–2154.

Brahma, S. 2015. Spherically symmetric canonical quantum gravity. *Phys. Rev.*, **D91**(12), 124003.

Brahma, S., Ronco, M., Amelino-Camelia, G., and Marciano, A. 2017. Linking loop quantum gravity quantization ambiguities with phenomenology. *Phys. Rev.*, **D95**(4), 044005.

Cailleteau, T., Mielczarek, J., Barrau, A., and Grain, J. 2012. Anomaly-free scalar perturbations with holonomy corrections in loop quantum cosmology. *Class. Quant. Grav.*, **29**, 095010.

Collins, J. C. 1986. *Renormalization*. Cambridge Monographs on Mathematical Physics, vol. 26. Cambridge: Cambridge University Press.

Corichi, A., Olmedo, J., and Rastgoo, S. 2016. Vacuum CGHS in loop quantum gravity and singularity resolution. *Phys. Rev.*, **D94**, 084050.

de Blas, D. M., Olmedo, J., and Pawlowski, T. 2015. Loop quantization of the Gowdy model with local rotational symmetry. ArXiv e-prints.

Deruelle, N., Sasaki, M., Sendouda, Y., and Yamauchi, D. 2010. Hamiltonian formulation of f(Riemann) theories of gravity. *Prog. Theor. Phys.*, **123**, 169–185.

Dirac, P. A. M. 1958. Generalized Hamiltonian dynamics and the theory of gravitation in Hamiltonian form. *Proc. R. Soc. Lond.*, **A**(246), 333.

Gambini, R., and Pullin, J. 2013. Loop quantization of the Schwarzschild black hole. *Phys. Rev. Lett.*, **110**(21), 211301.

Gomes, H., and Shyam, V. 2016. Extending the rigidity of general relativity. *J. Math. Phys.*, **57**(11), 112503.

Green, M. B., Schwarz, J. H., and Witten, E. 2012. *Superstring Theory 25th Anniversary Edition*. Cambridge: Cambridge University Press.

Han, M., and Hung, L. Y. 2017. Loop quantum gravity, exact holographic mapping, and holographic entanglement entropy. *Phys. Rev.*, **D95**(2), 024011.

Hartle, J. B., and Hawking, S. W. 1983. Wave function of the universe. *Phys. Rev.*, **D28**, 2960–2975.

Hawking, S. W., and Ellis, G. F. R. 2011. *The Large Scale Structure of Space-Time*. Cambridge Monographs on Mathematical Physics. Cambridge: Cambridge University Press.

Hojman, S. A., Kuchar, K., and Teitelboim, C. 1976. Geometrodynamics regained. *Annals Phys.*, **96**, 88–135.

Kuchar, K. 1974. Geometrodynamics regained—A Lagrangian approach. *J. Math. Phys.*, **15**, 708–715.

Laddha, A., and Varadarajan, M. 2011. The diffeomorphism constraint operator in loop quantum gravity. *Class. Quant. Grav.*, **28**, 195010.

Lin, J., Marcolli, M., Ooguri, H., and Stoica, B. 2015. Locality of gravitational systems from entanglement of conformal field theories. *Phys. Rev. Lett.*, **114**, 221601.

Mielczarek, J. 2012. Asymptotic silence in loop quantum cosmology. *AIP Conf. Proc.*, 1514, 81.

Mielczarek, J., and Trześniewski, T. 2016. Spectral dimension with deformed spacetime signature. *Phys. Rev.*, **D96**(2), 024012.

Polchinski, J. 2007a. *String Theory Vol. 1: An Introduction to the Bosonic String.* Cambridge: Cambridge University Press.

Polchinski, J. 2007b. *String Theory Vol. 2: Superstring Theory and Beyond.* Cambridge: Cambridge University Press.

Rastelli, L. 2005. String field theory. In: *The Encyclopedia of Mathematical Physics. arXiv e-print hep-th/0509129.* Ithaca, NY: Cornell University.

Reuter, M., and Saueressig, F. 2012. Quantum Einstein gravity. *New J. Phys.*, **14**, 055022.

Thiemann, T. 2008. *Modern Canonical Quantum General Relativity.* Cambridge: Cambridge University Press.

Verlinde, E. P. 2011. On the origin of gravity and the laws of Newton. *JHEP*, **04**, 029.

Vilenkin, A. 1984. Quantum creation of universes. *Phys. Rev.*, **D30**, 509–511.

4

Beyond Standard Inflationary Cosmology

ROBERT H. BRANDENBERGER

4.1 Introduction

4.1.1 Goals of Early Universe Cosmology

Explaining the origin and early evolution of the universe has been a goal of cosmology for millennia. However, it is only over the past few decades that cosmology has moved from being a branch of philosophy and theology to being a mainstream area of physics research. This change is due to an explosion of data about the structure of our universe that experimentalists and observers have gathered.

The prime example of data is the cosmic microwave background (CMB). The CMB was discovered serendipitously in the 1960s (Penzias and Wilson, 1965). Only in the early 1990s detailed measurements of the black body nature of the spectrum became available (Gush, Halpern, and Wishnow, 1990; Mather et al., 1990), and the first anisotropies were discovered (Smoot et al., 1992). Over the following two decades, rapidly improving measurements of the angular power spectrum[1] of the CMB anisotropies were made, culminating with the results from the WMAP (Hinshaw et al., 2000)[2] and Planck (Ade et al., 2013) satellites. To better than 1 part in 10^4, the temperature of the CMB is isotropic. At a slightly lower level, anisotropies appear. Figure 4.1 depicts the map of CMB anisotropies from the WMAP satellite. In this figure, the sky is projected onto a plane in the same way that the surface of the earth is sometimes projected onto a plane. To quantify the anisotropies, we can expand the map in spherical harmonics and compute the angular power spectrum. The results are shown in Figure 4.2 where the horizontal

---

[1] The CMB temperature map of the sky can be expanded in terms of spherical harmonics (the analog of Fourier expansion in Euclidean spaces). The spherical harmonics are labeled by two integers l and m with $-l \le m \le l$. The coefficient of the (l, m) spherical harmonic gives the amplitude of the CMB anisotropies on an angular scale proportional to l^{-1} and in a direction given by m. If we average the square of the coefficients for fixed l over the allowed values of m, we obtain what is called the *angular power spectrum* of CMB anisotropies.

[2] WMAP stands for Wilkinson Microwave Anisotropy Probe.

Robert H. Brandenberger

Figure 4.1 Map of the microwave background temperature anisotropies. What is shown is a projection of the sky onto a plane. The color scale indicates temperature. The temperature differences between hot and cold areas is of the order 10^{-5} of the average temperature. Credit: NASA/WMAP Science Team.

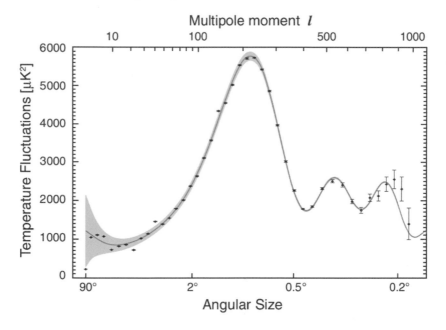

Figure 4.2 Angular power spectrum of the data shown in the previous figure. The horizontal axis is an inverse angular scale. The vertical axis gives the mean magnitude of fluctuations on the corresponding angular scale. Credit: NASA/WMAP Science Team.

axis is the value of l (or equivalently the angular scale), and the vertical axis gives the amplitude. The figure shows several interesting features. First of all, there are oscillations in the spectrum with a first peak at an angular scale of about $1°$. Second, at large angular scales the spectrum is quite flat. Finally, on small angular scales the spectrum is suppressed. One of the main goals of early universe cosmology is to provide an explanation for these data.

The CMB is not the only window we have to probe the structure of the universe. Using optical and infrared telescopes we can study the distribution of galaxies and galaxy clusters. X-ray telescopes allow us to map out the distribution of hot gas (gas with a temperature much higher than the surface temperature of the sun) in the universe. Radio telescopes allow the exploration of cold gas (gas with a temperature lower than the surface temperature of the sun). A new window to probe the universe is emerging: the 21 cm window (Furlanetto et al., 2006) (radiation at a frequency of 21 cm), a window that allows us to map out the distribution of neutral hydrogen, the atom that dominated matter content, up to very large distances and early times. These windows have the added advantage (compared to the CMB) of providing us with 3-dimensional maps as opposed to only 2-dimensional ones.

The second main goal of early universe cosmology is to provide explanations for these data based on physical models that obey the principles of causality. As we will see, standard big bang cosmology, the paradigm of cosmology until 1980, cannot provide a causal explanation (an explanation from a theory that obeys the principles of special relativity, namely that no information can travel faster than the speed of light) of the data, and we have to go back to the very early universe if we want to find explanations.

An important aspect of cosmology as a branch of physics is that, once we have a model that can address the data, we can make predictions for future observations. As will be discussed later, the current paradigm of early universe cosmology made predictions concerning data that were not yet available at the time when the model was formulated. As mentioned later, standard big bang cosmology predicted the existence and black body nature of the CMB, and inflationary cosmology predicted an almost spatially flat universe with a spectrum of CMB anisotropies with specific properties. These predictions were confirmed much later once observations became available. Any alternative to the whatever the current cosmological paradigm might be must make specific predictions for future observations with which it can be distinguished from the current paradigm. I regard this falsifiability aspect as an important challenge for modern cosmology.

Summarizing, the goals of early universe cosmology are:

- Explain the origin and early evolution of the universe.
- Explain the currently available data on the large-scale structure of the universe based on causal physics.
- Make predictions for future observations.

A few words on the notation. We will mostly consider homogeneous and isotropic space-times in which the line element is given by:

$$ds^2 = dt^2 - a(t)^2 d\mathbf{x}^2, \tag{4.1}$$

where t is physical time (time measured by our physical clocks), $\mathbf{x}$ are comoving spatial coordinates (a coordinate system that expands as space expands), and $a(t)$ is the cosmological *scale factor*. Particles at rest have time-independent comoving coordinates, and the change in $a(t)$ describes the expansion (or contraction) of space. Light travels on curves $\mathbf{x}(t)$ for which $ds^2 = 0$. It is often useful to work with *conformal time* τ in terms of which light travels at $45°$ lines when drawing space-time diagrams in terms of conformal time and comoving spatial coordinates. The Hubble expansion rate $H(t)$ is given by:

$$H(t) = \frac{\dot{a}}{a}, \tag{4.2}$$

where an overdot represents the derivative with respect to time.

We are using units in which the speed of light, Planck's constant, and Boltzmann's constant are all set to 1. These are units that are generally used in the field of high energy physics.

4.1.2 Problems of Standard Big Bang Cosmology

Cosmology deals with space, time and matter. Standard big bang (SBB) cosmology is based on describing space and time using Einstein's classical theory of general relativity, and matter in terms classical perfect fluids. At the present time (which will be denoted by t_0 in the following), matter is dominated by a fluid with vanishing pressure, which in cosmology is called *cold matter*, while the CMB provides a component that is subdominant at the present time.

The equations of motion of general relativity imply that in the presence of homogeneously distributed matter, the scale factor $a(t)$ cannot be constant. Space is either expanding or contracting. Observations tell us that we live in an expanding universe. Since the energy density in radiation increases faster (namely as $a(t)^{-4}$) as we go back in time than the energy density in cold matter (which increases as $a(t)^{-3}$), there is a time t_{eq} when the energy densities of the two components are equal. Before t_{eq} the universe is dominated by radiation, afterward by cold matter. It turns out that t_{eq} is the time before which inhomogeneities do not increase in time (up to factors that are logarithmic in time).[3]

---

[3] The dominant radiation causes a homogeneous gravitational potential that impedes the growth of density fluctuations in the cold matter.

There is another time that is important in the evolution of the late universe. This is the time t_{rec} of recombination before which the energy density was larger than the ionization energy of hydrogen. Hence, for $t < t_{rec}$ the universe was an electromagnetic plasma, and only after which it becomes neutral. The photons (particles of light) that are present at the time t_{rec} can then travel unimpeded to us. This is in fact the origin of the CMB.

For $t < t_{eq}$ the scale factor increases as:

$$a(t) \sim \left(\frac{t}{t_0}\right)^{1/2},\tag{4.3}$$

and vanishes at the time $t = 0$. At this time, the curvature and energy densities become infinite. This is the *big bang* singularity of standard cosmology.

The most impressive success of standard big bang cosmology is that it predicted the existence and black body nature of the CMB. However, the scenario is not able to explain the near isotropy of this radiation. This is the so-called horizon problem and it is illustrated in Figure 4.3. Here, the horizontal axis indicates comoving spatial coordinates, the vertical axis in conformal time. The lines at 45° represent light rays. The speed of propagation of causation is limited by the speed of light. As indicated in the figure, the region of causal contact between $t = 0$ and the time of recombination is smaller than the region of the last scattering surface (the intersection of our past light cone l_p with the surface at $t = t_{rec}$) over which

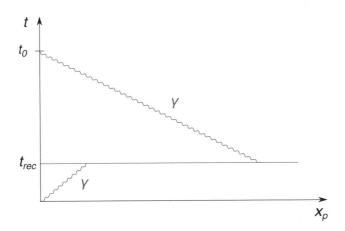

Figure 4.3 Illustration of the horizon problem. The vertical axis is time, the horizontal axis is distance—implicitly using coordinates in which light travels at 45° angles. We see light rays (wavy lines) starting at the time t_{rec} of recombination and reaching us today (time t_0). As shown, the distance over which we see the microwave light to be isotropic is much larger than the distance that could have been in causal contact starting at the initial time $t = 0$.

the microwave background is observed to be isotropic. The maximal angular scale where isotropy can in principle be explained by causal physics is the angular scale of the first peak in the CMB angular power spectrum.

As Figure 4.2 shows, there is nontrivial structure in the CMB on angular scales that, according to SBB cosmology, could never have been in causal contact. Hence SBB cannot explain the origin of structure on these large scales. This is the *formation of structure problem* from which the SBB scenario suffers.

In addition, SBB cosmology cannot explain the degree of spatial flatness of the universe that is currently observed. According to SBB cosmology, a spatially flat universe is unstable to the development of curvature in the expanding phase. The near spatial flatness that is currently observed requires a tuning of the relative contribution of the spatial curvature to the total energy density, which is of the order of 10^{-50} at energy scales corresponding to particle physics grand unification.

Since the SBB scenario is well tested at late cosmological times, any solution of the above-mentioned problems will require new physics during the stages of the very early universe.

4.1.3 Preview

In the following sections I discuss different scenarios of very early universe cosmology that can provide solutions to the horizon, flatness, and structure formation problems. I compare the current paradigm, the inflationary universe scenario, with a couple of alternatives. The main message will be that there are a number of early universe scenarios, and not just inflation, that are compatible with current observations.

I then ask the question, What kind of picture of the early universe emerges if superstring theory is the correct theory that unifies all forces of nature at high energies? Hints will be discussed indicating that superstring cosmology will be nonsingular and may not include a phase of inflation.

4.2 Theory of Cosmological Perturbations

4.2.1 Basics of Cosmological Perturbation Theory

Observations of CMB anisotropies and large-scale structure concern linear fluctuations about the cosmological background (4.1). Hence we must start with a brief summary of how these inhomogeneities are described and how they evolve (see, e.g., Mukhanov, Feldman, and Brandenberger [1992] and Brandenberger [2004] for details). We here work in the context of Einstein gravity with a matter source. For simplicity, matter is modeled in terms of a scalar field φ with a nontrivial background dynamics $\varphi_0(t)$. Since the universe is observed to converge to homogeneity on large scales, the fluctuations can be described in linear theory. This means that any fluctuating field can be expanded in plane waves (Fourier modes) and each such mode evolves independently. The Fourier modes can be labeled by the comoving wavenumber k.

Linear fluctuations of geometry and matter can be classified according to how they transform under spatial rotations. Out of the 10 degrees of freedom of the metric, four are scalars, four vectors, and two tensors (the two polarization states of gravitational waves). The physics is independent of which coordinates are used, and hence there are four coordinate modes (gauge modes) that can be factored out, leaving only two scalars and two vectors, plus the gravitational waves. For simple matter models, such as a scalar field with a homogeneous component that is evolving in time (a *rolling* scalar field), the vector modes are not sourced at linear order in perturbation theory, and hence I will not consider them. I focus on the scalar modes that are sourced by matter. One of the scalar modes vanishes for matter without anisotropic stress, and the remaining mode is determined by the matter fluctuations.

We can choose coordinates in which the scalar metric fluctuations are diagonal, and the metric is:

$$ds^2 = a^2 \left\{ (1 + 2\Phi) \, d\eta^2 - \left[(1 - 2\Phi) \, \delta_{ij} + h_{ij} \right] dx^i dx^j \right\}, \tag{4.4}$$

where Φ is a function of space and time and represents the scalar metric fluctuations. The matter field including linear fluctuations is:

$$\varphi(\boldsymbol{x},t) = \varphi_0(t) + \delta\varphi(\boldsymbol{x},t). \tag{4.5}$$

The equations of motion for the fluctuations can be obtained by expanding the action of matter plus gravity to second order about the background. Since the background satisfies the equations of motion, terms linear in cosmological fluctuations cancel out in the action, leaving the quadratic terms as the leading fluctuation terms. There is only one canonical fluctuation variable $v(\boldsymbol{x},t)$ (variable in terms of which the action has a canonical kinetic term). As shown in (Sasaki, 1986; Mukhanov, 1988), this variable is:

$$v = a \left(\delta\varphi + \frac{z}{a}\Phi \right), \tag{4.6}$$

where:

$$z = \frac{a\varphi_0'}{\mathcal{H}}, \tag{4.7}$$

and obeys the equation of motion:

$$v_k'' + \left(k^2 - \frac{z''}{z} \right) v_k = 0, \tag{4.8}$$

where a prime denotes the derivative with respect to conformal time τ. Here we see the crucial role which the Hubble radius H^{-1} plays. If the equation of state of matter is constant in time, then the factor z''/z is of the order H^2. Hence on sub-Hubble scales ($k > H$) fluctuations oscillate, while on super-Hubble scales ($k < H$) they are *frozen in* and are *squeezed*, i.e., in an expanding universe the amplitude of the dominant mode of the above equation grows as $z \sim a$.

The curvature fluctuation ζ is in fact given by v/z. What is measured is the power spectrum $P_\zeta(k)$ of ζ, which gives the mean square fluctuation of ζ on a length scale k^{-1} and is given in terms of the Fourier modes $\zeta(k)$ of ζ by:

$$P_\zeta(k) = k^3 |\zeta(k)|^2. \tag{4.9}$$

If this quantity is independent of k, one has a *scale-invariant* spectrum. More generally, one has $P(k) \sim k^{n_s-1}$, where n_s is called the scalar spectral index. The observed spectrum is nearly scale-invariant with a small red tilt (Ade et al., 2014).

An initial vacuum spectrum at time t_i is:

$$v_k(t_i) = \frac{1}{\sqrt{2k}}. \tag{4.10}$$

This is a deep blue power spectrum, i.e., there is more power on short wavelengths. If fluctuations originate as vacuum perturbations (which they are assumed to in inflationary cosmology and in a number of alternatives), then the squeezing of the fluctuations on super-Hubble scales must have exactly the right features to convert the initial vacuum spectrum to a scale-invariant one. As discussed in the next section, this happens both for inflationary cosmology and for the matter bounce scenario.

Note that the equation of motion for the Fourier mode of the amplitude of gravitational waves is the same as (4.8) except that the function $z(\eta)$ is replaced by the scale factor $a(\eta)$. If the equation of state of matter is constant, then $z \sim a$ and hence the gravitational waves evolve as the curvature fluctuations.

4.2.2 Criteria for a Successful Early Universe Cosmology

The presence of acoustic oscillations in the angular power spectrum of the CMB and in the power spectrum of matter density fluctuations (the so-called baryon oscillation peak) were predicted (Peebles and Yu, 1970; Sunyaev and Zeldovich, 1970) a decade before the development of inflationary cosmology. In these works it was realized that a roughly scale-invariant spectrum of standing wave fluctuations on super-Hubble scales at cosmological times before the time t_{rec} of recombination would, as the scales enter the Hubble radius and begin to oscillate as described in the previous section, be in good agreement with the data on the distribution of

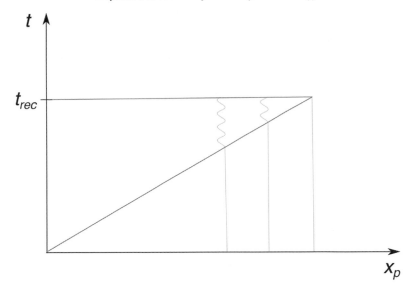

Figure 4.4 Origin of acoustic oscillations in the CMB angular power spectrum. Assuming that fluctuations originate as standing waves on scales that are super-Hubble at early times, then different modes perform a different number of oscillations between when they enter the Hubble radius and start to oscillate and the time of recombination. Those that perform $n = 0, 2, 4, \ldots$ quarter oscillations are standing waves with maximal amplitude at t_{rec} and yield peaks in the anisotropy spectrum, whereas those that perform $n = 1, 3, \ldots$ quarter oscillations have velocity inhomogeneities only at t_{rec} and yield minima in the angular power spectrum (see Sunyaev and Zeldovich [1970] from where the figure is adapted).

galaxies that were available at the time when the papers were written and give rise to oscillations in the angular power spectrum of the CMB and in the power spectrum of density fluctuations. The dynamics is illustrated in Figure 4.4. At that time, however, no models based on causal physics were known for how to produce such primordial fluctuations.

Here we outline the criteria that a successful early universe scenario must satisfy. Let us emphasize again the difference between the horizon and the Hubble radius. The horizon is the forward light cone of a point at the initial time and carries causality information. Beginning at the initial time, it is possible to have causal contact within the horizon. The Hubble radius $H(t)^{1}$, on the other hand, is a local concept. It separates scales where fluctuations oscillate (sub-Hubble modes) from scales where the inhomogeneities are frozen in and get squeezed (super-Hubble modes).

Since the Hubble radius at t_{rec} is smaller than the radius of the part of the universe that we probe with the CMB, the condition for an early universe scenario to solve the horizon problem is that the horizon at t_{rec} be much larger than the Hubble radius at that time (at least two orders of magnitude since the Hubble radius at t_{rec}

corresponds to an angular scale of about $1°$ and we need to explain the absence of order one anisotropies on the angular scales of the full sky). Thus the first requirement on a successful early universe scenario is that the horizon at late times be much larger than the Hubble radius.

In order to be able to have a causal mechanism of structure formation, scales that are currently observed in cosmology must originate in the early universe with a wavelength smaller than the Hubble radius. This is the second requirement.

Third, the scales that we observe today must propagate for a long time at super-Hubble lengths. This is in order that the fluctuations are squeezed and enter the Hubble radius at late times as standing waves. This is required in order to provide an explanation for the observed acoustic oscillations in the angular power spectrum of the CMB.

Finally, the generation mechanism that acts on sub-Hubble scales must yield a nearly scale-invariant spectrum of cosmological perturbations.

The inflationary universe scenario provides a simple realization of these four criteria. However, there are alternative realizations as explained in the following section.

4.3 Paradigms of Early Universe Cosmology

4.3.1 Cosmological Inflation

Cosmological inflation (Brout, Englert, and Gunzig, 1978; Fang, 1980; Guth, 1981; Sato, 1981) assumes that space underwent a period of almost exponential expansion during a finite time interval in the very early universe, starting at some time t_i and ending at a time t_R.[4] The first model of inflation was based on a modified gravitational action (Starobinsky, 1980) where higher curvature terms dominate at early times and lead to accelerated expansion. Later models assumed that the accelerated expansion of space is realized in the context of Einstein gravity, but assuming the presence of a slowly rolling (Albrecht and Steinhardt, 1982; Linde, 1982, 1983) scalar field whose energy-momentum tensor is dominated by the scalar field potential energy. After the end of inflation at time t_R, the evolution of the universe is like in SBB cosmology.

Figure 4.5 represents a spacetime sketch of inflationary cosmology. The vertical axis is time, the horizontal axis denotes physical spatial distance. The solid curve labeled H^{-1} represents the Hubble radius, the dashed line is the horizon, and the curve labeled k represents the physical wavelength of a fluctuation mode.

During the phase of accelerated expansion for $[t_i < t < t_R]$, the horizon is growing very fast while the Hubble radius increases only slowly. For almost exponential

---

[4] The subscript R stands for *reheating* since at this time the conditions of a hot early universe must be created.

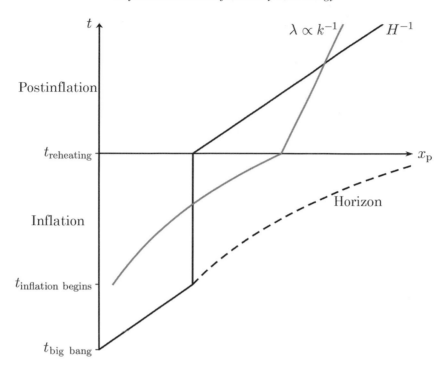

Figure 4.5 Space-time sketch of inflationary cosmology. The vertical axis is time, the horizontal axis is physical distance. The curve labeled H^{-1} is the Hubble radius, and the curve labeled k is the physical wavelength of the fluctuation on a particular (comoving) scale. Note that during the period inflation both the wavelength of a fluctuation and the horizon grow exponentially (for exponential inflation), which is indicated by the curvature of the lines. After the end of inflation, the wavelength grows as $a(t) \sim t^{1/2}$ which is indicated by the near linearity of the curve.

expansion the horizon increases nearly exponentially while the Hubble radius is almost constant. Only a very short interval of accelerated expansion is needed to solve the horizon problem (the horizon needs to increase by a factor of greater than 100 relative to the Hubble radius).

If the energy scale of inflation corresponds to the scale of particle physics Grand Unification (an energy scale of about 10^{16} GeV), then about 60 e-foldings of exponential expansion are required in order that scales that we observe now in cosmology had a wavelength smaller than the Hubble radius at the beginning of the inflationary period. In this case, the second of the general criteria for a successful early universe cosmology is satisfied.

As is evident from Figure 4.5, the wavelength of fluctuations is larger than the Hubble radius for a long time. Hence the squeezing of the fluctuations that is required in order to obtain the acoustic oscillations in the CMB angular power spectrum is realized.

Finally, as initially argued in Starobinsky (1979) for gravitational waves and in Mukhanov and Chibisov (1981) for the curvature perturbations, the power spectrum of fluctuations will be approximately invariant of scale. This is a consequence of the time-translation invariance of the phase of exponential expansion (Press, 1980; Sato, 1981). It is usually assumed that fluctuations originate as quantum vacuum perturbations at early times during the inflationary phase. This assumption can be justified since, in the absence of interactions with the agent driving the accelerated expansion, any initial fluctuations are redshifted away, leaving the matter in a vacuum state. If the agent driving inflation interacts with matter, it is possible to realize inflation while maintaining the dominance of thermal fluctuations. This is the *warm inflation* scenario (Berera, 1995).

The scale invariance of the spectrum of curvature fluctuations can also be seen using the general formalism described in the previous section. The curvature fluctuation variable, starting out with a deep blue vacuum spectrum on sub-Hubble scales, gets squeezed on super-Hubble scales. Large wavelength fluctuations exit the Hubble radius earlier and are thus squeezed more. For almost exponential inflation the squeezing is just right to turn the initial vacuum spectrum into a scale-invariant one. To see this, consider first the Hubble crossing condition:

$$a^{-1}(t_H t(k))k = H. \tag{4.11}$$

Before $t_H(k)$, the amplitude of v_k is constant, afterwards it increases as z. Thus the power spectrum of ζ at some late time t is:

$$P_\zeta(k,\eta) = k^3 z^{-2}(\eta) \left(\frac{z(\eta)}{z((\eta_H(k)))} \right)^2 |v_k(t_i)|^2$$

$$\simeq \frac{1}{2} \left(\frac{a(t_H(k))}{z(t_H(k))} \right)^2 H^2. \tag{4.12}$$

The k-dependence has canceled out, and we thus obtain a scale-invariant spectrum.

Gravitational waves evolve as curvature fluctuations, with the function $z(\eta)$ being replaced by $a(\eta)$. Hence inflationary cosmology predicts a scale invariant spectrum of gravitational waves whose amplitude is suppressed compared to that of curvature fluctuations by the slow-roll parameter ϵ, which gives the ratio between z and a during the inflationary phase.

Of the early universe scenarios discussed here, inflationary cosmology is the only one that predicted (as opposed to postdicted) various observational results, e.g., the spatial flatness of the universe, the almost scale invariance of the spectrum of cosmological perturbations, and the near Gaussianity of this spectrum. In fact, inflation predicted (Mukhanov and Chibisov, 1981) a slight red tilt of the spectrum, a tilt that has now been established by the recent CMB observations (see, e.g., Ade et al. [2014]). Inflation predicts a nearly scale-invariant spectrum of gravitational waves (Starobinsky, 1979) with a slight red tilt. Note that, whereas the tilt of the spectrum of cosmological perturbations can be made blue by complicating the scalar field model of inflation, the spectrum of gravitational waves has a red tilt unless the matter that is responsible for inflation violates standard energy conditions.[5]

A drawback of the inflationary scenario is the *trans-Planckian problem* for cosmological perturbations (Brandenberger and Martin, 2001; Martin and Brandenberger, 2001). (See e.g., Brandenberger and Martin [2013] for a review with an extended list of references.) If the period of inflation lasts only slightly longer than the minimal amount of time it needs to in order to solve the flatness and structure formation problems of SBB cosmology, then the wavelengths of the fluctuation modes that we observe today are smaller than the Planck length at the beginning of inflation. We do not understand the physics on these wavelengths, and hence it is unclear if the initial conditions for the fluctuations that are used are well justified. Some other problems of inflationary cosmology are discussed in later sections.

4.3.2 Matter Bounce

Another paradigm to obtain successful structure formation is the *matter bounce* scenario (Wands, 1999; Finelli and Brandenberger, 2002). It is assumed that there is new physics[6] that resolves the cosmological singularity. Time runs from $-\infty$ to ∞. The universe begins in a contracting phase that is the mirror inverse of the expanding standard big bang cosmology evolution, which is to say that at very early times during contraction, a dark energy component may dominate, followed by a period of matter-dominated contraction, followed by a radiation-dominated phase, and then a nonsingular bounce. The bounce point (minimal value of the scale factor) is $t = 0$.

Figure 4.6 presents a sketch of the resulting spacetime diagram. The vertical axis is time, the horizontal axis represents comoving spatial distance. The Hubble radius is symmetric about $t = 0$. The perpendicular curve indicates the wavelength of a fluctuation. Since there is no origin of time, the horizon is infinite and there is no horizon problem. As obvious from the figure, all scales originate inside the Hubble

---

[5] At this point, we have not yet observed a stochastic background of gravitational waves, and observing the small tilts predicted by various early universe models will be very challenging.

[6] The new physics could be string theory, loop quantum gravity, or some effective field theory that corresponds to Einstein gravity coupled to some matter that violated the usual Hawking–Penrose energy conditions. For a review of bouncing cosmologies see, e.g., Brandenberger and Peter (2017). I am emphasizing here ways to obtain a spectrum of cosmological fluctuations in agreement with observations, and the specific dynamics of the bounce phase has little to say about this issue.

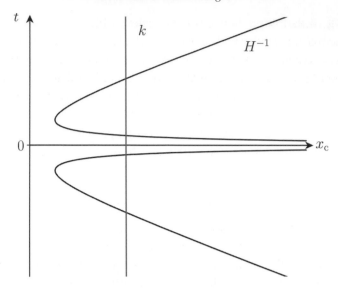

Figure 4.6 Spacetime sketch of the matter bounce cosmology. The vertical axis is conformal time, the horizontal axis is comoving distance. The curve labeled H^{-1} is the Hubble radius, and the vertical line represents the wavelength of a fluctuation mode. The bounce phase (for which new physics is required) lasts from the time when the Hubble radius takes on its minimal value in the contracting phase to the corresponding time during expansion.

radius, and hence a causal structure formation scenario is possible. In fact, for scales that exit the Hubble radius during the matter-dominated phase of contraction, an initial vacuum spectrum on sub-Hubble scales evolves (Wands, 1999; Finelli and Brandenberger, 2002) into a scale-invariant spectrum on super-Hubble scales. If we take into account the dark energy component in the contracting phase, a slight red tilt of the spectrum results, as in inflationary cosmology (Cai and Wilson-Ewing, 2015).

The reason why a scale-invariant spectrum of curvature fluctuations is obtained from initial vacuum perturbations is that the squeezing function in (4.8) has the same dependence on η as in the case of inflation. Hence the conversion of a vacuum spectrum to a scale-invariant way proceeds as in the case of inflation.

Note that a scale-invariant spectrum of gravitational waves is also generated in the matter bounce scenario. This demonstrates that the often stated claim that a detection of primordial gravitational waves on cosmological scales will confirm inflation is false (see Brandenberger [2011a] for a detailed discussion of this point). In fact, in the matter bounce scenario the analog of the inflationary slow-roll parameter ϵ is of order one, and hence there is no suppression of the amplitude of gravitational waves.

The matter bounce scenario faces significant problems. For one, the contracting phase is unstable against anisotropies (Cai et al., 2013). In addition, there is no suppression of gravitational waves compared to cosmological perturbations, and hence the amplitude of gravitational waves is predicted to be in excess of the observational bounds. If the cosmological perturbations are boosted in the bounce phase, then non-Gaussianities are induced, which are in excess of observational bounds (Quintin et al., 2015). Postulating a large graviton mass in the contracting phase can solve both of these problems (Lin, Quintin, and Breadenberger, 2018).

4.3.3 Pre–Big Bang and Ekpyrotic Scenarios

The pre–big bang (PBB) (Gasperini and Veneziano, 1993, 2003) and ekpyrotic scenarios (Khoury et al., 2001) are bouncing cosmologies that avoid the anisotropy and overproduction of gravitational wave problems of the matter bounce scenario. In both of these cosmologies, the contracting phase is dominated by a form of matter whose energy density increases as fast (in the case of the PBB scenario) or faster (in the case of the ekpyrotic scenario) than the contribution of anisotropy. In the case of the PBB scenario, it is the kinetic energy of the dilaton field (one of the massless degrees of freedom of string theory) that dominates in the contracting phase; in the case of the ekpyrotic scenario it is the energy density of a new scalar field with negative exponential potential. Both of these scenarios were initially proposed based on ideas in superstring theory, but they can also be viewed as effective field theories involving a new scalar field with some special features.

The spacetime sketches of the PBB and ekpyrotic scenarios are similar to that of the matter bounce paradigm (see Figure 4.6), except that the bounce is not symmetric. In particular, the horizon problem of standard big bang cosmology is solved in the same way as in the matter bounce, and in the same way fluctuations on all scales observed today originate inside the Hubble radius at early times, thus allowing a causal structure formation scenario. However, unlike what happens in the matter bounce scenario, the growth of fluctuations on super-Hubble scales in the contracting phase is too weak to convert an initial vacuum spectrum into a scale-invariant one. The resulting spectrum of curvature fluctuations and gravitational waves is blue (see, e.g., Brandenberger and Finelli [2001] and Lyth [2002]), thus not allowing initial vacuum perturbations to explain the observed structures in the universe, and predicting a negligible amplitude of gravitational waves on cosmological scales. As studied in Copeland, Easther, and Wands (1997a) and Copeland, Lidsey, and Wands (1997b) in the case of the PBB scenario and in Notari and Riotto (2002), Finelli (2002), Di Marco, Finelli, and Brandenberger (2003), Lehners et al. (2007), Buchbinder, Khoury, and Ovrut (2007), and Creminelli and Senatore (2007) in the case of the ekpyrotic scenario, a scale-invariant spectrum of curvature fluctuations can be obtained by using primordial vacuum fluctuations in a second scalar field.

4.3.4 Emergent String Gas Cosmology

Another alternative to cosmological inflation as a theory for the origin of structure in the universe is the *emergent scenario* as realized in string gas cosmology (SGC) (Brandenberger and Vafa, 1989). This scenario (see, e.g., Brandenberger [2011b, 2009] and Battefeld and Watson [2006] for reviews) is based, as discussed in the following section, on the idea that there was a long quasi-static phase in the early universe dominated by a thermal gas of fundamental strings. This phase may be past-eternal, or it may be preceded by a previous phase of contraction. At the end of this phase there is a transition to the expanding radiation phase of SBB cosmology.

Figure 4.7 is a spacetime sketch of SGC. The vertical axis is time, with the time t_R denoting the end of the quasi-static phase, the horizontal axis is physical distance. The curve labeled H^{-1} is the Hubble radius which is infinite in the Hagedorn phase, falls to a microscopic value at t_R and then increases linearly in time as in SBB cosmology. Since time goes back to $-\infty$, there is no horizon problem, as in the

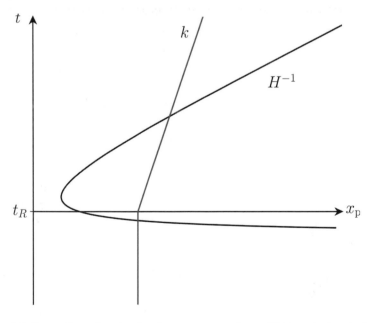

Figure 4.7 Spacetime sketch of string gas cosmology. The vertical axis is time, the horizontal axis is physical distance. The time t_R is the end of the Hagedorn phase and beginning of the radiation phase of SBB cosmology. The curve labeled k represents the wavelength of a fluctuation mode. The Hubble radius is labeled H^{-1}.

bouncing models discussed in previous sections. As is obvious from Figure 4.7, all scales originate from inside the Hubble radius at early times, thus allowing for a causal structure formation scenario.

As first realized in Nayeri, Brandenberger, and Vafa (2006), Brandenberger et al. (2007), and Brandenberger, Nayeri, and Patil (2014), thermal fluctuations of a gas of closed strings on a compact space with the topology of a torus lead to a scale-invariant spectrum of curvature fluctuations and gravitational waves. The tilt of the spectrum of curvature fluctuations is predicted to be red (as in the case of inflationary cosmology), but that of the gravitational waves is slightly blue, in contradistinction to what is obtained in inflation. The amplitude of gravitational waves is suppressed compared to that of curvature fluctuations by the equation of state parameter $w = p/\rho$ (Brandenberger et al., 2007, 2014).

The emergent string gas cosmology model is an example where the fluctuations are of thermal origin, not of quantum origin, as they are in most inflationary models (warm inflation being the exception) and the ekpyrotic scenario.

4.3.5 *Comparisons*

The main messages from this section is that there are a number of different early universe scenarios that can solve the horizon problem of SBB cosmology and that can generate a spectrum of curvature fluctuations consistent with observations. Inflationary cosmology is not the only scenario, although it was the first one and the only one that can claim to have made successful *predictions* (as opposed to postdictions). Inflation is also the only scenario that is at the present time self-consistent from the point of view of low energy effective field theory (Einstein gravity coupled to low energy quantum field matter). All the other scenarios mentioned earlier need new physics to obtain essential aspected of the dynamics of the cosmological background, be it a cosmological bounce or a quasi-static phase. Another nice feature that inflation has is that—at least in the case of large-field inflation—the slow roll trajectory in field space that leads to inflation is a local attractor in initial condition space (Brandenberger and Kung, 1990; Brandenberger, 2016). This feature is shared by the ekpyrotic scenario, but not some of the other ones.

On the other hand, in all of the alternatives to inflation mentioned, the physical wavelength of the fluctuations that are currently observed are much larger than the Planck length (as long as the maximal temperature is smaller than the Planck scale) at all times. Hence the trans-Planckian problem that the theory of cosmological fluctuations in inflationary cosmology suffers from is not present.

Unless model parameters are finely tuned, the energy scale at which inflation takes place is close to the particle physics scale of Grand Unification. This is close to the Planck scale and even closer to the preferred string scale (Green, Schwarz, and Witten, 1987). Hence the extrapolation of low energy physics to the scale of inflation and (to the bounce scale in bouncing cosmologies) needs to be justified. In spite of a large body of work, there have so far not been any convincing realizations of inflation in the context of superstring theory. Hence there are good reasons to look beyond standard inflationary cosmology.

Note that the structure formation scenarios described in this chapter are not the only ones. The main goal was to present a few very different scenarios and to show how the general criteria for a successful early universe model can be realized.

4.4 Hints from Superstring Theory

4.4.1 Challenges

The goal of superstring theory is to provide a quantum theory that unifies all four forces of nature, including gravity (see, e.g., Green et al. [1987] and Polchinski [1998a, 1998b] for textbook treatments of string theory). If nature is indeed described by superstring theory, then string theory should play a crucial role at the high energy densities that were present in the very early universe. Whichever of the scenarios for structure formation described in the previous section is in fact realized in nature should then be determined by string theory.

It has been shown to be very challenging to obtain an inflationary phase from string theory (see, e.g., Baumann and McAllister [2014] for a detailed discussion). The problem is that in order to obtain a period of slow-roll inflation from simple scalar field potentials, field values in excess of the Planck mass m_{pl} are required (see, e.g., the review in Brandenberger [2016]). However, for such large field values, string effects on the shape of the potential need to be considered and tend to destroy the required flatness of the potential unless special symmetries of the field theory (e.g., shift symmetry [Kawasaki, Yamaguchi, and Yanagida, 2000]) are considered. But even in this case, string theory arguments such as the weak gravity conjecture (Arkani-Hamed et al., 2007; Cheung and Remmen, 2014) tend to invalidate the effective field theory constructions.

4.4.2 String Thermodynamics

There are indications that the description of the very early universe in string theory will look very different from what can be obtained by models based on point particle theories (which includes all existing "string-derived" low energy effective field theories). The first evidence for this comes from string thermodynamics. It has been known (Hagedorn, 1965) from the earliest days of string theory that there is a maximal temperature T_H of a gas of closed string in thermal equilibrium, the so-called Hagedorn temperature. The value of this limiting temperature is given by the string scale. Assuming that space is a torus of radius R, then the evolution of the temperature T as a function of R is sketched in Figure 4.8. The length of the plateau region of $T(R)$ depends on the total entropy of the system: the larger the entropy, the larger the plateau region.

The origin of the $T(R)$ curve of Figure 4.8 is easy to understand. On a compact space, there are three types of states of strings: momentum modes (which correspond to the center of mass motion of the string), winding modes (which count the number of times the string wraps the torus), and the string oscillatory modes (whose energies are independent of R). The number of oscillatory states increases exponentially with the energy of the state. It is this fact that leads to the existence of the limiting temperature. As the energy density in a system is increased, then the energy will go into the excitation of new oscillatory states as opposed to the increase in the temperature.

The energies of the momentum and winding modes are quantized in units of $1/R$ and R, respectively. For momentum modes we have:

$$E_n = n\frac{1}{R}, \tag{4.13}$$

and for winding modes:

$$E_m = mR \tag{4.14}$$

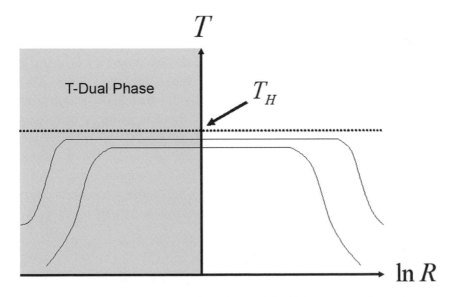

Figure 4.8 Temperature (vertical axis) as a function of the radius (horizontal axis) of a box of strings in thermal equilibrium.

(in string units), where n and m are integers. In fact, there is one momentum and one winding quantum number for each spatial dimension.

The above is the mass spectrum of free string states. It reflects an important symmetry of string theory, the T-duality symmetry, which implies that under the transformation:

$$R \to \frac{1}{R} \qquad (4.15)$$

(in string units), the spectrum of states is unchanged. This symmetry corresponds to an interchange between momentum and winding quantum numbers. It is a symmetry of string interactions, and it is assumed to be a symmetry of nonperturbative string theory, giving rise to the existence of D-branes (Polchinski, 1998b; see also Boehm and Brandenberger, 2003). This symmetry allows us to understand the symmetry of the $T(R)$ curve of Figure 4.8 about $\ln R = 0$: for large values of R, the energy wants to be in the modes that are light for large values of R, namely the momentum modes; while for $R \ll 1$ in string units, the energy will drift into the winding modes, which are the light ones in this range of R values.

Figure 4.8 immediately (Brandenberger and Vafa, 1989) leads one to expect that in the context of string theory the cosmological singularity can be avoided,[7] while in the context of Einstein gravity coupled to particle matter a temperature singularity as $R \to 0$ is unavoidable.

4.4.3 T-Duality Symmetry and Doubled Space

What is missing from the previous discussion is the dynamics of the background spacetime. This can obviously not be described by the Einstein–Hilbert action since the this action is incompatible with the T-duality symmetry of string theory. Dilaton gravity as described in Gasperini and Veneziano (1993, 2003) is a better starting point. In this case the T-duality symmetry of string theory is reflected in the so-called scale factor duality, which involves a transformation of both the metric and the dilaton field.

However, there may be a modified background description that is more useful for superstring cosmology. The starting point of this description is the following: in quantum mechanics the position eigenstates $|x>$ are dual to momentum eigenstates $|p>$. In a toroidal background, the momenta are discrete, labeled by integers n, and hence:

$$|x> = \sum_n e^{inx} |n>, \qquad (4.16)$$

where $|n>$ are the momentum eigenstates with momentum quantum numbers n. As discussed earlier, in string theory on a torus the windings are T-dual to momenta,

---

[7] The logic here is that if the temperature remains finite, the curvature should not be able to blow up.

and it is possible (Brandenberger and Vafa, 1989) to define a T-dual position operator:

$$|\tilde{x} >= \sum_{m} e^{im\tilde{x}}|m >, \qquad (4.17)$$

where $|m >$ are the eigenstates of winding, labeled by an integer m.

String states with both momenta and windings can be viewed as point particles propagating on a *doubled space*, which is spanned by both the position and the dual position eigenstates. The coordinates of this space are X^i:

$$X^i = (x^i, \tilde{x}^i). \qquad (4.18)$$

If the underlying target space of the strings has d spatial dimensions, the number of spatial dimensions of the doubled space is $2d$. This is the same space that is used in the double field theory (for reviews, see Siegel [1993a, 1993b], Hull and Zweibach [2009], and Aldazabal, Marques, and Nunez [2013]) approximation to string theory.

Let us consider a torus with radius R. If R is large in string units, then the light states are the momentum modes and a physical apparatus to measure length will be built from momentum modes. However, if R is small compared to the string length, then it is the winding modes that are light, and hence a physical apparatus will be constructed from the winding modes. Hence an apparatus measuring the physical length $l_p(R)$ will be measuring (see Figure 4.9):

$$l_p(R) = R \quad R \gg 1 \qquad (4.19)$$

$$l_p(R) = \frac{1}{R} \quad R \ll 1,$$

when R is expressed in units of the string length (Brandenberger and Vafa, 1989).

This yields a new interpretation of a dynamical evolution $R(t)$. The physical interpretation of R decreasing toward $R = 0$ is that the dual radius (which is also the physical radius) is increasing to $R = \infty$ in the dual directions. In double space we can express this dynamics by taking a cosmological metric in double space, which is:

$$ds^2 = dt^2 - a(t)dx^2 - a(t)^{-2}d\tilde{x}^2. \qquad (4.20)$$

As mentioned, physical measuring sticks measure length in terms of the coordinates related to the light string modes. Hence, for $R > 1$ (in string units), length will be measured in terms of x, while for $R < 1$ it is measured in terms of $\tilde{x}$. Thus a physical device will see space as contracting as R decreases toward $R = 1$, but for $R < 1$ it will be seen as increasing. Thus a physical observer will see no singularity. This argument is elaborated on in a recent paper (Brandenberger et al., 2018).

Robert H. Brandenberger

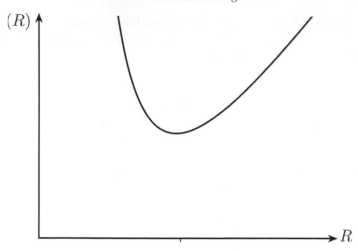

Figure 4.9 Physical length (vertical axis) as a function of the coordinate length (horizontal axis).

4.5 Discussion

In spite of the fact that the inflationary scenario has been a model of early universe cosmology with many successes, we may have to look beyond the standard inflationary scenario in order to understand the complete evolution of the early universe. All inflationary practitioners would admit that standard inflation cannot explain the evolution all the way back to the big bang as a consequence of the cosmological singularities that it does not avoid (Borde and Vilenkin, 1994). Although it is possible that the correct picture of the very early universe will involve a new phase followed by a period of inflation like we understand inflation today, this needs not be the case: there are a number of alternative early universe scenarios that lead to cosmological structure formation consistent with current observations.

Of all of the structure formation paradigms, inflation may be the only one that is self-consistent at the level of effective point particle field theories. However, if nature is described by string theory, then it may be difficult to embed standard inflation into the model, and an alternative such as string gas cosmology may emerge in a more natural way, as already argued in Brandenberger and Vafa (1989).

Acknowledgments

The final section of this article is based on collaborative research with Renato Costa, Guilherme Franzmann, and Amanda Weltman, research that is supported by the IRC—South Africa—Canada Research Chairs Mobility Initiative Grant No. 109684. I would like to express my thanks for all of my collaborators on the research reported on here. In particular, I thank Elisa Ferreira for producing many of the figures. I also thank the Canadian NSERC and the Canada Research Chair program for partial financial support.

References

Ade, P. A. R., Aghanim, N., Armitage-Caplan, C., et al. [Planck Collaboration]. (2013). Planck 2013 results. XV. CMB power spectra and likelihood. *Astron. Astrophys.*, 571, A15.

Ade, P. A. R., Aghanim, N., Armitage-Caplan, C., et al. [Planck Collaboration]. (2014). Planck 2013 results. XVI. Cosmological parameters. *Astron. Astrophys.*, 571, A16.

Albrecht, A. and Steinhardt, P. J. (1982). Cosmology for grand unified theories with radiatively induced symmetry breaking. *Phys. Rev. Lett.*, 48, 1220–1223.

Aldazabal, G., Marques, D. and Nunez, C. (2013). Double field theory: A pedagogical review. *Class. Quant. Grav.*, 30, 163001.

Arkani-Hamed, N., Motl, L., Nicolis, A. and Vafa, C. (2007). The string landscape, black holes and gravity as the weakest force. *JHEP*, 0706, 060.

Battefeld, T. and Watson, S. (2006). String gas cosmology. *Rev. Mod. Phys.*, 78, 435–454.

Baumann, D. and McAllister, L. (2014). Inflation and string theory. arXiv:1404.2601 [hep-th].

Berera, A. (1995). Warm inflation. *Phys. Rev. Lett.*, 75, 3218–3221.

Boehm, T. and Brandenberger, R. (2003). On T duality in brane gas cosmology. *JCAP*, 0306, 008.

Borde, A. and Vilenkin, A. (1994). Eternal inflation and the initial singularity. *Phys. Rev. Lett.*, 72, 3305–3309.

Brandenberger, R. H. (2004). Lectures on the theory of cosmological perturbations. *Lect. Notes Phys.*, 646, 127–167.

Brandenberger, R. H. (2009). String gas cosmology. Pages 193–230 of: Erdmenger, J. (ed.). *String Cosmology*. Hoboken, NJ: Wiley.

Brandenberger, R. H. (2011a). Is the spectrum of gravitational waves the 'holy grail' of inflation? arXiv:1104.3581 [astro-ph.CO].

Brandenberger, R. H. (2011b). String gas cosmology: Progress and problems. *Class. Quant. Grav.*, 28, 204005.

Brandenberger, R. (2016). Initial conditions for inflation—A short review. arXiv:1601.01918 [hep-th].

Brandenberger, R., Costa, R., Franzmann, G. and Weltman, A. (2018). Dual space-time and nonsingular string cosmology. arXiv:1805.06321 [hep-th]. Ithaca, NY: Cornell University.

Brandenberger, R. and Finelli, F. (2001). On the spectrum of fluctuations in an effective field theory of the ekpyrotic universe. *JHEP*, 0111, 056.

Brandenberger, R. H. and Kung, J. H. (1990). Chaotic inflation as an attractor in initial condition space. *Phys. Rev. D*, 42, 1008–1015.

Brandenberger, R. H. and Martin, J. (2001). The robustness of inflation to changes in superPlanck scale physics. *Mod. Phys. Lett. A*, 16, 999–1006.

Brandenberger, R. H. and Martin, J. (2013). Trans-Planckian issues for inflationary cosmology. *Class. Quant. Grav.*, 30, 113001.

Brandenberger, R. H., Nayeri, A., Patil, S. P. (2014). Closed string thermodynamics and a blue tensor spectrum. *Phys. Rev. D*, 90(6), 067301.

Brandenberger, R. H., Nayeri, A., Patil, S. P. and Vafa, C. (2007). Tensor modes from a primordial Hagedorn phase of string cosmology. *Phys. Rev. Lett.*, 98, 231302.

Brandenberger, R. and Peter, P. (2017). Bouncing cosmologies: Progress and problems. *Found. Phys.*, 47(6), 797–850.

Brandenberger, R. H. and Vafa, C. (1989). Superstrings in the early universe. *Nucl. Phys. B*, 316, 391–410.

Brout, R., Englert, F. and Gunzig, E. (1978). The creation of the universe as a quantum phenomenon. *Ann. Phys.*, 115, 78–106.

Buchbinder, E. I., Khoury, J. and Ovrut, B. A. (2007). New Ekpyrotic cosmology. *Phys. Rev. D*, 76, 123503.

Cai, Y. F., Brandenberger, R. and Peter, P. (2013). Anisotropy in a nonsingular bounce. *Class. Quant. Grav.*, 30, 075019.

Cai, Y. F. and Wilson-Ewing, E. (2015). A ΛCDM bounce scenario. *JCAP*, 1503(03), 006.

Cheung, C. and Remmen, G. N. (2014). Naturalness and the weak gravity conjecture. *Phys. Rev. Lett.*, 113, 051601.

Copeland, E. J., Easther R. and Wands, D. (1997a). Vacuum fluctuations in axion-dilaton cosmologies. *Phys. Rev. D*, 56, 874–888.

Copeland, E. J., Lidsey, J. E. and Wands, D. (1997b). S duality invariant perturbations in string cosmology. *Nucl. Phys. B*, 506, 407–420.

Creminelli, P. and Senatore, L. (2007). A smooth bouncing cosmology with scale invariant spectrum. *JCAP*, 0711, 010.

Di Marco, F., Finelli, F. and Brandenberger, R. (2003). Adiabatic and isocurvature perturbations for multifield generalized Einstein models. *Phys. Rev. D*, 67, 063512.

Fang, L. Z. (1980). Entropy generation in the early universe by dissipative processes near the Higgs' phase transitions. *Phys. Lett. B*, 95, 154–156.

Finelli, F. (2002). Assisted contraction. *Phys. Lett. B*, 545, 1–7.

Finelli, F. and Brandenberger, R. (2002). On the generation of a scale-invariant spectrum of adiabatic fluctuations in cosmological models with a contracting phase. *Phys. Rev. D*, 65, 103522.

Furlanetto, S., Oh, S. P. and Briggs F. (2006). Cosmology at low frequencies: The 21 cm transition and the high-redshift universe. *Phys. Rept.*, 433, 181–301.

Gasperini, M. and Veneziano, G. (1993). Pre–big bang in string cosmology. *Astropart. Phys.*, 1, 317–339.

Gasperini, M. and Veneziano, G. (2003). The pre–big bang scenario in string cosmology. *Phys. Rept.*, 373, 1–212.

Green, M. B., Schwarz, J. H. and Witten, E. (1987). *Superstring Theory*. Vols. 1 & 2. Cambridge: Cambridge University Press.

Gush, H. P., Halpern, M. and Wishnow, E. H. (1990). Rocket measurement of the cosmic-background-radiation mm-wave spectrum. *Phys. Rev. Lett.*, 65, 537–540.

Guth A. H. (1981). The inflationary universe: A possible solution to the horizon and flatness problems. *Phys. Rev. D*, 23, 347.

Hagedorn, R. (1965). Statistical thermodynamics of strong interactions at high-energies. *Nuovo Cim. Suppl.*, 3, 147.

Hinshaw, G., Spergel, D. N., Verde, L., et al. [WMAP Collaboration]. (2000). First year Wilkinson Microwave Anisotropy Probe (WMAP) observations: The angular power spectrum. *Astrophys. J. Suppl.*, 148, 135.

Hull, C. and Zwiebach, B. (2009). Double field theory. *JHEP*, 0909, 099.

Kawasaki, M., Yamaguchi, M. and Yanagida, T. (2000). Natural chaotic inflation in supergravity. *Phys. Rev. Lett.*, 85, 3572.

Khoury, J., Ovrut, B. A., Steinhardt, P. J. and Turok, N. (2001). The ekpyrotic universe: Colliding branes and the origin of the hot big bang. *Phys. Rev. D*, 64, 123522.

Lehners, J. L., McFadden, P., Turok, N. and Steinhardt, P. J. (2007). Generating ekpyrotic curvature perturbations before the big bang. *Phys. Rev. D*, 76, 103501.

Lin, C., Quintin, J. and Brandenberger, R. H. (2018). Massive gravity and the suppression of anisotropies and gravitational waves in a matter-dominated contracting universe. *JCAP*, 1801, 011.

Linde, A. D. (1982). A new inflationary universe scenario: A possible solution of the horizon, flatness, homogeneity, isotropy and primordial monopole problems. *Phys. Lett.*, 108B, 389–393.

Linde, A. D. (1983). Chaotic inflation. *Phys. Lett.*, 129B, 177–181.

Lyth, D. H. (2002). The primordial curvature perturbation in the ekpyrotic universe. *Phys. Lett. B*, 524, 1–4.

Martin, J. and Brandenberger, R. H. (2001). The TransPlanckian problem of inflationary cosmology. *Phys. Rev. D*, 63, 123501.

Mather, J. C., Cheng, E. D., Eplee, R. E., et al. (1990). A preliminary measurement of the cosmic microwave background spectrum by the Cosmic Background Explorer (COBE) satellite. *Astrophys. J.*, 354, L37–L40.

Mukhanov, V. F. (1988). Quantum theory of gauge invariant cosmological perturbations. *Sov. Phys. JETP*, 67, 1297–1302. [*Zh. Eksp. Teor. Fiz.*, 94N7, 1].

Mukhanov, V. F. and Chibisov, G. (1981). Quantum fluctuation and nonsingular universe. (In Russian). *JETP Lett.*, 33, 532–535. [*Pisma Zh. Eksp. Teor. Fiz.*, 33, 549].

Mukhanov, V. F., Feldman, H. A. and Brandenberger, R. H. (1992). Theory of cosmological perturbations. Part 1. Classical perturbations. Part 2. Quantum theory of perturbations. Part 3. Extensions. *Phys. Rept.*, 215, 203–333.

Nayeri, A., Brandenberger, R. H. and Vafa, C. (2006). Producing a scale-invariant spectrum of perturbations in a Hagedorn phase of string cosmology. *Phys. Rev. Lett.*, 97, 021302.

Notari A. and Riotto, A. (2002). Isocurvature perturbations in the ekpyrotic universe. *Nucl. Phys. B*, 644, 371–382.

Peebles, P. J. E. and Yu, J. T. (1970). Primeval adiabatic perturbation in an expanding universe. *Astrophys. J.*, 162, 815–836.

Penzias, A. A. and Wilson, R. W. (1965). A measurement of excess antenna temperature at 4080-Mc/s. *Astrophys. J.*, 142, 419–421.

Polchinski, J. (1998a). *String Theory*. Vol. 1: *An Introduction to the Bosonic String*. Cambridge: Cambridge University Press.

Polchinski, J. (1998b). *String Theory*. Vol. 2: *Superstring Theory and Beyond*. Cambridge: Cambridge University Press.

Press, W. H. (1980). Spontaneous production of the Zel'dovich spectrum of cosmological fluctuations. *Phys. Scripta*, 21, 702.

Quintin, J., Sherkatghanad, Z., Cai, Y. F. and Brandenberger, R. H. (2015). Evolution of cosmological perturbations and the production of non-Gaussianities through a nonsingular bounce: Indications for a no-go theorem in single field matter bounce cosmologies. *Phys. Rev. D*, 92(6), 063532.

Sasaki, M. (1986). Large scale quantum fluctuations in the inflationary universe. *Prog. Theor. Phys.*, 76, 1036–1046.

Sato, K. (1981). First order phase transition of a vacuum and expansion of the universe. *Mon. Not. Roy. Astron. Soc.*, 195, 467–479.

Siegel, W. (1993a). Superspace duality in low-energy superstrings. *Phys. Rev. D*, 48, 2826.

Siegel, W. (1993b). Two vierbein formalism for string inspired axionic gravity. *Phys. Rev. D*, 47, 5453.

Smoot, G. F., Bennett, C. L., Kogut, A., et al. [COBE Collaboration] (1992). Structure in the COBE differential microwave radiometer first year maps. *Astrophys. J.*, 396, L1–L5.

Starobinsky, A. A. (1979). Spectrum of relict gravitational radiation and the early state of the universe. *JETP Lett.*, 30, 682. [*Pisma Zh. Eksp. Teor. Fiz.*, 30, 719].

Starobinsky, A. A. (1980). A new type of isotropic cosmological models without singularity. *Phys. Lett. B*, 91, 99.

Sunyaev, R. A. and Zeldovich, Y. B. (1970). Small scale fluctuations of relic radiation. *Astrophys. Space Sci.*, 7, 3–19.

Wands, D. (1999). Duality invariance of cosmological perturbation spectra. *Phys. Rev. D*, 60, 023507.

5

What Black Holes Have Taught Us about Quantum Gravity

DANIEL HARLOW

Formulating a theory of quantum gravity has been a major goal of theoretical physics for almost a century. Early efforts to 'quantize' general relativity using the techniques that had succeeded for electrodynamics did not succeed for gravity, despite considerable efforts by some of the greatest theoretical physicists of the twentieth century (Feynman, 1963; DeWitt, 1967; 't Hooft and Veltman, 1974). In more precise terms, they were unable to fit general relativity within the framework of relativistic quantum field theory. There are two reasons that are sometimes given (see, e.g., the introduction of Becker, Becker, and Schwarz [2006]) for this failure:

- **Gravity is not renormalizable:** If one attempts to construct a perturbative description of gravitational interactions starting from the Einstein–Hilbert action, as one does in the standard model of particle physics starting from the Yang–Mills/matter Lagrangian, then, unlike in the latter case, one encounters an infinite number of short-distance divergences as one goes to higher and higher orders in perturbation theory.
- **Gravity is diffeomorphism invariant:** In general relativity the causal structure of spacetime depends on the metric, which is a dynamical variable. There is no physical meaning to any particular set of coordinates we use to describe physics, and as a result there is no notion of strictly local observables.

These statements are both true, but in fact neither one necessarily prevents a complete formulation of quantum gravity based on a local Lagrangian in spacetime. For example the Fermi theory of weak interactions and the chiral Lagrangian description of pions interacting with nucleons are both nonrenormalizable field theories, but both have long ago been 'UV completed' by the renormalizable standard model of particle physics. So far this has not been done for gravity, but the program of 'asymptotic safety' (see, e.g., Benedetti, Machado, and Saueressig [2009]), seeks to do precisely this. Similarly, there are several low-dimensional examples of diffeomorphism-invariant theories that have perfectly satisfactory formulations based on local Lagrangians: two especially simple ones are the worldline action for a free relativistic particle and the Nambu–Goto action for a free relativistic string (see, e.g., Polchinski [2007] for both).[1] If nonrenormalizability and diffeomorphism invariance are not to blame for the difficulty in quantizing gravity however, what is? The goal of this essay is to argue that ultimately it is black holes that are to blame.

The existence of black holes arises from the combination of relativistic invariance and the universality of gravitational interactions. There could be no such thing as a 'Newtonian black hole', since we could always escape any gravitational field by moving fast enough, and there could also be no such thing as an 'electromagnetic black hole', because no matter how strong the electric field in a region of space, a particle that does not carry electric charge can simply fly through unaffected. By contrast, relativity tells us that no object can move faster than the speed of light, and the universality of gravity tells us that there is no such thing as a particle that is 'neutral under gravity'. It is therefore possible to have a region of spacetime where the gravitational field is so strong that nothing can escape it.

Black holes are important in quantum gravity because they put strong restrictions on what kind of experiments probing the structure of spacetime are possible. In particular, they prevent construction of the kind of local observables that are the basic currency of quantum field theory. One way to understand this is to imagine how we might operationally measure a local observable in quantum gravity. For example we could try to construct a cubic network of rods of length ℓ filling a region of space, and then we could do experiments like 'walk 12 rods in the x direction, 5 rods in the y direction, and 17 rods in the z direction, and then measure the electric field.' In order for this experiment to make sense, however, there are two basic constraints it needs to satisfy. The first is that our collection of rods should not be so heavy that it collapses to form a black hole, and the second is that the rods should be heavy enough that quantum fluctuations in their position are small compared to the distances we are trying to resolve. I'll now show that these constraints are not compatible for networks that are fine enough to probe quantum gravity.

Indeed say the linear size of the collection of rods is L, and that each rod has mass m. To avoid collapse to a black hole, L needs to be much larger than the Schwarzschild radius of the whole collection:

$$L \gg \frac{2GmL^3}{c^2\ell^3}. \tag{5.1}$$

---

[1] It is true that theories with diffeomorphism invariance cannot quite be quantum field theories in the precise sense, since they lack an energy momentum tensor and local operators, but these examples show that they can nonetheless have local path integral formulations, which is all we could have asked for from gravity in the first place. The 'local observables' I describe in this chapter always include 'gravitational dressing' to make them diffeomorphism invariant.

There are various ways to rewrite this constraint, one useful way is as

$$\frac{m}{m_p} \ll \left(\frac{\ell}{L}\right)^2 \frac{\ell}{\ell_p}, \qquad (5.2)$$

where m_p and ℓ_p are the Planck mass and length. In order to have a useful set of rods we need $L \gg \ell$, so the first factor on the right-hand side will be small. In daily situations however ℓ/ℓ_p is very large, so this bound will not be very constraining on m: this is a good thing, since, for example, it is what prevents the Laser Interferometer Gravitational Wave Observatory (LIGO) detector from collapsing into a black hole despite the fact that $m_p \sim 10^{-8}$ kg. In quantum gravity, however, we are indeed interested in distances of order $\ell_p \sim 10^{-35}$ m, so we need to take $\ell \sim \ell_p$. Our 'no black hole' inequality (5.2) then tells us that to probe Planck-scale physics, we need

$$m \ll m_p. \qquad (5.3)$$

To see how quantum fluctuations constrain the mass in the other direction, we can use the uncertainty principle. Indeed we need

$$\ell \gg \delta x > \frac{\hbar}{\delta p} \approx \frac{\hbar}{m \delta v} \gg \frac{\hbar}{mc}, \qquad (5.4)$$

where I've also required that the rod velocities are nonrelativistic. We can rewrite this as

$$\frac{m}{m_p} \gg \frac{\ell_p}{\ell}, \qquad (5.5)$$

so in daily situations the right-hand side is very small and we again don't have much to worry about. In studying quantum gravity, however, we again need to take $\ell \sim \ell_p$, in which case we have

$$m \gg m_p, \qquad (5.6)$$

which is obviously inconsistent with (5.3). Thus the combination of black holes and quantum mechanics forbids the operational definition of local observables, which are sharp enough to probe quantum gravity at the Planck scale.

Another way to get at the same physics is via the famous Bekenstein–Hawking formula for the entropy of a black hole (Bekenstein, 1973; Hawking, 1975),

$$S_{BH} = \frac{c^3 A}{4G\hbar}. \qquad (5.7)$$

Here A is the horizon area of the black hole. This formula tells us something rather remarkable about quantum gravity: the number of independent degrees of freedom in a spacetime region is *subextensive*, or in other words, it grows with the size of the region less quickly than its volume. This again is telling us that the quantum gravity does not have all the local observables that a quantum field theory would.

In the early 1990s, the Bekenstein–Hawking formula (5.7) led 't Hooft and Susskind to the idea that if a theory of quantum gravity is local, it has to be local in a *lower number of spacetime dimensions* ('t Hooft, 1993; Susskind, 1995). Susskind christened this idea the *holographic principle*. At first this was viewed with some skepticism by the theoretical physics community, but this changed immediately in 1998 with the discovery by Maldacena of a set of explicit examples of holographic theories of quantum gravity: the anti–de Sitter spacetime conformal field theory (AdS/CFT) correspondence (Maldacena, 1999). Today this correspondence remains our most complete and well understood example of a theory of quantum gravity, and for 20 years it has been perhaps the most prolific source of new ideas in theoretical physics (see Harlow [2018] for a recent overview of the correspondence as well as the ideas described in the remainder of this essay).

The AdS/CFT correspondence says that any theory of quantum gravity in $d + 1$ dimensional asymptotically anti–de Sitter space is defined nonperturbatively by a d-dimensional conformal field theory living on the boundary of that space. I illustrate this setup in Figure 5.1: the CFT is the 'can', while the AdS is the 'soup'. This situation is usually described by saying that the 'bulk' quantum gravity theory is *dual* to the 'boundary' conformal field theory, and the radial direction, and therefore the bulk spacetime, is described as *emergent*. This necessary emergence of the bulk spacetime is perhaps the main thing black holes have taught us about quantum gravity.

In the philosophy literature there has been some confusion over the meaning of the words *dual* and *emergent* in this context (Rickles, 2013; De Haro, 2015; Dieks, van Dongen, and De Haro, 2015). This confusion arises because usually something that is emergent is only approximate, while AdS/CFT is usually claimed to be an

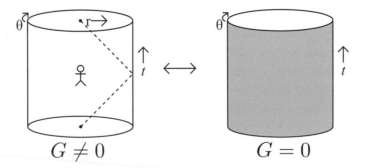

Figure 5.1 Three-dimensional AdS/CFT. On the left we have AdS_3, with gravity, while on the right we have CFT_2, without gravity. The t and θ coordinates are manifest in the boundary CFT, while the radial r coordinate is emergent.

exact correspondence. The explanation of this terminology is that in AdS/CFT there are really *three* theories being considered: the boundary conformal field theory, nonperturbative quantum gravity in the bulk, and low-energy effective field theory in the bulk. The first two are claimed to be exactly dual/equivalent to each other, while the third emerges approximately from either of the first two in appropriate situations. This is illustrated for the best-known example of AdS/CFT, IIB string theory on $AdS_5 \times \mathbb{S}^5$, in Figure 5.2. Since we do not currently have an independent formulation of nonperturbative string theory, it is perhaps safest to view the duality as using the boundary conformal field theory to define it. We may then ask to what extent this is a good definition, with the criterion for success being that low-energy effective field theory coupled to gravity emerges in appropriate situations.

How can a conformal field theory in d spacetime dimensions moonlight as a theory of quantum gravity in $d+1$ spacetime dimensions with an emergent effective field theory description? In recent years considerable progress has been made on this question. One way of understanding the basic problem is as follows (Almheiri, Dong, and Harlow, 2015): consider the algebra of an operator $\phi(x)$ in the center of a time slice of AdS space with the set of operators $\mathcal{O}(X)$ at the boundary of the slice. The basic setup is shown in Figure 5.3. Since $\phi(x)$ and $\mathcal{O}(X)$ are spacelike separated in the radial direction r, locality naively suggests that they should be commuting operators. But this is in tension with a basic property of any quantum field theory, including the boundary conformal field theory: any operator that commutes with all local operators at a fixed time must be proportional to the identity (Streater and Wightman, 1989). This observation formalizes the notion that a given theory cannot really be local with two different spacetime dimensionalities. But how then is AdS/CFT possible? What was realized in (Almheiri et al., 2015) is

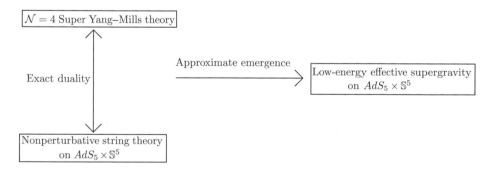

Figure 5.2 The logic of AdS/CFT. The exact duality for this example was established by the D-brane decoupling argument of Maldacena (1999), *assuming* that nonperturbative string theory exists. The emergence of low-energy supergravity is obvious in string perturbation theory, but is quite mysterious from the point of view of the boundary conformal field theory.

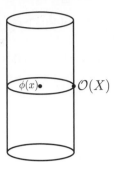

Figure 5.3 Locality in the radial direction requires operators near the center of AdS commute with operators near the boundary.

that this problem had actually already been solved, but by quantum information theorists!

Quantum information theory is a relatively young field, which grew up around the idea of figuring out what might be possible with a quantum computer (see Nielsen and Chuang [2010] for an introduction and many more references). One of the most basic problems in the theory of quantum information is to find a way to perform quantum computations with imperfect hardware. The standard way of doing this with classical computers is to do the computation redundantly: make extra copies of everything and do the steps multiple times to make sure you get the right answer. In quantum computing, however, we are constrained by the famous no-cloning theorem, which says that quantum information cannot be copied (Wootters and Zurek, 1982). This might seem fatal for the feasibility of quantum computation, but in fact while quantum mechanics taketh away, it also giveth: quantum entanglement gives a new way to introduce redundancy without running afoul of the no-cloning theorem, and fault-tolerant quantum computation is indeed possible. The essential idea is to encode the computation using a *quantum error correcting code*, which stores the relevant quantum information nonlocally in the entanglement between many degrees of freedom. This nonlocality offers the encoded information a degree of protection against outside interference: what was pointed out in (Almheiri et al., 2015) is that this protection is precisely analogous to how $\phi(x)$ and $\mathcal{O}(X)$ are able commute in AdS/CFT as required by radial locality. The physical qubits of the quantum computer are analogous to the boundary conformal field theory, and the encoded computation is analogous to the quantum gravity that it is simulating.

This idea can be studied in considerable generality, but here I will just give a simple example to illustrate the basic point. The example is illustrated in Figure 5.4: the CFT Hilbert space is the tensor product of three qutrits, which are shown as solid dots. The bulk theory consists of a single qutrit, sitting at the hollow dot in the

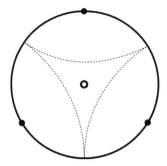

Figure 5.4 The simplest model of emergent spacetime: the three qutrit code.

center of the bulk. The state of the bulk qutrit is encoded in the boundary qutrits as follows:

$$|\widetilde{0}\rangle = \frac{1}{\sqrt{3}} (|000\rangle + |111\rangle + |222\rangle)$$

$$|\widetilde{1}\rangle = \frac{1}{\sqrt{3}} (|012\rangle + |120\rangle + |210\rangle)$$

$$|\widetilde{2}\rangle = \frac{1}{\sqrt{3}} (|021\rangle + |102\rangle + |210\rangle). \tag{5.8}$$

The subspace of states spanned by $|\widetilde{i}\rangle$ is called the *code subspace*, it is a 3-dimensional subspace of the 27-dimensional Hilbert space of the three boundary qutrits. Bulk operators like $\phi(x)$ are represented in this model as operators that act within the code subspace, and the magic of the subspace (5.8) is that any such operator ϕ has the rather remarkable property that

$$\langle \widetilde{\psi} | [\phi, X] | \widetilde{\chi} \rangle = 0 \tag{5.9}$$

for any states $|\widetilde{\psi}\rangle$ and $|\widetilde{\chi}\rangle$ in the code subspace and X any operator with support on only one of the boundary qutrits. In other words, any operator on the hollow dot commutes with any operator on any one of the solid dots: this is precisely locality in the radial direction! The reason we avoid the contradiction of Figure 5.3 is that we did *not* demand that $[\phi, X] = 0$ as an operator equation: we only have its matrix elements in the code subspace vanish.

But what are we to make of the states in the orthogonal complement of the code subspace? If we are to interpret the AdS/CFT correspondence as a duality, then all states in the CFT Hilbert space must bulk interpretations, while in our model so far we have given a bulk interpretation only to states in the code subspace spanned by the states (5.8). I claim that the remaining 24-dimensional set of states are nothing but the microstates of a black hole that in those states has swallowed our bulk point!

Daniel Harlow

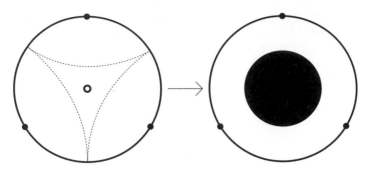

Figure 5.5 Leaving the code subspace leads to black hole formation.

This is shown in Figure 5.5. This may seem a bit ad hoc in this model, but study of further examples has shown that this is the correct interpretation. In particular, in AdS/CFT a rather general theorem of quantum error correction can be used to show that pushing the holographic error correcting code to the point where boundary locality will no longer allow locality in the emergent radial direction corresponds precisely in the gravitational description to the formation of a black hole (Almheiri et al., 2015; see also Harlow [2018]). Thus we return to our initial discussion: black holes tell us that the local structure of the degrees of freedom in bulk gravitational effective field theory is an emergent notion, valid only in a subset of all physical states, and that the real microscopic degrees of freedom are organized in a quite different manner. In recent years we have begun to get a rather detailed quantitative understanding of how this works in our best-understood theory of quantum gravity, the AdS/CFT correspondence. Surely there is more excitement ahead.

References

Almheiri, A., Dong, X., and Harlow, D. 2015. Bulk locality and quantum error correction in AdS/CFT. *JHEP*, **04**, 163, 1–33.
Becker, K., Becker, M., and Schwarz, J. H. 2006. *String Theory and M-Theory: A Modern Introduction*. Cambridge: Cambridge University Press.
Bekenstein, J. D. 1973. Black holes and entropy. *Phys. Rev.*, **D7**, 2333–2346.
Benedetti, D., Machado, P. F., and Saueressig, F. 2009. Asymptotic safety in higher-derivative gravity. *Mod. Phys. Lett.*, **A24**, 2233–2241.
De Haro, S. 2015. Dualities and emergent gravity: Gauge/gravity duality. *Stud. Hist. Philos. Sci. B Stud. Hist. Philos. Mod. Phys.*, **59**, 109–125.
DeWitt, B. S. 1967. Quantum theory of gravity. 1. The canonical theory. *Phys. Rev.*, **160**, 1113–1148.
Dieks, D., van Dongen, J., and De Haro, S. 2015. Emergence in holographic scenarios for gravity. *Stud. Hist. Philos. Sci. B Stud. Hist. Philos. Mod. Phys.*, **52**, 203–216.
Feynman, R. P. 1963. Quantum theory of gravitation. *Acta Phys. Polon.*, **24**, 697–722.
Harlow, D. 2018. TASI Lectures on the Emergence of the Bulk in AdS/CFT. asXiv:1802.01040 [hep-th]. Ithaca, NY: Cornell University.

Hawking, S. W. 1975. Particle creation by black holes. *Commun. Math. Phys.*, **43**, 199–220.

Maldacena, J. M. 1999. The Large N limit of superconformal field theories and supergravity. *Int. J. Theor. Phys.*, **38**, 1113–1133.

Nielsen, M. A., and Chuang, I. L. 2010. *Quantum Computation and Quantum Information*. Cambridge: Cambridge University Press.

Polchinski, J. 2007. *String Theory*. Vol. 1: *An Introduction to the Bosonic String*. Cambridge: Cambridge University Press.

Rickles, D. 2013. AdS/CFT duality and the emergence of spacetime. *Stud. Hist. Philos. Sci. B Stud. Hist. Philos. Mod. Phys.*, **44**(3), 312–320.

Streater, R. F., and Wightman, A. S. 1989. *PCT, Spin and Statistics, and All That*. Princeton, NJ: Princeton University Press.

Susskind, L. 1995. The world as a hologram. *J. Math. Phys.*, **36**, 6377–6396.

't Hooft, G. 1993. Dimensional reduction in quantum gravity. *Proceedings of the Conference on Highlights of Particle and Condensed Matter Physics (SALAMFEST)*, **C930308**, 284–296.

't Hooft, G., and Veltman, M. J. G. 1974. One loop divergencies in the theory of gravitation. *Ann. Inst. H. Poincare Phys. Theor.*, **A20**, 69–94.

Wootters, W. K., and Zurek, W. H. 1982. A single quantum cannot be cloned. *Nature*, **299**(5886), 802–803.

Part II

Time in Quantum Theories of Gravity

6

Space and Time in Loop Quantum Gravity

CARLO ROVELLI

6.1 Introduction

Newton's success sharpened our understanding of the nature of space and time in the seventeenth century. Einstein's special and general relativity improved this understanding in the twentieth century. Quantum gravity is expected to take a step further, deepening our understanding of space and time, by grasping the implications for space and time of the quantum nature of the physical world.

The best way to see what happens to space and time when their quantum traits cannot be disregarded is to look how this actually happens in a concrete theory of quantum gravity. Loop quantum gravity (LQG) (Gambini and Pullin, 1996, 2010; Rovelli, 2004; Theimann, 2007; Ashtekar, 2011; Perez, 2012; Rovelli and Vidotto, 2014) is among the few current theories sufficiently developed to provide a complete and clear-cut answer to this question.

Here I discuss the role(s) that space and time play in LQG and the version of these notions required to make sense of a quantum gravitational world. For a detailed discussion, see the first part of Rovelli (2004).

A brief summary of the structure of LQG is given in the Appendix to this chapter, for the reader unfamiliar with this theory.

6.2 Space

Confusion about the nature of space—even more so for time—originates from failing to recognize that these are stratified, multilayered concepts. They are charged with a multiplicity of attributes and there is no agreement on a terminology to designate spatial or temporal notions lacking some of these attributes. When we say *space* or *time* we indicate different things in different contexts.

The only route to clarify the role of space and time in quantum gravity is to ask what we mean *in general* when we say *space* or *time* (van Fraassen, 1985). There are distinct answers to this question; each defines a different notion of space or time. Let's disentangle them. I start with space, and move to time, which is more complex, later on.

6.2.1 Relational Space

Space is the relation we use when we locate things. We talk about space when we ask "*Where* is Andorra?" and answer "Between Spain and France." Location is established in relation to something else (Andorra is located by Spain and France). Used in this sense space is a relation between things. It does not require metric connotations. It is the notion of space Aristotle refers to in his *Physics*, Descartes founds on contiguity, and so on. In mathematics it is studied by topology. This is a very general notion of space, equally present in ancient, Cartesian, Newtonian, and relativistic physics.

This notion of space is equally present in LQG. In LQG, in fact, we can say that something is in a certain location with respect to something else. A particle can be at the same location as a certain quantum of gravity. We can also say that two quanta are *adjacent*. The network of adjacency of the elementary quanta of the gravitational field is captured by the graph of a *spin network* (see Appendix). The links of the graph are the elementary adjacency relations. Spin networks describe relative *spatial* arrangements of dynamical entities: the elementary quanta.

6.2.2 Newtonian Space

In the seventeenth century, in the *Principia*, Newton (1934) introduced a *distinction* between two notions of space. The first, which he called the "common" one, is the one illustrated in the previous item. The second, which he called the "true" one, is what has been later called Newtonian space. Newtonian space is not a relation between objects: it is assumed by Newton to exist also in the absence of objects. It is an entity with no dynamics, with a metric structure: that of a 3-dimensional Euclidean manifold. It is postulated by Newton on the basis of suggestions from ancient Democritean physics and is essential for his theoretical construction.[1] Special relativity alters this ontology substantially by merging Newtonian space and time into Minkowski's spacetime, but retains the Newtonian logic of a (background)

---

[1] During the nineteenth century, certain awkward aspects of this Newtonian hypostasis led to the development of the notion of "physical reference system": the idea that Newtonian space captures the properties of preferred systems of bodies not subjected to forces. This is correct but already presupposes the essential ingredient: a fixed metric space, permitting to locate things with respect to distant references bodies. Thus the notion of reference system does not add much to the novelty of the Newtonian ontology.

entity assumed to exist also in the absence of objects. It is an entity with no dynamics, with a metric structure.

In quantum gravity, Minkowski spacetime and hence Newtonian space *appear only as an approximations*, as we shall see later. They have no role at all in the foundation of the theory.

6.2.3 General Relativistic Space

Our understanding of the true physical nature of Newtonian space (and spacetime) underwent a radical sharpening with general relativity (GR). The empirical success of GR, which slowly accumulated for a century before booming recently, adds strong credibility to the effectiveness of this step. GR shows that Newtonian space is in fact an entity as Newton postulated—it exists even where nothing else exists, has properties, affects the behavior of objects—but it is not nondynamical as Newton assumed. Rather, it is a *dynamical* entity, very much akin to the electromagnetic field: a "gravitational field." Therefore there are two distinct notions of space in GR. The first is the concept of relational space, the simple fact that dynamical entities (all entities in the theory are dynamical) are localized with respect to one another ("This gravitational wave pulse is inside the black hole"). The second is a modification of the Newtonian concept of space: one specific dynamical entity, the gravitational field, which we keep calling space (or spacetime) as a leftover habit from Newtonian logic. There is nothing wrong in doing so, provided that the strong difference between these three notions of space (order of localization, Newtonian nondynamical space, gravitational field) remains clear.

LQG treats space (in this sense) precisely as GR does: a dynamical entity that behaves as Newtonian space in a certain approximation. In addition, this entity has the usual properties of *quantum* entities. Thus GR is the classical field theory analogous to classical electromagnetism (EM), and LQG is the quantum theory that describes the quantum gravitational field, analogous to quantum electrodynamics.

The characteristic properties of a quantum entity are three: (1) Granularity: photons embody the quantum granularity of the EM field. They describe the fact that exchanges with the quantum field are in packets. The quantum gravitational field has the same granularity; it interacts in discrete packets. Its elementary quanta are given by the discrete nodes of a spin network (see Appendix). Spin network states form a basis in the Hilbert space of LQG, like photon states form a basis in the Hilbert state of quantum EM. (2) Indeterminism: the dynamics of such "quanta of space" is probabilistic like that of photons. (3) Quantum relationalism: in its basic interpretation, quantum theory describes *interactions* among systems (uglily called "measurements") where properties actualize (Rovelli, 2017). So is in

LQG: the theory describes properties of the gravitational field becoming actual in interactions.

Section 6.3, illustrates the remarkable outcome from combining such quantum relationalism with the relational nature of localization.

6.3 Time

The case with time is parallel to space, but with some additional levels of complexity.

6.3.1 Relational Time

Time is the relation we use when we locate events. We are talking about time when we ask "*When* shall we meet?" and answer "In three days." Location of events is given with respect to something else. (We shall meet after three sunrises.) Used in this sense time is a relation between events. This is the notion Aristotle refers to in his *Physics*,[2] and so on. It is a very general notion of time, equally present in ancient, Cartesian, Newtonian, and relativistic physics.

When used in this wide sense, time is definitely present in LQG. In LQG we can say that something happens *when* something else happens. For instance, a particle is emitted when two quanta of gravity join. Also, we can say that two events are temporally adjacent. A network of temporal adjacency of elementary processes of the gravitational field is captured by the *spinfoams* (see Appendix).

6.3.2 Newtonian Time

In the *Principia*, Newton distinguished between two notions of time. The first, which he called the common one, is the one in the previous section. The second, which he called the true one, is what has been later called Newtonian time. Newtonian time is assumed to be "flowing uniformly," even when nothing happens, with no influence from events, and to have a metric structure: we can say when two time intervals have equal duration. Special relativity shows that the clock time between two events depends on the clock's motion, but it modifies the basic structure of Newtonian ontology only by merging Newtonian space and time into Minkowski spacetime.

In LQG (Minkowski spacetime and hence) Newtonian time *appears only as an approximation*. It has no role at all in the foundation of the theory.

---

[2] The famous definition is: Time is ἀριθμός κινήσεως κατὰ τὸ πρότερον καὶ ὕστερον "The number of change with respect to before and after" (*Physics*, IV, 219 b 2; see also 232 b 22–23) (Aristotle, 1990).

6.3.3 General Relativistic Time

What GR has shown is that Newtonian time is indeed (part of) an entity as Newton postulated, but this entity is not nondynamical as Newton assumed. Rather, it is an aspect of a dynamical field, the gravitational field. What the reading T of a common clock tracks, for instance, is a function of the gravitational field $g_{\mu\nu}$,

$$T = \int \sqrt{g_{\mu\nu}\, dx^\mu dx^\nu}. \tag{6.1}$$

In GR, therefore, there are two distinct kinds of temporal notions. The first is the simple fact that all events are localized with respect to one another ("This gravity wave has been emitted *when* the two neutron stars have merged," "The binary pulsar emits 700 pulses *during* an orbit"). The second is a leftover habit from Newtonian logic: the habit of calling time (in spacetime) aspects of one specific dynamical entity: the gravitational field. Again, there is nothing wrong in doing so, provided that the difference between three notions of time (relative order of events, Newtonian nondynamical time, the gravitational field) are clear.

LQG treats time (in this sense) as GR does: there is no preferred clock time, but many clock times measured by different clocks. In addition, however, clock times undergo standard quantum fluctuations like any other dynamical variable. There can be quantum superpositions between different values of the same clock time variable T.

What I mean is not to downplay the novelties imposed by quantum gravity. What I mean is to place them where they belong. Confusion in the discussion about time in quantum gravity comes from mixing up the "natural" notion of time as a measure of change—which is common in our daily life and was the basis of the understanding of time from antiquity to Descartes—with the radically new Newtonian idea of a time that exists by itself independent from things. The second is altered by quantum gravity. The first is left untouched.

Our common intuition about time is profoundly marked by natural phenomena that are not *generally* present in fundamental physics. Unless we disentangle these from the aspects of time described earlier, confusion reigns (I have extensively discussed the multiple aspects of temporality in a recent book [Rovelli, 2018]). These fall into two classes.

6.3.4 Irreversible Time

When dealing with many degrees of freedom we recur to statistical and thermodynamical notions. In an environment with an entropy gradient, namely where in one time direction (the "past") entropy was low, there are irreversible phenomena. The existence of traces of the past versus the absence of traces of the future, or the apparent asymmetry of causation and agency, are consequences of this entropy gradient since this is the only source of time asymmetry, as all fundamental physics, including LQG, is time symmetric. Our common intuition about time is profoundly marked by these phenomena. We do not know why entropy was as low in the past universe (Earman, 2006). A possibility is that this is a perspectival effect due to the way the physical system to which we belong couples with the rest of the universe (Rovelli, 2015). Whatever the origin of the entropy gradient, it is a fact that all irreversible phenomena of our experience can be traced to (some version of) it (Reichenbach, 1958; Price, 1996; Albert, 2000). This has nothing to do with the role of time in classical or quantum mechanics, in relativistic physics or in quantum gravity. There is no compelling reason to confuse these phenomena with issues of time in quantum gravity.

Accordingly, nothing refers to "causation," "irreversibility," or similar in LQG. LQG describes physical happening, the way it happens, its probabilistic relations, the microphysics and not the statistics of many degrees of freedom, entropy gradients, or related irreversible phenomena.

To address these, and understand the source of features that make a time variable special, we need a general covariant quantum statistical mechanics. Key steps in this direction exist (see Connes and Rovelli [1994] and Rovelli [1993] on *thermal time*, and Chirco, Josset, and Rovelli [2016] and references therein) but are incomplete. They have no *direct* bearing on LQG.

6.3.5 *Experiential Time*

The second class of phenomena that profoundly affects our intuition of time are those following from the fact that our brain is a machine that (because of the entropy gradient) remembers the past and works constantly to anticipate the future (Buonomano, 2017). This working of our brain gives us a distinctive feeling about time: this is the feeling we call "flow," or the "clearing" that is our experiential time (Heidegger, 1977). This depends on the working of our brain, not on fundamental physics (James, 1890). It is a mistake to search for something pertaining to our feelings uniquely in fundamental physics. It would be like asking fundamental physics to directly justify the fact that a red frequency is more vivid to our eyes than a green one: a question asked in the wrong chapter of science.

Accordingly, nothing refers to flowing, passage, or the similar in LQG. LQG describes physical happenings (Dorato, 2013), the way they happen, their probabilistic relations and not idiosyncrasies of our brain (or our culture [Everett 2008]).

6.4 Presentism or Block Universe?
A False Alternative

An ongoing discussion on the nature of time is framed as an alternative between presentism and block universe (or eternalism). This is a false alternative. Let me get rid of this confusion before continuing.

Presentism is the idea of identifying what is *real* with what is present now, everywhere in the universe. Special relativity and GR make clear that an objective notion of *present* defined all over the universe cannot be defined in the physical world. Hence there can be no objective universal distinction between past, present, and future. Presentism is seriously questioned by this discovery, because to hold it we have to base it on a notion of present that lacks observable ground and appears theoretically ill-motivated, and this is unpalatable.

A common response states that (1) we must therefore identify what is *real* with the ensemble of all events of the universe, including past and future ones (Putnam, 1967), and (2) this implies that, since future and past are equally real, the passage of time is illusory, and there is no becoming in nature (McTaggart, 1908).[3]

I maintain that the claim that points 1 and 2 follow from the demise of presentism is wrong.

Point 1 is just a grammatical choice about how we decide to use the ambiguous adjective *real*; it has no content (Quine, 1948; Austin, 1962). We find no difficulty in distinguishing "real now" from "real in the past" or "real in the future." If the temporal structure of the world is richer than just the three alternatives "now / in the past / in the future, we can simply enrich the manner in which we say something can be real. It can be "real here and now," "real in the future of here and now," or "real in a region space-like separated from here and now," and so on. There is nothing that forces us to give up distinctions between these manners of being real.

Point 2 is mistaken because it treats time too rigidly, failing to realize that time can behave differently from our experience and still deserve to be called time. The arguments against the reality of time are all weak because they consider only two rigid alternatives: either there is in nature something that corresponds to our *entire* intuition about time or there is no time in nature. But this is silly alternative, because our intuition is multilayered and complex and *certainly* does not capture what happens in nature. But this does not imply that our intuition about time, stripped by some layers, cannot still be applied to some general aspects of nature.

---

[3] The expression *block universe* is sometimes used to refer to this view, but its meaning varies according to authors. I agree with those using it to say that fundamental physics does not require modification to be able to describe the universe of our experience.

Therefore the absence of a preferred objective present does not imply that temporality and becoming (in a sufficiently general sense) are illusions. Fundamental physics turns out to work much better in the language of becoming than in the language of being. Quantum theory is about *transitions*, general relativity about *events*. Events *happen*, rather than *are*, and this we call "becoming." They have temporal relations (in the sense of relational time) and these temporal relations form a structure richer than we previously thought. We better adapt our notion of becoming to what has been discovered, not discard it.

There are temporal relations, but these are local and not global; more precisely, there is a temporal ordering, but it is a partial ordering and not a complete one. The universe is an ensemble of processes that happen, and these are not organized in a unique global order. In the classical theory, they are organized in a nontrivial geometry. In the quantum theory, in possibly more complex patterns.

The expression *real here now* can still be used to denote an ensemble of events that sit on the portion of a common simultaneity surface for a group of observers in slow relative motion; the region it pertains to must be small enough for the effects of the finite speed of light to be smaller than the available time resolution. When these conditions are not met, the expression *real now* simply makes no sense.

Therefore the discovery of relativity does not imply that becoming or temporality are meaningless or illusory: it implies that they behave in a more subtle manner than in our prerelativistic intuition. In classical relativity, for example, they are not structured according to a complete order, but only in a (locally) partial order, in the mathematical sense. The best language for describing the universe remains a language of happening and becoming, not a language of being. Even more so when we fold quantum theory in. This is the language used in LQG.

LQG describes reality in terms of processes. The amplitudes of the theory determine probabilities for processes to happen. This is a language of becoming, not being. In a process, variables change value. The quantum states of the theory code the possible set of values that are transformed into each other in processes.

In simple words, the *now* is replaced by *here and now*, not by a frozen eternity.

Temporality in the sense of becoming is at the roots of the language of LQG. But in LQG there is no preferred time variable, as I discuss in the next section.

6.5 "Absence of Time" and Relative Evolution: Time Is Not Frozen

What is missing in LQG is not becoming. It is a (preferred) time variable, a fixed Newtonian time, a continuous Einstein's geometry.

Let me start by reviewing the (different) roles of the coordinates in Newtonian physics and GR. Newtonian space is a 3-dimensional Euclidean space and Newtonian time is a uniform 1-dimensional metric line. Euclidean space admits families of Cartesian coordinates $\vec{X}$ and the time line carries a natural (affine) metric coordinate T. These quantities are tracked by standard rods and clocks. Rods and clocks are not strictly needed for localization in time and space, because anything can be used for relative localization, but they are convenient in the presence of a rigid background metric structure such as the Newtonian, or the special relativistic one.

Rods and clocks are also useful in GR, but far less central. Einstein relayed on rods and clocks in the early days of the theory, but later realized that this was a mistake and repeatedly deemphasized their role at the foundation of his theory. In fact, he cautioned against giving excessive weight to the fact that the gravitational field defines a geometry (Lehmkuhl, 2014). He regarded this fact as a convenient mathematical feature and a useful tool to connect the theory to the geometry of Newtonian space (Einstein, 1921a), but the essential about GR is not that it describes gravitation as a manifestation of a Riemannian spacetime geometry; it is that it provides a field theoretical description of gravitation (Einstein, 1921b).

GR's general coordinates $\vec{x}, t$ are devoid of metrical meaning, unrelated to rods and clocks, and arbitrarily assigned to events. They do not have the direct physical interpretation of Newtonian and special relativistic coordinates, which are interpreted as the reading of rods and clocks. In GR, instead, behavior of rods and clocks is determined by interaction with the gravitational field, it is not expressed by the general coordinates $\vec{x}, t$. The arbitrariness of these coordinates implies that to compare the theory to reality we have to find coordinate-invariant quantities, as any experimental physicists working with GR knows well. This generates some technical complication but is never particularly hard in realistic applications. In particular, the relativistic t coordinates should not be confused with intuitive time, nor with clock time. Clock time is computed in the theory by the proper time (6.1) along a worldline. This quantity counts, say, the oscillations of a mechanism following the worldline. Contrary to what is sometimes stated, this is not a postulate of the theory: it is a consequence of the equations of motion of the mechanism.

Given two events in spacetime, the clock time separation between them depends on the worldline of the clock. Therefore there is no single meaning to the time separation between two events. This does not make the notion of time inconsistent: it reveals it to be richer than our naive intuition. It is a fact that two clocks separated and then taken back together in general do not indicate the same time. Accord of clocks is an approximate phenomenon due to the peculiar environment in which we conduct our usual business.

Due to the discrepancy between clocks, it makes no sense to interpret dynamics as evolution with respect to one particular clock, as Newton wanted.[4] Accordingly,

---

[4] Given two clocks that measure different time intervals between two events, it make no sense to ask which of the two is "true time": the theory simply allows us to compute the way each changes with respect to the other.

the dynamics of GR is not expressed in terms of evolution in a single clock time variable; it is expressed in terms of relative evolution between observable quantities (a detailed discussion is in chapter 3 of Rovelli [2004]). This fact makes it possible to get rid of the t variable altogether, and express the dynamical evolution directly in terms of the relative evolution of dynamical variables (chapter 3 of Rovelli [2004]). Thus special clocks or preferred spatial or temporal variables are *not* needed in relativistic physics.

A formulation of *classical* GR that does not employ the time variable t at all is the Hamilton–Jacobi formulation (Peres, 1962). It is expressed uniquely in terms of the 3-metric q_{ab} of a spacelike surfaces and defined by two equations:

$$D_a \frac{\delta S[q]}{\delta q_{ab}} = 0, \quad G_{abcd} \frac{\delta S[q]}{\delta q_{ab}} \frac{\delta S[q]}{\delta q_{cd}} + \det q \, R[q] = 0, \tag{6.2}$$

where $G_{abcd} = q_{ac}q_{bd} + q_{ad}q_{bc} - q_{ab}q_{cd}$, and R is the Ricci scalar of q. Notice the absence of any temporal coordinate t. In principle, knowing the solutions of these equations is equivalent to solving the Einstein equations. Here $S[q]$ is the Hamilton–Jacobi function of GR. When q is the 3-metric of the boundary of a compact region R of an Einstein space, $S[q]$ can be taken to be the action of a solution of the field equations in this region. It is the quantity connected to the LQG amplitudes as in (6.7).

Absence of a time variable does not mean that time is frozen or that the theory does not describe dynamics, as unfortunately is still heard.

Equations (6.2) indeed provide an equivalent formulation of standard GR and can describe the solar system dynamics, black holes, gravitational waves, and any other *dynamical* process, where things become, without any need of an independent t variable. In these phenomena many physical variables change together and no preferred clock or parameter is needed to track change.

The same happens in LQG. The quantum versions of (6.2) formally determine the transition amplitudes between quantum states of the gravitational field. These can be coupled to matter and clocks. Variables change together and no preferred clock variable is used in the theory.[5]

It is in this weak sense that it is sometimes said that time does not exist at the fundamental level in quantum gravity. This expression means that there is no time variable in the fundamental equations. It does not mean that there is no change in nature. The theory indeed is formulated in terms of probability amplitudes for *processes*.

---

[5] The canonical formulation of the theory leads to the same conclusion, but in a turned-around way: one first describes evolution in a single parameter time, but the Hamiltonian that generates this evolution vanishes, and only relations between quantities invariant under this evolution are predictable. So, the theory predicts only relative evolution, and once again there is no preferred observable independent time.

6.6 Quantum Theory without Schrödinger Equation

Quantum mechanics requires some cosmetic adaptations in order to deal with the way general relativistic physics treats becoming. General relativistic physics describes becoming as evolution of variables that change together, any of them can be used to track change. No preferred time variable is singled out. Quantum mechanics instead is commonly formulated in terms of a preferred independent clock variable T. Evolution in T is expressed either in the form of Schrödinger equation:

$$i\hbar\frac{\partial\psi}{\partial T} = H\psi,$$ (6.3)

or as a dynamical equation for the variables:

$$\frac{dA}{dT} = i\hbar[A, H],$$ (6.4)

where H is the Hamiltonian operator. Neither of these equations is adept to describe relativistic relative evolution. The extension of quantum theory to the relativistic evolution is, however, not very hard, and has been developed by many authors, starting from Dirac. See for instance chapter 5 of Rovelli (2004), or Rovelli and Vidotto (2014), or, from a slightly different perspective, the extensive work of Jim Hartle (1995) on this topic.

Like classical mechanics, quantum mechanics can be phrased as a theory of the probabilistic relations between the values of variables evolving together, rather than variables evolving with respect to a single time parameter. The Schrödinger equation is then replaced by a Wheeler–DeWitt equation:

$$C\psi = 0,$$ (6.5)

or as a dynamical equation for the variables:

$$[A, C] = 0,$$ (6.6)

for a suitable Wheeler–DeWitt operator C. Again: these equations do not mean that time is frozen or there is no dynamics. They mean that the dynamics is expressed as a joint evolution between variables, rather than evolution with respect to a single special variable. (On the issue of time in quantum gravity, see Huggett, Vistarini, and Wüthrich [2013] and Rickles [2006].)

Formally: the $\hbar \to 0$ limit of equation (6.5) is the second equation in (6.2); given boundary values, (6.5) is formally solved by the transition amplitudes W. These can be expressed as a path integral over fields in the region and in the $\hbar \to 0$ limit $W \sim e^{i\hbar\frac{S}{\hbar}}$, where S is a solution to (6.2). These are formal manipulations. LQG provides a finite and well-defined expression for W, at any order in a truncation in the number of degrees of freedom.

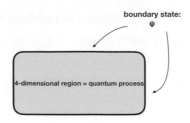

Figure 6.1 A compact spacetime region is identified with a quantum transition. The states of LQG sit on its boundary.

6.7 Quantum Process = Spacetime Region

Quantum theory does not describe how things *are*. It describes quantum events that happen when systems interact (Rovelli, 2017). We mentally separate a quantum system, for a certain time interval, from the rest of the world and describe the way this interacts with its surroundings. This peculiar conceptual structure at the foundations of quantum theory takes a surprising twist in quantum gravity.

In quantum gravity we identify the process of the quantum system with a *finite spacetime region*. This yields a remarkable dictionary between the relational structure of quantum theory and the relational structure of relativistic spacetime:

quantum transition	$\leftrightarrow$	4-dimensional spacetime region
initial and final states	$\leftrightarrow$	3-dimensional boundaries
interaction (measurement)	$\leftrightarrow$	contiguity

Thus the quantum states of LQG sit naturally on 3-dimensional boundaries of 4-dimensional regions (see Figure 6.1) (Oeckl, 2003). The quantum amplitudes are associated to what happens inside the regions. Intuitively, they can be understood as path integrals over all possible internal geometries, at fixed boundary data. For each set of boundary data, the theory gives an amplitude, that determines the probability for this process to happen, with respect to other processes.

Remarkably: the net of quantum interactions between systems *is the same thing* as the net of adjacent spacetime regions. This comes naturally from the application of quantum theory to finite spacetime regions.

6.8 Conclusion

Space and *time* are expressions that can mean many different things:

1. Space can refer to the **relative localization** of things, time can refer to the **becoming** that shapes nature. As such, they are present in LQG like in any other physical theory.

2. Spacetime is a name given to the **gravitational field** in classical GR. In LQG there is a gravitational field, but it is not a continuous metric manifold. It is a quantum field with the usual quantum properties of discreteness, indeterminism, and quantum relationality.

3. Space and time can refer to preferred variables used to locate things or to track change, in particular **reading of meters and clocks**. In LQG, rods and clocks and their (quantum) behavior can in principle be described, but play no role in the foundation of the theory. The equations of the theory do not have preferred spatial or temporal variables.

4. Thermal, causal, **flowing** aspects of temporality are grounded on chapters of science distinct from the elementary quantum mechanics of reality. They may involve, for example, thermal time, perspectival phenomena, statistics, and brain structures.

5. The universe described by quantum gravity is not flowing along a single time variable or organized into a smooth Einsteinian geometry. It is a network of quantum processes, related to one another, each of which obeys probabilistic laws that the theory captures. **The net of quantum interactions between systems is identified with the net of adjacent spacetime regions**.

These are the roles of space and time in loop quantum gravity. Much confusion about these notions in quantum gravity is confusion between these different meanings of space and time.

Appendix: Loop Quantum Gravity in a Nutshell

As any quantum theory, LQG can be defined by a Hilbert space, an algebra of operators, and a family of transition amplitudes. The Hilbert space $\mathcal{H}$ of the theory admits a basis called the *spin network basis*, whose states $|\Gamma, j_l, v_n\rangle$ are labeled by a (abstract, combinatorial) graph Γ, a discrete quantum number j_l for each link l of the graph, and a discrete quantum number v_n for each node n of the graph. The nodes of the graph are interpreted as elementary "quanta of gravity" or "quanta of space," whose adjacency is determined by the links, see Figure 6.2. These quanta do not live on some space: rather, they themselves build up physical space.

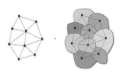

Figure 6.2 The graph of a spin network and an intuitive image of the quanta of space it represents.

Figure 6.3 Spinfoam: the time evolution of a spin network.

The volume of these quanta is discrete and determined by v_n. The area of the surfaces separating two nodes is also discrete, and determined by j_l. The elementary quanta of space do not have a sharp metrical geometry (volume and areas are not sufficient to determine geometry), but in the limit of large quantum numbers, there are states in $\mathcal{H}$ that approximate 3-dimensional geometries arbitrarily well, in the same sense in which linear combinations of photon states approximate a classical electromagnetic field. The spin network states are eigenstates of operators A_l and V_l in the operator algebra of the theory, respectively associated to nodes and links of the graph. In the classical limit, these operators become functions of the Einstein's gravitational field $g_{\mu\nu}$, determined by the standard relativistic formulas for area and volume. For instance, $V(R) = \int_R \sqrt{\det q}$, for the volume of a 3-dimensional spatial region R, where q is the 3-metric induced on R.

In the covariant formalism (see Rovelli and Vidotto [2014]), transition amplitudes are defined order by order in a truncation on the number of degrees of freedom. At each order, a transition amplitude is determined by a *spinfoam*: a combinatorial structure $\mathcal{C}$ defined by elementary faces joining on edges in turn joining on vertices (in turn, labeled by quantum numbers on faces and edges), as in Figure 6.3.

A spinfoam can be viewed as the Feynman graph of a history of a spin network; equivalently, as a (dual) discrete 4-dimensional geometry: a vertex corresponds to an elementary 4-dimensional region, an elementary process. The boundary of a spinfoam is a spin network. The theory associates an amplitude $W_{\mathcal{C}}(\Gamma, j_l, v_n)$ (a complex number) to spinfoams. These are ultraviolet finite. Several theorems relate them to the action (more precisely the Hamilton function S) of GR, in the limit of large quantum numbers (Freidel and Krasnov, 1998; Barrett et al., 2009, 2011; Fairbairn and Meusburger, 2010; Rovelli and Vidotto, 2014; Han, 2017). This is the expected formal relation between the quantum dynamics, expressed in terms of transition amplitudes W and its classical limit, expressed in terms of the action S:

$$W \sim e^{i\frac{S}{\hbar}}, \tag{6.7}$$

where W and S are both functions of the boundary data.

This concludes the sketch of the formal structure of (covariant) LQG. Notice that nowhere in the basic equations of the theory does a time coordinate t or a space coordinate x show up.

References

Albert, D. (2000). *Time and Change*. Cambridge, MA: Harvard University Press.

Aristotle. (1990). Physics. Pages 257–355 of: *The Works of Aristotle*. Vol. 1. Chicago: The University of Chicago Press.

Ashtekar, A. (2011). Introduction to loop quantum gravity. *PoS QGQGS2011*, 1.

Austin, J. (1962). *Sense and Sensibilia*. Oxford: Clarendon Press.

Barrett, J. W., Dowdall, R. J., Fairbairn, W. J., Hellmann, F., and Pereira, R. (2009). Lorentzian spin foam amplitudes: Graphical calculus and asymptotics. *Class. Quant. Grav.*, 27, 165009.

Barrett, J. W., Dowdall, R. J., Fairbairn, W. J., Hellmann, F., and Pereira, R. (2011). Asymptotic analysis of Lorentzian spin foam models. *PoS QGQGS2011*, 9.

Buonomano, D. (2017). *Your Brain Is a Time Machine: The Nueroscience and Physics of Time*. New York: W. W. Norton.

Chirco, G., Josset, T., and Rovelli, C. (2016). Statistical mechanics of reparametrization-invariant systems. It takes three to tango. *Class. Quant. Grav.*, 33, 4.

Connes, A., and Rovelli, C. (1994). Von Neumann algebra automorphisms and time thermodynamics relation in general covariant quantum theories. *Class. Quant. Grav.*, 11, 2899–2918.

Dorato, M. (2013). Rovelli's relational quantum mechanics, monism and quantum becoming. Pages 290–324 of: Marmodoro, A., and Yates, A. (eds.). *The Metaphysics of Relations*. Oxford: Oxford University Press.

Earman, J. (2006). The past hypothesis: Not even false. *Stud. Hist. Philos. Mod. Phys.*, 37, 399–430.

Einstein, A. (1921a). *Geometrie und Erfahrung*. Berlin: Julius Springer.

Einstein, A. (1921b). *The Meaning of Relativity*. Princeton, NJ: Princeton University Press.

Everett, D. (2008). *Don't Sleep, There Are Snakes*. New York: Pantheon Books.

Fairbairn, W. J., and Meusburger, C. (2010). Quantum deformation of two four-dimensional spin foam models. *J. Math. Phys.*, 53, 22501.

Freidel, L., and Krasnov, K. (1998). Spin foam models and the classical action principle. *Adv. Theor. Math. Phys.*, 2, 1183–1247.

Gambini, R., and Pullin, J. (1996). *Loops, Knots, Gauge Theories and Quantum Gravity*. Cambridge: Cambridge Monographs on Mathematical Physics. Cambridge University Press.

Gambini, R., and Pullin, J. (2010). *Introduction to Loop Quantum Gravity*. Oxford: Oxford University Press.

Han, M. (2017). Einstein equation from covariant loop quantum gravity in semiclassical continuum limit. *Phys. Rev. D*, 96, 2.

Hartle, J. B. (1995). Spacetime quantum mechanics and the quantum mechanics of spacetime. In: Julia, B., and Zinn-Justin, J. (eds.). *Gravitation and Quantizations*. Amsterdam: Elsevier.

Heidegger, M. (1977). Gesamtausgabe Vol. 5. *Holzwege*. Frankfurt am Main: Klostermann.

Huggett, N., Vistarini, T., and Wüthrich, C. (2013). Time in quantum gravity. Pages 242–261 of: Dyke, H., and Bardon, A. (eds.). *A Companion to the Philosophy of Time*. Chichester: Wiley-Blackwell.

James, W. (1890). *The Principles of Psychology*. New York: Henry Holt.

Lehmkuhl, D. (2014). Why Einstein did not believe that general relativity geometrizes gravity. *Stud. Hist. Philos. Sci.*, 46, 316–326.

McTaggart, J. (1908). The unreality of time. *Mind*, 17, 457–474.

Newton, I. (1934). *Scholium to the Definitions in Philosophiae Naturalis Principia Mathematica. Book 1: (1689)*. Trans. Andrew Motte (1729), rev. Florian Cajori. Berkeley: University of California Press.

Oeckl, R. (2003). A 'general boundary' formulation for quantum mechanics and quantum gravity. *Phys. Lett.*, B575, 318–324.

Peres, A. (1962). On Cauchy's problem in general relativity. *Nuovo Cimento*, 26, 53.

Perez, A. (2012). The spin-foam approach to quantum gravity. *Living Rev. Rel.*, 16.

Price, H. (1996). *Time's Arrow*. Oxford: Oxford University Press.

Putnam, H. (1967). Time and physical geometry. *J. Philos.*, 64, 240–247.

Quine, W. (1948). On what there is. *Rev. Metaphys.*, 2, 21–38.

Reichenbach, (1958). *The Philosophy of Space and Time*. New York: Dover Books.

Rickles, D. (2006). Time and structure in canonical gravity. Pages 152–196 of: Rickles, D., French, S., and Saatsi, J. (eds.). *The Structural Foundations of Quantum Gravity*. Oxford: Oxford University Press.

Rovelli, C. (1993). Statistical mechanics of gravity and the thermodynamical origin of time. *Class. Quant. Grav.*, 10, 1549–1566.

Rovelli, C. (2004). *Quantum Gravity*. Cambridge: Cambridge University Press.

Rovelli, C. (2015). Is time's arrow perspectival? In: Chamcham, K., Silk, J.,Barrow, J., and Saunders, S. (eds.). *The Philosophy of Cosmology*. Cambridge: Cambridge University Press.

Rovelli, C. (2017). Space is blue and birds fly through it. *Philosph. Trans. R. Soc. A Math. Phys. Eng. Sci.*, 376, 20170312.

Rovelli, C. (2018). *The Order of Time*. New York: Riverhead Books.

Rovelli, C., and Vidotto, F. (2014). *Covariant Loop Quantum Gravity*. Cambridge Monographs on Mathematical Physics. Cambridge: Cambridge University Press.

Thiemann, T. (2007). *Modern Canonical Quantum General Relativity*. Cambridge: Cambridge University Press.

van Fraassen, B. (1985). *An Introduction to the Philosophy of Time and Space*. New York: Columbia University Press.

7

Being and Becoming on the Road to Quantum Gravity; or, the Birth of a Baby Is Not a Baby

FAY DOWKER

7.1 A Persistent Lack of Consensus

The ancient dichotomy between Being and Becoming continues to find expression in the ongoing struggle to understand the nature of time. This struggle encompasses disagreement on exactly what the two positions in the Being vs. Becoming debate are.[1] For the purpose of this paper I take the two positions to be, "The universe is nothing more than a Block in which past, present and future events have the same physical status" and "The universe comes into being in a physical process that corresponds to the passage of time." Imprecise as these positions are, they are distinct enough and suggestive enough to be heuristics for the stimulation of research on the physics of spacetime in different directions. And progress in relevant physics will surely illuminate the terms of the debate. This interplay between the Being vs. Becoming dichotomy and research in fundamental physics forms the background to this article.

To set the scene, let us consider some statements supporting the two sides from modern thinkers. On the Becoming side, philosopher J. Norton states (2010),

Time really passes.... Our sense of passage is our largely passive experience of a fact about the way time truly is, objectively. The fact of passage obtains independently of us. Time would continue to pass for the smoldering ruins were we and all sentient beings in the universe suddenly to be snuffed out.... We have no good grounds for dismissing the passage of time as an illusion. It has none of the marks of an illusion. Rather, it has all the marks of an objective process whose existence is independent of the existence of we humans.

On the Being, or Block side, physicist P. Davies (2013) says,

The flow of time is an illusion, and I don't know very many scientists and philosophers who would disagree with that, to be perfectly honest. . . . And presumably the explanation for this illusion has to do with something up here (in your head) and is connected with

---

[1] Two representative works from the philosophical literature that set out some of the possibilities are Butterfield (2012) and Price (2011).

memory I guess – laying down of memories and so on. So it's a feeling we have, but it's not a property of time itself . . .. Time doesn't flow. That's part of psychology.

Arguments such as Davies's are strongest when they are grounded in our best current scientific theory of spacetime, general relativity, as described by physicist S. Carroll (2011),

Modern physics suggests that we can look at the entire history of the universe as a single four-dimensional thing. That includes our own personal path through it, which defines our world line. This seemingly conflicts with our intuitive idea that we exist at a moment, and move through time. Of course there is no real conflict, just two different ways of looking at the same thing. There is a four-dimensional universe that includes all of our world line, from birth to death, once and for all; and each moment along that world line defines an instantaneous person with the perception that they are growing older, advancing through time.

Davies and Carroll are in accord with the view of H. Weyl (1949),

The objective world simply *is*, it does not *happen*. Only to the gaze of my consciousness, crawling upward along the life line of my body, does a section of this world come to life as a fleeting image in space which continuously changes in time.

Setting aside the question of whether we really do experience "fleeting image[s] in space" or something more dynamic, or how to make sense of an "instantaneous person," let us look in more detail at the reason that general relativity draws many modern thinkers to adopt the Block Universe view.

7.2 Being and Becoming in General Relativity

In general relativity (GR) spacetime is a 4-dimensional manifold with a Lorentzian geometry. The Lorentzian geometry gives spacetime its lightcone structure and its, closely related, causal structure. These structures are fundamental to the physics of GR and mean that the spacetime in GR cannot be understood as any kind of combination of a spatial geometry and a temporal geometry.

Nowhere in GR does one find the concept of a 3-dimensional physical "Being": there is, fundamentally, no such thing. There are no 3-dimensional objects because there is no 3-dimensional space. In order to give physical meaning to physically meaningful *time coordinate*. This would contradict one of the basic principles of GR, general covariance. General covariance says that coordinates on spacetime have no physical significance, no more significance than a choice of coordinate grid on a map of Mexico City, say. The concept of "persisting physical object" is replaced, in GR, with a pattern of regularity in the values of the fields and distribution of worldlines on 4-dimensional spacetime.

The only concept of time that exists in GR is that of *proper time duration along timelike worldlines*. Physical time pertains to only worldlines and the time that elapses along a worldline is independent of the coordinates of that worldline. Time in GR cannot be thought of as a fourth dimension. Indeed, the difference between space and time in GR could hardly be more stark in the sense that there is time duration but there is no concept, fundamentally, of physical spatial distance. GR, then, denies the notion of Being, *if* Being is interpreted as having a 3-dimensional character. However, general relativity does contain a Being that is 4-dimensional: the physical world in GR is a 4-dimensional Lorentzian spacetime together with matter in the form of fields and worldlines on it.

Does GR force a Block view on us? Can it accommodate the concept of Becoming? To incorporate Becoming, while still doing justice to the 4-dimensional character of the world, a *growing* 4-dimensional spacetime seems to be called for, as suggested by C. D. Broad (1923). One obvious way to achieve this would be to postulate that spacetime in GR continually, physically grows by what has sometimes been called "Hypersurface Becoming." Such an incarnation of Becoming has been supported in recent times by G. Ellis and R. Goswami (2014) and also by L. Smolin (2013). However, such Hypersurface Becoming contradicts general covariance, providing as it does a physical foliation of spacetime into 3-dimensional spatial hypersurfaces. It picks out a special time coordinate labeling the leaves of the foliation and gives it physical significance. This return to the prerelativistic notions of a global time and of absolute physical simultaneity is unpalatable to many physicists. This aversion to violation of general covariance leads many workers to claim GR as support for a Block view of the physical world.

Despite this, the continuing arguments for Becoming, though not quantitative and mostly composed of appeals to intimate experience, are undeniably powerful. Nothing is more fundamental to our experience of the world than the temporal nature of that experience. And so, dissatisfaction with the Block view persists, and consensus has not been reached. The gist of the following, frustrated exchange may be familiar (*Blockhead* and *Broadhead* are J. Earman's [2008] terms):

Blockhead: Events happen in the Block. There in the Block, a tree is growing; there, a supernova is exploding; there, a person is experiencing time passing.

Broadhead: No. The Block is static. It corresponds to events that *have happened*, or that *will have happened*, not to events *happening*. There, in the Block, is the history of the growth of a tree; there, is a supernova having exploded; there, is a person having experienced time passing.

Blockhead: The Block *does* correspond to things happening.

Fay Dowker

Broadhead: No it doesn't.
Blockhead: Yes it does.

It's a pantomime of an argument, going nowhere.

7.3 Growing a Discrete Spacetime

Recently, however, novel possibilities for the physics of spacetime have been introduced that merit a reevaluation of the question of physical Becoming. This has taken place in the context of the causal set approach to the problem of quantum gravity, which is based on the hypothesis that spacetime is fundamentally atomic at the Planck scale.

Most workers on quantum gravity accept that a differentiable manifold will not be a good description of spacetime at length scales at the Planck length and smaller. Spacetime discreteness or atomicity at the Planck scale is perhaps the simplest hypothesis that can be made in response to this widespread expectation. The causal set program for quantum gravity is based on this atomic hypothesis (Bombelli et al., 1987; Sorkin, 1991, 2003). It proposes that our seemingly smooth, continuous, Lorentzian spacetime is an approximation at large scales to a discrete, partially ordered set or *causal set*.

The elements of the causal set are conceived of as the atoms of spacetime—note, atoms of *spacetime* and not atoms of *space*—and the number of spacetime atoms comprising a region manifests itself as the spacetime 4-volume of that region in (close to) Planck units. The observable universe has a spacetime volume of roughly 10^{240} Planck volumes, and this therefore is the order of magnitude of the number of spacetime atoms needed to describe what we observe. The order relation between the spacetime atoms in the causal set manifests itself as the spacetime causal order in the approximating Lorentzian spacetime.[2]

Figure 7.1 is an example of a small causal set illustrated as a graph.

The causal set hypothesis is grounded in work by Kronheimer and Penrose (1967), Hawking (Hawking, King, and McCarthy, 1976), and Malament (1977) showing that, for a continuum Lorentzian spacetime, the causal order and local scale information together are equivalent to the full geometry. This fact underpins the claim that a causal set can indeed be approximated by a Lorentzian geometry:

---

[2] In GR the spacetime causal order is a relation of *precedence*, of *before and after*, not of *causation*. The ambiguous term *causal order* is used because the spacetime *chronological relation* is defined by timelike curves and to distinguish this from the relation defined by timelike or null curves, the term *causal relation* is used for the latter relation. It is this spacetime causal relation that is hypothesized to be underpinned by the causal set's order relation. The name *causal set* has therefore inherited this confusion from continuum GR: the order of a causal set is an order of precedence and not of causation. "Temporal set" might have been a better name but it is too late to change it now.

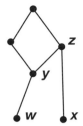

Figure 7.1 The Hasse diagram of a causal set with 6 elements. The vertices are the elements of the causal set. The edges represent the covering relations and the upward going direction on the page represents the direction of the order: given an edge joining two vertices, the lower vertex precedes the upper vertex. The relation is transitive. Therefore, w precedes y, y precedes z and so w precedes z. w and x are unordered.

the causal set's order relation provides the approximating continuum's causal order and the local physical scale is set by the causal set's discreteness.

In the causal set program for quantum gravity, an outstanding task is to construct the physical dynamics of causal sets. This dynamics is expected to be quantum mechanical in order to do justice to the unity of physics. At the present stage of development of the theory, we do not have a quantum dynamics for causal sets but we do have a family of *classically* stochastic models, the classical sequential growth (CSG) models of Rideout and Sorkin (2000).

Each model in the CSG family is a stochastic process in which a causal set grows by a process of continual, random births of new spacetime atoms. In order to define a CSG dynamical model, a "gauge time" is introduced: integers that label discrete stages of the growth process. The process starts with the birth of a single causal set element labeled 0. At the beginning of stage $n > 0$, a causal set with elements labeled $0, 1, 2, \ldots n - 1$ has already grown and the stage consists of the birth of a new element labeled n. This newborn chooses, with a certain probability given by the model parameters, a subset, S, of the already existing elements to be to the future of.[3] At the end of stage n, therefore, one more element has been added to the growing causal set. That the labeling generated by this staged process is a gauge and not physical is shown by the fact that the probability of growing a given causal set by stage n does not depend on the labeling. Figure 7.2 shows two labelings of the same 6-element causal set. The probabilities of growing these two causal sets are equal in a CSG model: the labeling has no physical significance.

---

[3] A particular CSG model is defined by a set of non-negative parameters, $\{t_0, t_1, t_3, \ldots\}$ and the relative probability of the newborn choosing to be to the future of a set of elements S is $t_{|S|}$, where $|S|$ is the cardinality of S.

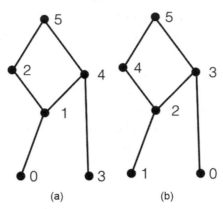

Figure 7.2 Two labelings of the same causal set.

The physical order in which the spacetime atoms are born in a CSG model is the partial order that the atoms possess in the growing causal set. For example, if the causal set in Figure 7.2(a) represents the causal set grown in a CSG model, atom labeled 1 is born before atom labeled 4 because the former precedes the latter in the causal set order. However there is no physical order between the birth of atom labeled 1 and that labeled 3 because there is no order relation between them in the causal set. Sorkin (2007) calls this partially ordered growth *Asynchronous Becoming* and argues that CSG models are counterexamples to the claim that relativity implies the Block view of spacetime (see also Dowker [2014]). A CSG model is a model of a spacetime that Becomes in a manner compatible with the absence of a physical global time.

7.4 Being and Becoming in Classical Sequential Growth

The physical world in a classical sequential growth model consists of two different types of things. One of these types is the *material* of the world consisting of the spacetime atoms and their order relations. The other type of thing is the asynchronous birth *process*. *Both* the atoms and the process are physical, and they are different: the birth of a baby is not a baby, the birth of a spacetime atom is not a spacetime atom.

The existence of CSG models allows the conversation between the Blockhead and the Broadhead to be more meaningful. The Broadhead has something concrete to hang their argument upon, a theory to compare to the Block.[4] A CSG model contains something new, something that is absent in the Block, namely the birth process. In a CSG model, an event corresponds to a collection of spacetime atoms

---

[4] Whether asynchronous becoming conforms to a Broadhead's intuitions and motivations is a question Callender and Wüthrich take up (Wüthrich and Callender, 2015).

and the causal relations between them. The *occurrence* of the event corresponds to the birth of these atoms.

> **Blockhead**: Events happen in the Block. There in the Block, a tree is growing; there, a supernova is exploding; there, a person is experiencing time passing.
>
> **Broadhead**: No. The Block is static. It corresponds to events that *have happened*, or that *will have happened*, not to events *happening*. There, in the Block, is the history of the growth of a tree; there, is a supernova having exploded; there, is a person having experienced time passing.
>
> **Blockhead**: The Block *does* correspond to things happening.
>
> **Broadhead**: Compare the Block to a CSG model, in which there are not just spacetime atoms but the birth process as well. Suppose, for the sake of argument, a CSG model produces a causal set, C, well approximated by a continuum spacetime like our universe, together with some decorations on the spacetime atoms, field values say, for the matter degrees of freedom. Suppose that a subcausal set, S, of that causal set corresponds to the history of a supernova explosion. It is the *births* of the spacetime atoms in S that correspond to the explosion *happening*. This birth process is missing in the Block; the Block is only the history of the process.

The Broadhead can further argue, following Sorkin (2007), that the CSG model "even provides an objective correlate of our subjective perception of 'time passing' in the unceasing cascade of birth events that build up the causal set, by 'accretion' as it were." This objective correlate is absent in the Block. Thus, if one accepts one's perception of time passing as empirical data, then a model with an asynchronous becoming process is empirically favored over a Block Universe model. The form of these data are peculiar, subjective, and nonquantitative. It seems therefore to be a matter of personal choice for a scientist whether to accept or reject the call of his or her own perception of time passing to be coordinated with some element of physical theory. However, one's perceptions change when one understands the world better. It is possible that to view the world through a theory with Becoming is necessary to be able to interrogate and describe one's perceptions systematically.

7.5 Other Dichotomies

Let us tie things together by considering, along with the dichotomy of Being vs. Becoming, two others: Objectivity vs. Subjectivity and Continuity vs. Atomicity.

Fay Dowker

There is Being and there is Becoming in a CSG model. The spacetime atoms and their relations realize Being[5] and the birth process realizes Becoming.[6]

The birth process is objective. It is physical and not dependent on any "observer." It would continue for the smoldering ruins were we and all sentient beings in the universe suddenly to be snuffed out. Moreover, the process is objective in the further sense that it is not subject to any atom or other being (as, for instance, the relative past is).[7]

And yet there is no single, objective picture of "the physical world that exists" in a CSG model, not even a world that grows and changes.[8] Instead, there are many 4-dimensional "Being"s, but they are subjective: each Being is subject to a spacetime atom's birth and is the *past* of that newborn atom, consisting of the set of spacetime atoms that precede the newborn. This subjectivity—subject to a birth event—leads to no contradictions or contrary inferences, since any two spacetime atoms will agree on the structure of the causal set in their common past.

Thus, in a CSG model, Becoming is objective and Being is subjective. Not only are the seemingly opposed concepts in each of the two dichotomies reconciled in the physics of CSG but the two dichotomies are intertwined.

As for the third dichotomy of Atomicity vs. Continuity, for myself I find it hard to conceive of Asynchronous Becoming except in the context of a discrete theory of spacetime. However, someone capable of conceiving of the Asynchronous Becoming of a continuum spacetime manifold could argue that Becoming can be realized even within GR.[9]

If causal set theory is to be successful as a theory of quantum gravity, classical sequential growth will have to be generalised to some kind of quantum sequential

[5] The Being in a CSG model is a spacetime, not a spatial Being, in accord with GR.
[6] CSG models are stochastic. However, I want to emphasize that it is not the stochasticity as such that means a CSG model embodies Becoming. Stochastic processes, including CSG models, can be given a Block interpretation and indeed this is the way they are treated formally mathematically. For example, a random walk on the integers is mathematically formalized as merely a "one-shot" random choice of one single infinite trajectory (aka Block) for the walker from a "sample space" of possible infinite trajectories (possible Blocks) with a probability measure on it. On the other hand, a deterministic becoming process is also conceivable. Consider a random walker on the integers who always steps to the right and never to the left. If the physical process of taking discrete steps is explicitly part of the model then the walker's trajectory grows physically and this model will embody Becoming, even though it is deterministic.
[7] I am trying to make it clear that, in the context of the heuristic dichotomy I am setting up here, "objective" means more than mind independence and here, conversely, "subjective" does not mean mind dependent.
[8] This fact is at the heart of Callender and Wüthrich's sceptical view on whether Becoming is manifested in a CSG model (Wüthrich and Callender, 2015).
[9] Note that it is essential to Becoming as conceived in this paper that the birth process is part of the theory as a physical element. In the continuum "Worldline Becoming" of Clifton and Hogarth (1995) there is no physical process: Becoming is, for Clifton and Hogarth, explicitly a *relation*, not a process. Stein (1991) also identifies Becoming with a relation and not a process, as do Callender and Wüthrich (Wüthrich and Callender, 2015). Dieks (2006) comes close to proposing an Asynchronous Becoming process for continuum spacetime, but muddies the proposal by claiming "There is no need to augment the block universe in any way." The Block is the history of the Becoming process, not the process itself so there *is* a need to augment the Block with the process in order to claim that Becoming is realized in the theory.

growth. The points presented here will have to be reassessed in the light of these future developments.

It is interesting to note that the Objectivity vs. Subjectivity dichotomy has played a central role in discussions of the interpretation of quantum theory. Disagreements about quantum theory often center on this dichotomy. On one view, a theory must be observer independent, or objective, to be satisfactory: it must provide an objective World Picture. On another view, it is acceptable, even necessary, for the role of subjects—observers or agents—to be privileged. CSG provides an example, albeit in a classical setting, of a theory in which there are both objective and subjective aspects of the physical world.

The broadest goal of those working on quantum gravity is to find a unified framework for the whole of fundamental physics. It is not surprising to find ancient, but persistent, disputes about the nature of the physical world being played out as we work toward that goal.

Acknowledgments

This paper is based on a seminar given to the Space and Time after Quantum Gravity Group, jointly at the University of Geneva and University of Illinois at Chicago, on March 8, 2017. I thank my colleagues there for enlightening discussion and useful comments. Supported by STFC Grant ST/P000762/1. I thank the Perimeter Institute for Theoretical Physics for hospitality and support during the writing of this paper. Research at Perimeter Institute is supported by the Government of Canada through the Department of Innovation, Science and Economic Development and by the Province of Ontario through the Ministry of Research and Innovation. Thanks to Chris Wüthrich, Nick Huggett, and an anonymous referee for helpful comments.

References

Bombelli, L., Lee, J. H., Meyer, D., and Sorkin, R. 1987. Space-time as a causal set. *Physical Review Letters*, **59**, 521.

Broad, C. 1923. *Scientific Thought*. London: Kegan Paul.

Butterfield, J. 2012. On time chez Dummett. *European Journal of Analytic Philosophy*, **8**, 77–102.

Carroll, S. 2011. The flow of time. Sean Carroll [blog]. www.preposterousuniverse.com/blog/2011/08/18/the-flow-of-time.

Clifton, R., and Hogarth, M. 1995. The definability of objective becoming in Minkowski spacetime. *Synthese*, **103**, 355–387.

Davies, P. 2013. Q & A with Paul Davies: What is time? https://fqxi.org/community/articles/display/187.

Dieks, D. 2006. Becoming, relativity and locality. Pages 157–176 of: Dieks, D. (ed.). *The Ontology of Spacetime*. Vol. 1. Amsterdam: Elsevier.

Dowker, F. 2014. The birth of spacetime atoms as the passage of time. *Annals of the New York Academy of Sciences*, **1326**, 18–25.

Earman, J. 2008. Reassessing the prospects for a growing block model of the universe. *International Studies in the Philosophy of Science*, **22**, 135–164.

Ellis, G. F., and Goswami, R. 2014. Spacetime and the passage of time. Pages 243–264 of: Ashtekar, A. and Petkov, V. (eds.). *Springer Handbook of Spacetime*. Berlin Heidelberg: Springer.

Hawking, S. W., King, A. R., and McCarthy, P. J. 1976. A new topology for curved spacetime which incorporates the causal, differential, and conformal structures. *Journal of Mathematical Physics*, **17**, 174–181.

Kronheimer, E. H., and Penrose, R. 1967. On the structure of causal spaces. *Mathematical Proceedings of the Cambridge Philosophical Society*, **63**, 481–501.

Malament, D. B. 1977. The class of continuous timelike curves determines the topology of spacetime. *Journal of Mathematical Physics*, **18**, 1399–1404.

Norton, J. 2010. Time really passes. *Humana Mente*, **13**, 23–34.

Price, H. 2011. The flow of time. Pages 276–311 of: *The Oxford Handbook of the Philosophy of Time*. Oxford: Oxford University Press.

Rideout, D. P., and Sorkin, R. D. 2000. newblock A classical sequential growth dynamics for causal sets. *Physical Review*, **D61**.

Smolin, L. 2013. Temporal naturalism. *Studies in History and Philosophy of Science Part B: Studies in History and Philosophy of Modern Physics*, **52**(Part A), 86–102.

Sorkin, R. D. 1991. Space-time and causal sets. In: *Relativity and Gravitation: Classical and Quantum*. Proceedings of the SILARG VII Conference. Singapore: World Scientific.

Sorkin, R. D. 2003. Causal sets: Discrete gravity (notes for the Valdivia Summer School). In: Gomberoff, A. and Marolf, D. (eds.). *Lectures on Quantum Gravity*. Proceedings of the Valdivia Summer School, Valdivia, Chile, January 2002. New York: Plenum.

Sorkin, R. D. 2007. Relativity theory does not imply that the future already exists: A counterexample. Pages 153–161 of: Petkov, V. (ed.). *Relativity and the Dimensionality of the World*. Fundamental Theories of Physics. Berlin Heidelberg: Springer.

Stein, H. 1991. On relativity theory and openness of the future. *Philosophy of Science*, **58**, 147–167.

Weyl, H. 1949. *Philosophy of Mathematics and Natural Sciences*. Princeton, NJ: Princeton University Press.

Wüthrich, C., and Callender, C. 2015. What becomes of a causal set. *The British Journal for the Philosophy of Science*, **68**(3), 907–925.

8

Temporal Relationalism

LEE SMOLIN

8.1 Introduction

I would like to introduce a research program aimed at solving the foundational issues in quantum mechanics in a way that also addresses the problem of quantum gravity.

There are, roughly speaking, two kinds of approaches to the measurement problem and the other issues in quantum foundations. The first take is that the principles of quantum mechanics are largely correct and therefore aim to make our thinking more compatible with the practice of quantum physics. The second kind of approach begins with the hypothesis that the foundational puzzles are consequences of an incompleteness of our understanding of nature, due primarily to the lack of a detailed description of individual systems. These approaches then aim to provide such a description by positing novel hypotheses about physics. Consequently these approaches are realist, whereas the first kind of approach are mostly anti-realist or operational. The aim is a deeper theory, inequivalent to quantum mechanics, which completes the partial description the standard theory now reveals. As it is a distinct theory, we may hope this completion will be testable. The two main examples of such completions of quantum mechanics that have been studied so far, pilot wave theory (Bohm, 1952) and dynamical collapse models (Pearle, 1976; Ghirardi, Grassi and Rimini, 1990), do make testable predictions, which differentiate them from quantum mechanics.[1]

The program I present here is in this second class. What distinguishes it from other realist completions of quantum theory is, first of all, the emphasis on relationalism, which the reader may recall is a way to characterize general relativity (GR).

---

[1] Pilot wave theory makes predications that differ from these of quantum mechanics because one can prepare an ensemble out of quantum equilibrium, in which Born's rule is not satisfied (Valentini, 1991, 2002; Valentini and Westman, 2005; Abraham, Colin and Valentini, 2014; Colin and Valentini, 2016; Underwood and Valentini, 2016).

That is, we connect quantum foundations to quantum gravity by seeking a relational completion of quantum mechanics, i.e., an extension of the theory in terms of "relational hidden variables."

A theory can be called relational if it satisfies the set of principles that I lay out in the next section. These basically dictate that the beables of the theory describe relationships between dynamical actors, as opposed to defining properties in terms of a fixed, unchanging background. But it is important to stress that there are different classes of relational theories, which differ by the choice of which of the basic elements of physics are to be regarded as fundamental and which as emergent. By *fundamental*, I mean irreducible, in the sense that it cannot be eliminated in a theory that aims to fully describe nature.

The program I advocate here regards time, causation, energy, and momentum as fundamental quantities. Among the things that are emergent, and hence reducible to more fundamental elements, are space and spacetime. Quantum mechanics itself will also be emergent from a deeper physical theory. In the next several sections I explain the motivation for these choices. They come from reflections on issues in quantum gravity, seen in the light of quantum foundations.

The first big divide among relationalist theories has to do with whether time is regarded as fundamental or emergent. Many relationalists, such as Julian Barbour and Carlo Rovelli, have advocated the view that time is emergent. I take the opposite view, that time is fundamental. My reasons will be only touched on here; the full argument is presented at length in two books (Smolin, 2013b; Unger and Smolin, 2014) and a number of supporting papers (Cortês and Smolin, 2014a; Smolin, 2015a). These arguments were developed in collaborations with Roberto Mangabeira Unger and Marina Cortês.

As Rovelli has emphasized, time is a complex phenomenon, and we should be precise and assert exactly which aspects are being claimed to be fundamental. I will be more precise in Section 8.3, but let me briefly say here that the aspect of time I assert is irreducible in its activity as the generator of novel events from present events (Cortês and Smolin, 2014a). This activity generates a *thick present,* by which is meant that two events in the present can be causally related with each other. This thick present is continually growing by the addition of novel events. At the same time other events in the thick present, having exhausted their potential to directly influence the future, slip from the present to join the always growing past. This continual construction of the future from the present, which then becomes past, makes the distinction among past, present, and future objective and universal.

I am aware of the arguments that claim that an objective observer-independent distinction among the past, present, and future conflicts with special relativity. This is addressed momentarily.

Because it is grounded on a version of relationalism that takes time to be primary, I propose to call this program *temporal relationalism*, as opposed to *timeless relationalism*, which supposes time to be emergent from a timeless fundamental theory.

Once one has decided that time is going to be fundamental or emergent, one must ask the same question in regard to space. I will explain why it is unlikely that, within a realist framework, time and space can both be fundamental. Since I choose time to be fundamental, I must choose space to be emergent. It is, by the way, the fact that in this construction time is fundamental, but space is not, that resolves the apparent conflict of having an objective present with special relativity. The relativity of inertial frames, and the consequent Lorentz invariance, is an emergent symmetry, which comes into effect only when space emerges and is not hence a symmetry of the fundamental laws, which govern a domain of events with causal relations, but no space.

The third choice one has to make concerns energy and momentum: are they fundamental or emergent? The more I reflect on the structure of our physical theories the more I realize that momentum and energy are at the heart of the foundations of physics and that this distinguishes physics, in ways that are sometimes under-appreciated, from other fields of science that describe systems that change in time. Only physics has the canonical structure indicated by the Poisson brackets, $\{x^a, p_b\} = \delta^a_b$, and the related principles of inertia and of the relativity of inertial frames. These lead to the most fundamental structure in quantum mechanics, the Heisenberg algebra, $[\hat{x}, \hat{p}] = \iota\hbar$, which cannot be expressed in a system constructed with a finite number of qubits.

This is one reason I find the claims to analogize nature to a computer, classical or quantum, to be inadequate: such claims neglect the fundamental roles energy and momentum each play in the structure of our physical theories, indeed, I would suggest, in nature. For reasons I describe later, I think it is interesting to develop the idea that energy and momentum are both fundamental.

Thus the class of theories I will develop treat time, energy, and momentum as fundamental. Space—and hence spacetime and its symmetries—are to be recovered as emergent. Toward the end of the chapter, I will describe a class of theories I constructed and studied with Marina Cortês that realize these ideas, called *energetic causal set models* (Cortês and Smolin 2014a, 2014b, 2016, 2017).

I learned about relationalism from Julian Barbour, from a conversation we had in the summer of 1980, in which he explained to me the debt that Einstein owed to Leibniz and Mach as well as his detailed understanding of the hole problem and the role of active diffeomorphisms. I realized immediately that I had always been an instinctive, if very naive, relationalist. Talking with Julian, I became a slightly less naive relationalist. My education has continued ever since, and I owe an enormous debt to philosophers since then.

Temporal relationalism is a part of a larger program of *temporal naturalism*, which Roberto Mangabeira Unger and I presented in a number of publications (Smolin, 2013b, 2015a; Unger and Smolin, 2014). The adjective *temporal* is again meant to emphasize the centrality of the hypothesis that time is fundamental and irreducible.

These temporal approaches are to be distinguished from timeless, or atemporal, forms of relationalism and naturalism, which hold that the most fundamental descriptions of the world are formulated without time, as time, in the sense of its flow or passage, is held not to exist fundamentally. According to this view, time is real but only in the sense that it is emergent, in the same sense that pressure and temperature are real. The block universe is an aspect of timeless naturalism, so is the notion that the laws of nature are fixed and unchanging.

In the next section I introduce principles for relationalism; the primacy of time and causation is the subject of Section 8.3. Section 8.4 is devoted to the question of whether energy and momentum should be treated as fundamental or emergent. Section 8.5 introduces a new notion of locality, which is entirely relational. The different strands of the argument come together in Section 8.6, where I introduce the full program of temporal relationalism.

8.2 Relationalism and Its Principles

We begin with the principle behind relationalism, which is Leibniz's *principle of sufficient reason*. In his *Monadology*, Leibniz (1714) states,

31. Our reasonings are based on two great principles, that of contradiction, in virtue of which we judge that which involves a contradiction to be false, and that which is opposed or contradictory to the false to be true.
32. And that of sufficient reason, by virtue of which we consider that we can find no true or existent fact, no true assertion, without there being a sufficient reason why it is thus and not otherwise, although most of the time these reasons cannot be known to us.

Leibniz (1717) also applied the principle to events: "[T]he principle of sufficient reason, namely, that nothing happens without a reason." I read this to say that every time we identify some aspect of the universe that seemingly may have been different, we will discover, on further examination, a rational reason why it is so and not otherwise.

I take this principle, not as metaphysics but, in a weak form, as methodological advice for physicists hoping to make progress with our understanding of nature.

• Seek to progress by making discoveries and inventing hypotheses and theories that lessen the arbitrary elements of our theories.

I call this the principle of *increasingly sufficient reason*, and it will be the form I employ here.

Here are some examples of steps in the history of physics that exemplify this advice.

- Eliminate references to unobservable absolute positions and motions and replace them by measurable relative positions and motions.
- Seek a theory of light in which its speed is not arbitrary but is computable in terms of known constants.
- Formulate the dynamics of general relativity and gauge theories directly in terms of gauge and active diffeomorphism invariant observables. This gives us unique dynamics to leading order in the derivative expansion.

And here are some further steps that we might yet be able to accomplish.

- Seek a theory that exists or makes sense only if the number of (macroscopic) spatial dimensions is three.
- Seek a completion of the standard model of particle physics that reduces the number of freely tuneable parameters.

I don't know whether or not there is an ultimate understanding in which sufficient reason is maximally satisfied and cannot be improved on. But each of these steps has been or would represent profound progress.

The principle of increasingly sufficient reason has a number of consequences that are each worthy principles in their own light. In each case they are to be read as advice, i.e., of the kinds of theories we ought to be aiming for, if we hope to progress fundamental physics.[2]

1. The principle of increased background independence.
2. The principle that properties that comprise or give rise to space, time, and motion are relational.
3. The principle of causal completeness.
4. The principle of reciprocity.
5. The principle of the identity of the indiscernible.

Each of these requires some elaboration.

8.2.1 Background Independence (Smolin, 2005)

All physical theories to date depend on structures that are fixed in time and have no justification; they are simply assumed and imposed. One example is the geometry of

---

[2] Notice that this advice, when followed, generates a dynamics for our theories to be challenged and improved, which is present in the absence of anomalies or conflicts with experiment. This does not mean experiment plays no role for, as Feyerabend pointed out, new theories may suggest new experiments, or give a new significance to old experiments, which can distinguish them from older theories.

space in all theories prior to general relativity. In Newtonian physics, the geometry of space is simply fixed to be Euclidean 3-dimensional geometry. It's arbitrary, it doesn't change in time, it can't be influenced by anything. Hence it is not subject to dynamical law. The principle of background independence requires that the choice is made not by the theorist but by nature, dynamically, as a part of solving the laws of physics.

The principle states that *a physical theory should depend on no structures that are fixed and do not evolve dynamically in interaction with other quantities.*

Nondynamical, fixed structures define a frozen background against which the system we are interested in evolves. I would maintain that these frozen external structures represent objects outside the system we are modeling, which influence the system but do not themselves change. (Or whose changes are too slow to be noticed.) Hence these fixed background structures are evidence that the theory in question is incomplete.

In many cases the observables of a background dependent theory describe how some quantity changes, or some body moves, with respect to those fixed external structures. This is the role of reference frames; they implicitly reflect a division of the universe into two parts—a dynamical system we aim to study and a part whose dynamics we neglect and fictionalize as fixed—for the purpose of pretending that relational quantities like relative position and relative motion have an absolute meaning.

It follows that any theory with fixed external structures can be improved if the external elements can be unfrozen, made dynamical, and brought inside the circle of mutually interacting physical degrees of freedom. This was the strategy that led Einstein to general relativity. The geometry of space and time is frozen in Newtonian physics, and it is also frozen in special relativity. In these theories, the spacetime geometry provides an absolute and fixed background against which measurements are defined. These are reflected in the role played by reference frames. Mach pointed out the fiction involved by identifying inertial frames with the "fixed stars." General relativity unfreezes geometry, making it dynamical. This freed the local inertial frames from an absolute and fixed dependence on the global mass distribution and made their relation dynamical, subject to solutions of the equations of motion.

This is rarely a one-step process. Typically, the new theory retains frozen elements, then it too is not the end of our search and will require further completion. The principle asserts we should seek to make progress by eliminating fixed background structures by identifying them as referring to external elements and making the choices involved subject to dynamical law.

It follows that the only complete theory of physics must be a cosmological theory, for the universe is the only system that has nothing outside of it. A theory of the whole universe must then be very different from theories of parts of the universe. It must have no fixed, frozen, timeless elements, as these refer to things outside the system described by the theory, but there is nothing outside the universe. A complete cosmological theory must be fully background independent.

It follows that quantum mechanics cannot be a theory of the whole universe because it has fixed elements. These include the algebra of observables and the geometry of Hilbert space, including the inner product.

This implies that there is no wave-function of the universe, because there is no observer outside the universe who could measure it. The quantum state is, and must remain, a description of part of the universe. Relational quantum mechanics develops this idea.

We then seek to complete quantum theory by eliminating background structures. We do this by exposing and then unfreezing the background and giving it dynamics. In other words, rather than quantizing gravity we seek to gravitize the quantum, that is, identifying and unfreezing those aspects of quantum theory that are arbitrary and fixed, making them subject to dynamical laws. Turning this around, we hope to understand the challenging features of quantum physics as consequences of separating the universe into two parts: the system we observe and the rest containing the observer and his or her measuring instruments. Closely related to background independence is another key idea: that the observables of physical theories should describe relationships.

8.2.2 Relational Space and Time

A relational observable, or property, is one that describes a relationship between two entities. In a theory without background structures, all properties that describe space, time, and motion should be relational. Background independent theories speak to us about nature through relational observables.

Leibniz, Mach, and Einstein taught us to distinguish absolute notions of space and time from relational notions and, more importantly, how to formulate physics entirely in terms of the latter.

A relational theory can have some beables that are intrinsic to individual events and processes. These describe properties of individual events, rather than relations between two or more events. I will suggest that energy is one such intrinsic property. I will argue that the structure of physics requires this.

8.2.3 Principle of Causal Completeness

We follow chains of causation back in time; the principle of causal completeness states that all chains of causation relate back to events and properties of our single universe.

If a theory is complete, everything that happens in the universe has a cause, which is one or more prior events. It is never the case that the chain of causes traces back to something outside the universe.

Consequently, our theory can contemplate only a single, causally connected universe.

8.2.4 Principle of Reciprocity

The principle of reciprocity was introduced by Einstein in his papers on general relativity and states that if an object, A, acts on a second object, B, then B acts back on A (Einstein, 1916).

8.2.5 Principle of the Identity of the Indiscernible

The principle of the identity of the indiscernible (PII) states that any two objects that have exactly the same properties are in fact the same object.

One important implication of the identity of the indiscernible is that there are no (global) symmetries. A symmetry of a physical theory is a group of transformations on the state space that take solutions to other solutions. Consider an arbitrary point in the state space a and consider also $T \circ a$, where T is a transformation. They have exactly the same properties, hence the PII requires that we identify them. The symmetry then acts as the identity.

An example is a translation in space. In the full description, there is a silent background, which includes the observers and the measuring instruments. The physical subsystem under study is being translated with respect to this silent and frozen audience. When we identify and unfreeze the silent background and bring it into the system of dynamical equations, we eliminate the symmetry.

This does not apply to gauge transformations, which take the mathematical description of a state to a different mathematical description of the same state.

Relationalism then offers an important piece of advice to the project of the unification of the elementary particles and their interactions, which is that the deeper a theory is the fewer global symmetries it should exhibit. This strikingly conflicts with the method that inspired the grand unified theories such as $SU(5)$, $SO(10)$, and $E(7)$, which aimed to unify by postulating larger and larger global as well as gauge symmetry groups. Of course, the program of grand unification failed experimentally because of proton decay.

On the other hand, this is something that general relativity gets right. As shown by Kuchar (1981), general relativity restricted to spatially compact solutions has no conformal Killing fields on its configuration space and hence no global symmetries. From general relativity we learn that global symmetries arise only as properties of certain very special, idealized solutions, where they may be regarded as signaling the existence of a frozen, decoupled background sector.

String theory aspires also to get this right. Particular perturbative, background-dependent string theories have global symmetries, but these are believed to represent expansions around solutions of a more fundamental, background independent string or $\mathcal{M}$ theory, which has no global symmetries.

The first time I heard the assertion that more fundamental theories should have fewer symmetries was in a talk by Roger Penrose in 1976 or 1977, in which he used it to argue that twistor theory was right to break parity and even CP. I have been thinking about it ever since. Thus, to me, one of the attractive features of loop quantum gravity, in the original formulation based on Ashtekar's original variables, is that there is evidence the quantum theory does break P and CP (Contaldi, Magueijo and Smolin, 2008).

8.2.6 Examples of Relationalism in Physical Theories

The importance of the methodological advice to "make our theories more relational by eliminating background structure" is shown in several instances where it was decisive. The role of relationalism in the passage from Newtonian dynamics to Einstein's general theory of relativity is paradigmatic, even iconic. So I will begin here with the earlier transition between Aristotle and Newton. I believe there is a way to tell that story that makes it also an instance where this advice was decisive. The importance is that the move to eliminate background structure—in particular the notions of absolute rest and absolute velocity—resulted in the discovery of energy and momentum and the central roles they have played in the structure of physical theories ever since.

- *Second-order equations of motion and the canonical dualities.* One of the most important instances of the transformation of background structure into dynamical degrees of freedom occurred during the transition from Aristotelian to Newtonian physics. For Aristotle and his followers dynamics is first order in time, i.e., a velocity is a response to a force. More generally, if x^a are the relevant dynamical quantities that live on a configuration space, $\mathcal{C}$, then the dynamics is given by a fixed vector field, $v^a[x]$ on $\mathcal{C}$:

$$\frac{dx^a(t)}{dt} = v^a[x(t)]. \tag{8.1}$$

This law made sense to Aristoteleans because they believed velocity had an absolute meaning. Hence its absence, rest, has an absolute meaning. Newton identified the absolute frame of reference needed to define rest as the Sun, which for some experiments could be approximated by the rest frame of the laboratory. He removed this background structure by replacing absolute velocity with respect to the laboratory with its relative counterpart. By making that subject to dynamical law he made the decisive step to the new dynamics. In retrospect the key insight was the discovery of momentum, as the dynamical quantity related to relative velocity.

Most applications of mathematics to science, such as models of economics, ecology, and biology are still Aristotelean. This is because they still have notions of absolute change and stasis. Consequently, they have no analog of the principles of inertia or of relativity. The dynamical equations in economics, biology, ecology, general dynamical systems, etc. are thus first order in time. Computers, and algorithms generally, such as cellular automata, have the same first-order structure, in which change is given by a first-order update rule:

$$X(n + 1) = F[X(n)]. \tag{8.2}$$

The fixed value of v^a or $F[X(n)]$ codes the knowledge we have of the causes of change in these systems. It is background structure.

This goes for computers as well. Whether classical or quantum, a computer has a well-defined notion of stasis; the update rule, or the identity operation, that changes nothing. So long as it does so, it misses a fundamental aspect of dynamics in physics, whose deepest principle is that stasis and change are, to first order, relative. This why I am not convinced by claims that physics could be modeled or envisioned as a bunch of interconnected qubits.

Newton made the background structure, implicit in the fixed vector field $v^a(x)$, dynamical. The clearest way to see this is in the Hamiltonian formalism, where the equations of motion are:

$$\dot{x}^a = \{x^a, H\} = \frac{1}{m} g^{ab} p_b. \tag{8.3}$$

But where p_b rather than being fixed, has been made to evolves from arbitrary initial conditions, subject to dynamical law:

$$\dot{p}_a = \{p_a, H\} = -\frac{\partial V}{\partial x^a}. \tag{8.4}$$

If we eliminate the p_a we get the usual form of Newton's laws, which are second order in time, so that,

$$\frac{d^2 x^a(t)}{dt^2} = G\left[x(t), \frac{dx(t)}{dt}\right], \tag{8.5}$$

so the initial velocity, $\frac{dx^a(t)}{dt}$, rather than being a fixed background structure, is a contingent initial condition, which can be varied so that the same law applies to a multitude of different circumstances.

The canonical first-order form of dynamics is elegantly defined on a phase space, $\Gamma = (x^a, p_b)$. This carries an imprint of the original second-order form of the dynamics in the canonical pairings:

$$\{x^a, p_b\} = \delta^a_b, \tag{8.6}$$

and in the form of the Hamiltonian:

$$H = g^{ab} p_a p_b + V(x). \tag{8.7}$$

There are still elements of background structure—they are coded into the Hamiltonian in the g^{ab} and $V(x)$. There is also the background time, which can be eliminated by making the Hamiltonian into a constraint. One way to do this was pioneered by Julian Barbour together with Bruno Bertotti. Instead of the usual action:

$$S = \int dt \sqrt{g} \left(\frac{1}{2} g_{ab} \dot{x}^a \dot{x}^b - V(x) \right), \tag{8.8}$$

we write the time reparameterization invariant action:

$$S = \int dt \sqrt{\frac{1}{2} g_{ab} \dot{x}^a \dot{x}^b V(x)}. \tag{8.9}$$

I want to emphasize that the canonical form of dynamics, based on the dualism of a pair (x^a, p_b) tells us something fundamental about physics and distinguishes motion in space from all other forms of change studied by the different sciences.

A key consequence of this structure is Noether's theorem, which relates symmetries in the x^a's to conservation laws involving total values of the p_a's. These symmetries arise from invariances of H, which reflect the principles of relationalism, i.e., that the forces are functions of relative coordinates.

All of this structure is special to the laws of motion and field equations of physical systems.

- *Mach's principle* was a step in articulating how space, time, and motion in a relational world would differ from Newtonian physics, with its dependence on absolute space and time. It gives a criteria for a relational theory, which is that the response to an acceleration or rotation of a body, measured against "the fixed stars" must be the same as if that body were fixed and the universe accelerated or rotated in the opposite sense.

  General relativity with spatially closed topology satisfies a form of Mach's principle. But general relativity with asymptotically flat boundary conditions does not, because asymptotic infinity defines a fixed inertial frame of reference.

- *Connections and gauge invariance.* One of the most basic things we know about nature is that the laws of physics are second order in time. This means that the law of motion prescribes accelerations as a function of positions and velocity. To define such a law we have to be able to compute the rate of change of a rate of change or, equivalently, compare two velocity vectors at slightly different times.

  To compare two vectors at different events in spacetime, we need to be able to bring them together at the same point. To do this requires a notion of parallel transport. If we give a fixed, nondynamical rule for comparing vectors at different events, we impose background structure and hence violate the principle of background independence. To satisfy that principle, we must introduce a notion of parallel transport that can be contingent and subject to dynamics. The structure needed to do so is a connection.

  The proof that a connection implements background independence is that one can rotate the tangent space at two nearby points independently, without affecting the laws of motion. This is an example of how background independence implies an important consequence:

- *Symmetries should all be local.* Indeed, it follows that a relational theory should have no global symmetries. This is true of general relativity with cosmological boundary conditions (Kuchar, 1981). Another way to say this is that only proper subsystems of the universe can have nonvanishing values of conserved charges. The values of any such charges code relations between the subsystem that is modeled and the rest of the universe. The symmetries such charges generate translate or transform the subsystem with respect to the rest of the universe.

- *General relativity, with compact spatial slices* is a relational theory of space and time. It is a story of events, their causal relations and their measure, but it is crucial to understand that physical events do not correspond to mathematical points in a differential manifold. Instead physical events correspond to equivalence classes of those points under active spacetime diffeomorphisms. This means that events have no intrinsic nondynamical labels. What distinguishes, and hence labels, a physical event is a conjunction of degrees of freedom. One way to say this is that physical events are labeled relationally in terms of what can be seen from there.

  Temporal relationalists disagree with timeless relationalists on one point of the interpretation of general relativity. Timeless relationalists take the diffeomorphisms discussed earlier to include the whole group of diffeomorphisms of the 4-dimensional manifold. This is in tension with some aspects of the primacy of time, in particular with the insistence on an objective notion of the present moment and an objective distinction among past, present, and future.

  In classical general relativity this amounts to the claim that there is a physically preferred slicing. Temporal relationalists then are interested in a reformulation of general relativity known as shape dynamics (Gomes, Gryb, and Koslowski, 2011; Mercati, 2018).

- *Shape dynamics* is a classical gravitational theory that invokes a preferred slicing of spacetime, which arises from constant mean curvature gauge of general relativity (Gomes et al., 2011; Mercati, 2018). But the many-fingered time symmetry is not broken, rather it is traded for local scale invariance of the Hamiltonian 3-dimensional theory. It has the same massless spin two degrees of freedom of general relativity and is equivalent to GR outside of horizons, but can differ within horizons.

  Shape dynamics is important because it shows us that all the empirical successes of special and general relativity can be made compatible with the postulation of a global time, which would be a consequence of an objective distinction among past, (a thick) present, and future.

- *Entanglement* reveals that a pair of quantum systems can share properties that are not properties of either. A completely relational description of a quantum system would be one that constructed all its properties from such shared properties. This is the case with the Einstein–Podolsky–Rosen (EPR) state, which has no definite value of any component of the individual spins.

- *Spin networks* are models of a discrete or quantum geometry invented by Penrose in the early 1960s. A Penrose spin network (Penrose, unpublished notes) is a trivalent graph, whose edges are labeled by half-integers (denoting representations of $SU(2)$), such that the triangle inequality is satisfied at each node. This condition means that there is an invariant or singlet in the product of the three spin representations at each node.

  Penrose was interested in the idea that the shared properties of entangled quantum systems provided a realization of Mach's principle, in the sense that the properties of a local system are determined by its coupling to the larger system in which it was embedded. Mach's particular suggestion was that local initial frames are determined to be those that don't rotate or accelerate relative to the large scale distribution of matter (which he referred to as the *fixed stars*).

  Penrose realized a version of Mach's principle in his spin networks. An edge in the network is chosen to have its spin measured along an axis that is determined by its coupling into the network as a whole. Through a combinatorial calculation on the graph Penrose defines an angle between a pair of spins, each on an edge. He then shows that in the limit of a large and complex graph, this reproduces the geometry of a two sphere, which is the space of directions in R^3.

- *Relational hidden variable theories.* One way to complete quantum mechanics would be by the introduction of additional degrees of freedom, whose statistical fluctuations are responsible for the fluctuations and uncertainties in quantum physics. These are called *hidden variables*, because it was presumed that they are not measured by the procedures that measure quantum observables. Given the ubiquity of nonlocality in quantum systems, due to entanglement, it is natural to hypothesize that the hidden variables are relational, in the sense that they describe shared properties of pairs of quantum systems.

We then might expect to have a hidden degree of freedom for every pair of ordinary degrees of freedom. These could be arranged as a matrix or a graph.

Several such relational hidden variables have been constructed (Smolin, 1985, 2002; Markopoulou and Smolin, 2004). These were formulated by constructing an explicit stochastic process in which, in a certain large N limit, the assumptions Nelson (1966, 1985) imposes on a stochastic process to yield a solution to the Schrödinger equation would be realized. This required a fine-tuning to keep the dynamics time reversible, as discussed in Smolin (2006).

- *Two-dimensional many body system (Smolin, 1985).* The degrees of freedom are complex numbers Z_{ij}, which are entries in an $N \times N$ matrix, Z. The N complex eigenvalues of Z, labeled λ_i, give the positions of N particles in the 2-dimensional space $\mathcal{C} = \mathcal{R}^2$.

  A dynamics was imposed on the Z_{ij}, including an arbitrary potential energy that is a function of differences, $V[\lambda_j - \lambda_k]$, such that in the limit of large N the eigenvalues λ move according to Newton's laws in a potential V. There are corrections that can be computed to order $\frac{1}{\sqrt{N}}$, and it is found that the probability distribution evolves according to the Schrödinger equation.

  A similar relational hidden variables model was constructed for a system of N particles in D dimensions, where the hidden degrees of freedom are now M $N \times N$ anti-Hermitian matrices (Smolin, 2002). The eigenvalues again give the positions of the particles in R^D.

  The dynamics is chosen to be the standard matrix model dynamics used in dimensionally reduced models of $\mathcal{M}$ theory and Yang–Mills theory, and it is again shown that the Schrödinger equation is recovered to first order in an expansion around the large N limit.

  A relational hidden variable whose hidden variables are a graph was constructed in Markopoulou and Smolin (2004); matrix hidden variables theories were also constructed by Steven Adler (2004) and Artem Starodubtsev (2003).

- *Relational approaches to quantum gravity.* The search for the correct quantum theory of gravity has in my view reached an unexpected situation, which is that several different approaches have achieved results that may be construed as descriptions of plausible, but distinct, regimes of quantum gravitational phenomenon (Smolin, 2017a, 2017b). At the same time, most have also encountered significant barriers that make it seem unlikely that it by itself is a complete theory.

  Several of these approaches reflect one or more aspects of the principles of relationalism, typically by having background-independent kinematics and dynamics, in the narrow sense that their formulations do not require a fixed, classical background spacetime. The background-independent approaches include causal set models (Bombelli et al., 1987), causal dynamical triangulations (Ambjorn et al., 2013), group field theory (Oriti, 2017), and loop quantum gravity, including both its Hamiltonian and path integral (or spin foam) formulations (Rovelli and Vidotto, 2015).

Non-background-independent formulations include perturbative string theory (Conlon, 2016), asymptotic safety (Eichhorn, 2018), and other perturbative approaches.

However, recent very promising explorations of the idea that space may be emergent from entanglement (an idea that at least roughly goes back to Penrose's early work on spin networks) certainly have a relational flavor (Lee, Kim, and Lee, 2013).

8.3 The Primacy of Time and Causation

There are several ways of expressing the idea that time and causation are fundamental and primary.[3]

- The present moment is fundamental, meaning that there is no deeper level of description that lacks a present moment, which is one of a flow of such moments. The reason our experience is structured as a flow of moments is because that is how the world is structured.
- The distinction among the past, present, and future is objective and universal. But the present may be thick, which means it contains events that are causally related.
- The process that brings further novel events from present events is real and fundamental. This defines a thick present of events that have yet to give rise to all their future events.
- Once an event has given rise to all the future events it is ever going to, it has used up its whole capability to influence the future. It drops out of the thick present and becomes part of the past. The past was real and is no longer real.[4]
- The future is not real and is, to some extent, open. There are no present facts of the matter concerning future events.

A concrete illustration of a thick present is given in the energetic causal set models I constructed with Marina Cortês (Cortês and Smolin, 2014a, 2014b). The history of the universe is a set of events with causal relations and other properties; for example, each event is endowed with energy and momentum. I can illustrate the thick present with just the causal relations.

---

[3] For more discussion on these, see Smolin (2013b, 2014a), Unger and Smolin (2014), and Cortês and Smolin (2014a).

[4] So an event is real and present by virtue of its capacity to directly influence the future.

Each event, *e* has a causal past, $\mathcal{P}(e)$, which has some mathematical representation. There is a measure of the distinctiveness of two pasts, which we can indicate symbolically:

$$\mathcal{D}(e, f) = g[\mathcal{P}(e), \mathcal{P}(f)]. \tag{8.10}$$

The causal set grows by a series of distinct moves. In each move a new event, *e*, is created, which has two parents, lets say *a* and *b*. These parent events are the immediate causal past of *e*, the full causal past is:

$$\mathcal{P}(f) = \mathcal{P}(a) \cup \mathcal{P}(b), \tag{8.11}$$

including their causal relations.

Moreover each parent can have up to two children. At every stage in the construction, the set of events with fewer than two children makes up the thick present.

There is an algorithm that determines the next event born. We used the rule that the two parents are the members of the thick present that have the largest distinctiveness, $\mathcal{D}(e, f)$, among all pairs in the thick present (Cortês and Smolin, 2014a). Thus, at each step, one new event is created, which has at first no children and so is part of the thick present. One or both of the parents may leave the thick present to become part of the past, if this is their second child.

8.3.1 The Nature of Laws

This view has strong implications for how we think about physical laws.

If time is to be truly fundamental and irreducible, then there cannot be absolute, unchanging deterministic laws. One reason is that if there are such laws, then any property of the state of the world at time t', $P(t')$ is a function of other properties at an earlier or later time t,

$$P(t') = \mathcal{F}[t', t; P_a(t)]. \tag{8.12}$$

But, if this is so, then time has been eliminated, as any property of the system at a time other than t can be expressed in terms of properties only at t.

Another way to say this is that, *if* there is a mathematical object, $\mathcal{M}$, that is a perfect mirror of the history of the universe, in the sense that every true property of the universe, and its history, corresponds to a true theorem about $\mathcal{M}$, *then* each instance of causation, i.e., of the generation of a causal relation, is equivalent to an instance of logical implication. But this means that causation, the generation of new events from present events, is equivalent to a timeless logical implication. So causation, the essence of time, is reducible to logical implication, which is timeless.

So, when we assert that time is irreducible, we are asserting that there exists no mathematical object, $\mathcal{M}$, that is equivalent to the history of the universe, in the sense specified. Indeed we can name at least one property of the universe, which is not shared by any mathematical object, which is that here in the universe it is always some present moment, which appears uniquely, i.e., at most once, in a series of such moments.

Several of our basic principles imply that laws cannot be absolute. The usual notion of law contradicts the principle of reciprocity because laws affect the motion of matter, but how matter moves or changes has no effect on what the laws are. Similarly, the principle of causal completion is violated because a law that affects[5] the motion of bodies, but cannot be affected by them, is a cause outside of the universe of causes.

Instead, we propose that laws are meaningful for only limited ranges of space and time, and that on cosmological scales, laws evolve. I have proposed, several mechanisms by which this might happen (Smolin, 1992, 1997, 2012a). These are briefly mentioned in the following discussion.

It is usually presumed that the fundamental laws are symmetric under time reversal. If indeed, time is inessential and emergent, there can be no absolute difference between the past and the future. Since we do have such a distinction, we are free to make the hypothesis that the fundamental laws are irreversible. Indeed we mean this in two senses, first that there is no symmetry of the laws that reverses the direction of time; second that once an event happens, it cannot be made to unhappen (Cortês and Smolin, 2014a).[6]

Penrose (1979) posited that a time asymmetry of the fundamental laws could be responsible for the arrow of time. The basic idea is that a time reversible law emerges at late times from a time irreversible law, but the former is subject to a restricted set of initial conditions, which are compatible with the time asymmetry of the fundamental law. We endorse this hypothesis, and we have found some results that support it (Cortês and Smolin, 2014a).

8.3.2 The Nature of Space

What about space? If time is fundamental, what is space? For a relationalist, it is assumed that space manifests a network of relationships, which define relative position and relative motion. But are these spatial relationships fundamental or emergent from a more elementary set of relations?

In general relativity, fixing a notion of time gives us a time slicing, $M = \Sigma \times R$. On each slice, a relational notion of space is given by the spatial metric,

---

[5] The law affects, but does not cause, the event. Only events cause other events.
[6] Suppose an event, E, among other actions, exchanges two distinguishable particles, A and B. This may be followed by a second event, F, which returns them to their original positions. F may reverse the action of E, but it does not nullify E. The result is a history with two events.

Lee Smolin

h_{ab}, modulo $Diff(\Sigma)$. If we don't fix a time gauge, we instead view the spacetime geometry as relational, corresponding to the spacetime metric, $g_{\mu\nu}$, modulo $Diff(M)$.

In background-independent approaches to quantum gravity, these spatial, or spacetime, relations are found to be emergent from a more fundamental network of relations. In the simplest theory, which is causal set theory, both causal and metric relations are emergent from a network of discrete causal relations. In other background-dependent approaches, such as causal dynamical triangulations, loop quantum gravity, spin foam models, and group field theories, the fundamental discrete structure is more complicated, but it is still the case that the spatial or spacetime geometry (modulo diffeomorphisms) is held to be emergent from a more fundamental discrete structure of dynamically evolving relationships.

Thus the lesson from quantum gravity is that space or spacetime is emergent, not fundamental.

It is also important to note that there are clues indicating that both space and time cannot be both fundamental. These include:

1. Both existing realist completions to quantum mechanics, namely pilot wave theory and dynamical collapse models, require a preferred simultaneity, and hence are in tension with special relativity. This is the case even in pilot wave models of relativistic quantum field theory, which reproduce the Lorentz invariance in the quantum statistics. When one puts these models out of quantum equilibrium, one sees instantaneous and nonlocal transmission of information that takes place in a preferred simultaneity.

2. The state of the art concerning models in which classical spacetime emerges from an underlying dynamical quantum geometry is that those that succeed either have a fixed boundary or assume a prior temporal or causal order.

This suggests that the fundamental level of description should have a global temporal order as well as a partial causal order and that space (and hence spacetime) should emerge from the network of relations existing in the fundamental description.

At this point we may note that the work of Sorkin and collaborators has given us a well-studied model of a fundamental spacetime that has only causal relations, this is causal set theory (Bombelli et al., 1987).[7] As candidates for a fundamental theory, these models must be credited with a number of unique successes, including the only genuine prediction of a cosmological constant of the right order of magnitude, from a theory of quantum gravity (Sorkin, 2007).

---

[7] Causal sets built out of intrinsic structures was developed by Furey (2006), Criscuolo and Waelbroeck (1999), and Markopoulou (2000).

Another important result is that causal sets can be produced by sampling Lorentzian spacetimes. However it is important to note that there is an inverse problem, which is the fact that almost no causal set arises from sampling a low (by which I mean, say, less than 11) dimensional spacetime. The problem is then to choose a dynamics on causal sets that induces a low-dimensional and weakly curved Lorentizian spacetime to emerge. Because of the inverse problem such a dynamics must suppress the typical contributions that fail to correspond to any spacetime.

8.4 Energy Is Fundamental

The next (and last) choice we have to make to construct a relational theory of fundamental physics has to do with the status of energy and momentum. We might take these to be emergent quantities that arise as a consequence of Noether's theorem. When the emergent spacetime has symmetries, these are generated by an emergent energy and momentum.

The other choice is to presume them fundamental. The conservation laws of momentum and energy are then to be posited ab initio. We conjecture that there is an inverse Noether effect whereby the conservation of energy and momentum imply the emergence of spacetime, with global translation symmetries. We see this occur in the energetic causal set models (Cortês and Smolin, 2014a, 2014b), where this solves the inverse problem for causal sets.

Another reason to presume energy is fundamental is that the concept is a necessary part of Jacobson's proof of the Einstein equations from the thermodynamics of a more fundamental theory (Jacobson, 1995).

8.4.1 Emergent Locality

If space and spacetime are emergent concepts, so is locality. To illustrate this important point, I'd like to present a class of theories in which locality is explicitly emergent, as a consequence of the classical equations of motion. Moreover, since locality is a consequence of dynamics, it can be modified; to be precise, it is relativized, in an exact sense I will describe. This is then the story called *relative locality*.

One way to approach this subject, which was the original way we discovered it, is to think like a phenomenologist. Imagine that we have a quantum theory of gravity; whatever it is it depends on *four* (not three) dimensional parameters, Newton's gravitational constant, G, $\hbar$, c, and the cosmological constant Λ. Various regimes of quantum gravity are expressible as limits of combinations of these constants. (These include the well-known limits, such as $G \rightarrow 0$ that gives quantum field theory on a fixed spacetime background, but there are a number of others that are not often discussed; these are discussed in Smolin [2017b].)

Consider the particular limit where we hold c fixed and take $\hbar$ and G both to zero,

$$\hbar \to 0, \quad G \to 0, \tag{8.13}$$

but with the ratio that defines the Planck energy fixed:

$$E_p = \sqrt{\frac{\hbar}{Gc^3}}. \tag{8.14}$$

This is a domain of experiment that should preserve the principle of the relativity of inertial frames that c marks, but it also has a fixed (and therefore, we expect) invariant energy scale, E_p. Because this is an extension of special relativity with two invariant scales the name of this story used to be *doubly* or *deformed* special relativity.

The same limit puts the Planck length to zero:

$$l_p = \sqrt{\frac{\hbar G}{c^3}} \to 0, \tag{8.15}$$

so there is no quantum geometry. We then expect that the new physics will appear first as corrections to the standard equations of relativistic physics, expressed as ratios $\frac{E}{E_p}$, where E is an energy or momentum, and will thus be naturally described in terms of modifications of momentum space.

But before we look at those modifications let us note that there is already, with $E_p \to \infty$, a formulation of relativistic particle dynamics that takes momentum space to be fundamental, while spacetime is an emergent or derivative concept.

Let us start with a single free relativistic particle. It has a 4-momentum, p_a, which lives on a flat momentum space, with flat metric η^{ab} and mass m. It has an action principle, based on the extremization of:

$$S^{free} = \int ds \left(-x^a \dot{p}_a - \mathcal{N}\mathcal{C}\right). \tag{8.16}$$

The action is invariant under reparametrizations of s; as a result there is an Hamiltonian constraint, which is the result of varying by the Lagrange multiplier $\mathcal{N}$:

$$\mathcal{C} = \eta^{ab} p_a p_b + m^2 c^4 = 0. \tag{8.17}$$

Notice that the only geometry that is invoked is that of momentum space; $x^a(s)$ is a momentum to the momentum (notice also the switch from $p_a \dot{x}^a$ to $-x^a \dot{p}_a$, this is not a typo.

Varying x^a yields:

$$\dot{p}_a = 0, \tag{8.18}$$

while varying p_a gets us to:

$$\dot{x}^a = -2\mathcal{N}\eta^{ab}p_b. \tag{8.19}$$

So far this describes a free particle, which means it is at rest in momentum space.

We can add more particles trivially, by adding their actions, which are so far independent. To begin to describe physics we introduce an interaction. The one thing we know about interactions is that they conserve energy and momentum. So let us consider an interaction where particles $A + B \to C$. We write $p_a^A(1)$ for the end of particle A's worldline and $p_a^C(0)$ for the beginning of particle C's worldline. The conservation law says:

$$\mathcal{P}_a = p_a^C(0) - p_a^A(1) - p_a^B(1) = 0. \tag{8.20}$$

We add this conservation law to the action, using a Lagrange multiplier, z^a:

$$S = \Sigma_{\text{world lines}} S^{free} + z^a \mathcal{P}_a. \tag{8.21}$$

Now let us consider the equation that comes from varying $p_a^C(0)$. Unlike the other points of the worldlines, this end-point variation picks up two terms, and we get:

$$x_C^a(0) = z^a; \tag{8.22}$$

varying at the other two end points we find they are also at z^a:

$$x_A^a(1) = x_B^a(1) = x_C^a(0) = z^a. \tag{8.23}$$

Hence the Lagrange multiplier z^a becomes identified with the interaction point, the event where all three worldlines meet.

We see that, as promised, the interaction is local, i.e., takes place at a single event and is emergent as a consequence of equations of motion. Now let us call that event e, which is embedded at point z_e^a. There is another event at the end of worldline C, which we will call f; it is embedded at z_f^a. Now on C the momentum is constant, so we will call it p_a^C. By integrating (8.19) we have:

$$\Delta z_{ef}^a = z_f^a - z_e^a = N\eta^{ab}p_b^C, \qquad N = \int_C \mathcal{N}. \tag{8.24}$$

Let us consider the case of massless particles, $m_C = 0$. Then we note that the inverse of the metric we defined on momentum space gives an induced metric on the space of z^a's. We write:

$$\Delta z_{ef}^a \Delta z_{ef}^b \eta_{ab} = N^2 p_c^C p_d^C \eta^{ce} \eta^{df} \eta_{ef} = N^2 p_c^C p_d^C \eta^{cd} = 0. \tag{8.25}$$

The lesson of this story is that we start out with a theory defined on momentum space, and the classical equations of motion induce an emergent spacetime, together with an embedding of the events of our process in it, on which emerges a metric. In the case of massless particles, what we induce is actually a conformal metric.

8.4.2 *Relative Locality*

Now that we are used to the idea that energy and momentum are primary and spacetime and its geometry are emergent from the classical equations of motion, we can go back to the original point, which was to represent quantum gravity effects that might arise in the limit of Equations (8.13) and (8.14).

We want terms in the ratio of energies to E_p to appear. How are we to do this? We can get these from modifying the geometry of momentum space (Amelino-Camelia et al., 2011; Freidel and Smolin, 2011).

A continuous geometry can depart from that of flat spacetime in two independent ways. The metric can be changed. The metric comes into the constraint $\mathcal{C}$, where the norm of the momentum p_a must be written as $|p|^2 = D^2(p)$, the geodesic distance from the point p_a of phase space to the origin.

The other structure that can be changed is the connection or parallel transport. This is given by a derivative operator or connection and is generally independent of the metric. The parallel transport comes into the conservation law. The momentum space is no longer a vector space, and we have to think about how we define the sum of two or more momenta, which we need to state the conservation law. We showed that a nonlinear combination operator is equivalent to a connection (Amelino-Camelia et al., 2011; Freidel and Smolin, 2011).

An important point is that we want to preserve the relativity of inertial frames, hence we want to preserve the fact that there is a symmetry group on momentum space, which has 10 generators in the case of $3 + 1$ dimensions. This means we want a homogeneous geometry on momentum space.

There are three independent, invariant measures that can characterize the parallel transport in a homogeneous geometry. These are torsion, nonmetricity, and curvature. To give leading-order quantum gravity effects, these should be set proportional to $\frac{1}{E_p}$ and $\frac{1}{E_p^2}$.

The effects of torsion and nonmetricity have been worked out. A typical result is the following (Freidel and Smolin, 2011). Consider an observer who measures a distant event, such as the one we just discussed, involving a photon of energy E, a distance d from them. This event we will assume is local from the point of an observer local to it. Then the distant observer will describe it to be nonlocal. For example, if the event is the absorption of a photon by an atom, and the local observer sees the photon disappear at the spacetime point where the photon encounters the atom, the distant observer will describe the absorption as happening when the photon was still a distance x from the atom, where in rough numbers,

$$x = dEN = d\frac{E}{E_p}, \qquad (8.26)$$

where $N \sim \frac{1}{E_p}$ is a component of the nonmetricity tensor.

This has observational consequences. Consider a gamma-ray burst a distance d away and consider two photons that are emitted simultaneously, according to the local observer. Then they arrive at the Earth with distinct times of arrivals different by:

$$\Delta t = \frac{d}{c}\frac{\Delta E}{E_p}, \qquad (8.27)$$

where ΔE is the energy difference of the two photons (Freidel and Smolin, 2011).

In special relativity simultaneity is relativized, but locality becomes absolute. If onc observer sees two events to coincide at the same time and place, that is the way all observers will see it. As we have just seen, an analysis of the possible limits of quantum gravity phenomena, shows us that this notion of locality also gets relativized.

The relativity of simultaneity is usually taken to mean that the notion of a spatial manifold of space, in other words, has no place in a world described by special relativity. It is a fiction created by the procedure that local observers use to construct a picture of what is happening around them. Similarly, we interpret the phenomenon of relative locality as a signal that spacetime itself is an observer-dependent notion that falls apart when one looks too closely. Just like the surface of simultaneity is a fiction tied to the motion of an observer, the notion of a local spacetime is also a fiction, which depends on the position and motion of the observer as well as on the energies of the probes she uses to observe the world distant from ourselves.

This relativization of the notion of locality was uncovered as the answer to paradoxes pointed out by Hossenfelder (2010) and Schutzhold and Unruh (2003) in several models with an invariant, observer-independent, energy scale (Amelino-Camelia, 2001, 2002; Kowalski-Glikman, 2002; Magueijo and Smolin, 2002).

8.5 The Dynamics of Difference: Replacing Locality with Similarity of Views

In a relational account of space, there is no intrinsic property of an object associated to its location in space. Instead, an object's position in space is a consequence of its place in a network of relationships with other objects. What is fundamental is that network of relationships from which position in space is emergent and, very possibly, approximate.

The network of relationships that describes a system may be illustrated by a graph where the objects are nodes and two nodes are connected by an edge if they are involved in a relationship. The edges may be labeled by properties of the relationship. There is a metric on the graph (one of several which exist; we will see another very different one shortly), g_{IJ}, that counts the minimal number of steps to walk from I to J on the graph.

An object has a view of the system it is a part of, which represents the knowledge it may have of the rest of the universe; all such knowledge is a function of the relationships that tie the object to the rest of the system. It is very useful to describe the knowledge in terms of neighborhoods. The first neighborhood of an object, A, consists of all objects in the network one step away on the graph, plus any edges that may join them. Similarly, the nth neighborhood of J, $\mathcal{N}_J^n$, consists of all nodes K such that:

$$g_{J,K} \leq n, \tag{8.28}$$

together with all the links in the graph between pairs of nodes in neighborhood. The graphs representing neighborhoods of J have an origin, or marked point, that is the object, J.

One way to represent the view of an object is by the sets of neighborhoods together with their embedding maps into each other:

$$\mathcal{V}_J = \{\mathcal{N}_J^0, \mathcal{N}_J^1, \ldots, \mathcal{N}_J^n, \ldots\}. \tag{8.29}$$

I want to suggest that similarities and differences of views are fundamental, whereas distance in space is emergent and approximate.

The usual idea of locality is that two objects will interact more often, or more strongly, the closer they are in space. But notice that two nearby objects have similar views of the rest of the universe. Here, by *your view* I mean, informally, what you see when you look around, i.e., the sky from your point of view.

We can also think of this in terms of the view of an event V_e. Think of the pattern of stars seen in the sky from a particular event's perspective, i.e., the pattern of incoming radiation on the sphere that is the space of directions on your backward light cone. Clearly there is at least a rough relationship between the distance between two events:

$$d(e, f) = \sqrt{|e - f|^2}, \tag{8.30}$$

and the similarity of their views:

$$d(e, f) \approx g(V_e, V_f), \tag{8.31}$$

where $g(V_e, V_f)$ is a metric on the space of views.

What if distance in spacetime is only a proxy for difference of views? What if the locality that matters fundamentally is the distance in the space of views? This means that two events are more likely to interact when their views are similar.

There are two ways that two events can have similar views. One is if they are events in the history of two macroscopic bodies and are close to each other in spacetime. This is the conventional case, which I've mentioned. The second is if the two events arise in the histories of two atoms or molecules. These systems have few degrees of freedom, so the space of possible views is going to be in some sense small. But these are also systems that exist in vast numbers of copies spread throughout the universe. So the view of an atom or molecule can have many neighbors in the space of views. These neighbors define ensembles of microscopic systems, which all interact with each other in spite of being spread through the universe. I would like to suggest that it is these ensembles that quantum states refer to. I would also suggest that the peculiarities of quantum mechanics arises from the fact that it is a course-grained description of these ensembles.

This is the basic idea behind the real ensemble formulation of quantum mechanics (Smolin, 2012b, 2016a, 2018).

To make this suggestion precise, we need to define the distinctiveness of the views of two objects, J and K, which we will call $\mathcal{D}_{JK}$. There are several ways this can be defined. The simplest is to define:

$$\mathcal{D}_{JK} = \frac{1}{n_{JK}^{p}}, \tag{8.32}$$

where p is some fixed power and n_{JK} is the smallest n such that $\mathcal{N}_J^n$ is not isomorphic to $\mathcal{N}_K^n$ under all maps that preserve the origin.

Alternatively, define a metric on views, μ_{JK}, which measures their distinctiveness, modulo all isomorphisms of the views that preserve the marked points.

Given a metric on the space of views, which measures their distinctiveness, we may define the variety of the set of relations (Barbour and Smolin, 1992). This has the general form:

$$\mathcal{V} = \frac{1}{N^2} \sum_{J \neq K} \mathcal{D}_{JK}. \tag{8.33}$$

This is, in general, an interaction among three systems, because for each pair J and K the distinctiveness $\mathcal{D}_{JK}$ involves a comparison of the views J and K each have to third bodies. Very remarkably, this turns out to yield exactly Bohm's quantum potential (Smolin, 2016a).

This is the core of how Schrödinger quantum mechanics emerges from a dynamics that involves comparisons among similar systems. I leave the details to available papers (Smolin, 2012b, 2016a, 2018).

8.5.1 *Causal Sets and Causal Neighborhoods*

Given the emphasis of the primacy of time and causation, we will be interested in models of fundamental physics based on causal structures. A set of events, $\{A, B, C, \ldots,\ \in\ \mathcal{C}$ together with a causal relation, $>$, is a causal set if for any two events, A and B, only one of the following is true:

$$A > B, \quad B > A, \quad \text{or, } A \text{ and } B \text{ are causally unrelated.} \tag{8.34}$$

We also require that the set be *locally finite*, i.e., for all A and B, the set of C such that $B > C > A$ is finite.

We require that causal relations are transitive, so that:

$$A < B, \text{ and } B < C \text{ implies } A < C. \tag{8.35}$$

We will also want to denote the primary causal relations, A ¡– B, which are irreducible and generate the rest.

The view of an event is its causal past:

$$B \in \mathcal{P}(A) \forall B \quad \text{such that} \quad A > B. \tag{8.36}$$

We can define the first, second and nth causal neighborhoods as the subsets of $\mathcal{P}(A)$ that are N or fewer causal steps to the past of A. These are denoted $\mathcal{P}_n(A)$.

We can define the causal versions of distinctiveness:

$$\mathcal{D}_{AB} = \frac{1}{n_{AB}^p}, \tag{8.37}$$

where p is some fixed power and n_{AB} is the smallest n such that $\mathcal{P}_A^n$ is not isomorphic to $\mathcal{P}_B^n$.

We then define the causal version of variety:

$$\mathcal{V} = \frac{1}{N^2} \sum_{A \neq B} \mathcal{D}_{AB}. \tag{8.38}$$

We can then consider theories based on an energetic causal set, in which the variety defined in this way acts as a potential energy. The dynamics of such a theory aims to maximize the variety of the system. This acts preferentially on pairs with small distinctiveness and changes them so as to increase the overall variety. In other words, the more similar two events are, the more likely they are to interact.

Under appropriate conditions, this leads also to a derivation of Schrödinger quantum mechanics, from a theory whose dynamics involves extremizing the variety. I leave the details to the original paper (Smolin, 2018).

8.6 Temporal Relationalism

As I said in Section 8.1, the program of temporal relationalism aims to complete quantum mechanics in a way that would simultaneously construct a quantum theory of gravity and spacetime. It can aspire to do both because it is based on a series of closely interconnected hypotheses about the nature of space and time. We discussed each of these in the preceding pages; it is now time to put them together and summarize the resulting picture.

1. Time, in the sense of causation is fundamental, by which is meant that it is irreducible (Smolin, 2013b, 2015a; Unger and Smolin, 2014). The activity of time is the unceasing and irreversible generation of novel events from present events; this generates a continually growing network of causal relations (Cortês and Smolin, 2014a).
2. Energy and momentum are also fundamental and irreducible (Amelino-Camelia et al., 2011; Freidel and Smolin, 2011; Cortês and Smolin, 2014a, 2014b). Events are endowed with energy and momentum and are transmitted by their causal relations (Cortês and Smolin, 2014a, 2014b).
3. Space is not present fundamentally but is emergent from the underlying, dynamically evolving network of causal relations (Cortês and Smolin, 2014a, 2014b). Locality is emergent, as therefore is nonlocality.
4. Lorentzian spacetime geometry is an emergent, macroscopic coarse graining of the fundamental causal structure, and the Einstein equations are consequences of the statistical thermodynamics that traces the flow of energy through the fundamental network of causal relations (Jacobson 1995; Smolin, 2014, 2017a, 2017c).
5. Locality is disordered, in the sense that there are mismatches between the emergent metric causal structure and the fundamental causal structure (Markopoulou and Smolin, 2007). These may be responsible for quantum nonlocality (Markopoulou and Smolin, 2004).
6. Locality in space is a consequence of locality in a space of views, where a view of an event is defined to be the information it receives from its causal past (Smolin, 2018).
7. The most fundamental law is the one that generates new events and is irreversible (Cortês and Smolin, 2014a). Other laws represent regularities, mostly emergent, and are local and evolving (Smolin, 2013b; Unger and Smolin, 2014).
8. Local, reversible laws emerge from global irreversible laws (Cortês and Smolin, 2017). This explains the existence of arrows of time.

The main results of this research program to date include the following.

- *Relative locality.* By studying a limit of quantum gravity defined by (8.13, 8.14), we find an extension of special relativity where causation, energy, and momentum are fundamental and space and spacetime are emergent (Amelino-Camelia et al., 2011; Freidel and Smolin, 2011). Locality is relativized, so that whether two events are observed to coincide depends on their energies and distance from the observer .

- *Energetic causal set models* (Cortês and Smolin, 2014a, 2014b, 2016, 2017). These were constructed with Marina Cortês and describe a world in which time, causation, energy, and momentum are fundamental and all take place in a space-less world. Some of the results include:

  - The emergence of a Lorentzian spacetime together with an embedding of the causal set into it (Cortês and Smolin, 2014a, 2014b).
  - The evolution passes through an initial phase, which is chaotic and irreversible, which gives way to an ordered phase where there emerge pseudo-particle-like excitations with quasi-reversible dynamics (Cortês and Smolin, 2014a). We now understand this to be very similar to the behavior of deterministic finite state dynamical systems, in which the initial phase is dominated by a basin of attraction that leads to a phase dominated by limit cycles (Cortês and Smolin, 2017). This mechanism can explain the emergence of time reversible effective laws from a fundamental theory that is irreversible.
  - In the limit of many events there emerges the dynamics of relativistic particles moving and interacting in the emergent Lorenzian spacetime (Cortês and Smolin, 2014a, 2014b).
  - A class of spin foam models invented by Wolfgang Wieland (2015) can be understood as energetic causal set models (Cortês and Smolin, 2016).
  - There appears to be a disordering of causality in the sense that the causal ordering in the fundamental law generating the causal set is not entirely consistent with the causal structure of the emergent spacetime in which the causal set is embedded (Cortês and Smolin, in preparation).

- *Time irreversible extensions of general relativity* (Cortês, Gomes, and Smolin, 2015; Smolin, 2016b). If general relativity emerges from a time irreversible theory early in the history of the universe, there ought to be an extension of general relativity that is not symmetric under time reversal. This could serve as an effective theory to describe the transition from the time asymmetric to the time symmetric dominated regime. We constructed two classes of such theories (Cortês et al., 2015; Smolin, 2016b) and studied their implications for the physics of the early universe (Cortês, Liddle, and Smolin, 2016).

- *The real ensemble formulation of quantum mechanics* (Smolin, 2012b, 2016a, 2018). This is a completion of quantum mechanics in which the wave-function of a microscopic system is derived from an ensemble consisting of all the similar systems spread throughout the universe. We posit a new and highly nonlocal interaction among the members of such ensembles, which extremizes the variety of the system, where by *variety* we mean a measure of the diversity of the views of the different subsystems (Barbour and Smolin, 1992). We find that the variety is closely related to the Bohmian quantum potential (Smolin, 2016a, 2018).

- *Relational hidden variables theories, in which the hidden variables are shared properties, associated to pairs of particles, were constructed* (Smolin, 1985, 2002, 2006; Markopoulou and Smolin, 2004).
- *Evolving laws.* One of the big questions of fundamental physics is how the universe chooses its laws from a reservoir of many possible laws. A natural solution is that the laws evolve in time (Smolin, 1992, 1997). This gives rise to the meta-law dilemma (Unger and Smolin, 2014), which starts by asking whether there is a law that governs the evolution of laws and, if so, how that was chosen. We proposed so far three approaches to this question.

  1. *Cosmological natural selection* (Smolin, 1992, 1997, 2013a). This is a proposal I made in the late 1980s of a cosmological scenario in which laws and their parameters evolve on a landscape of possible laws (analogous to the fitness landscape of population biology).
  2. *The Principle of Precedence* (Smolin, 2012a) is a proposal for laws to the dynamical laws of quantum mechanics to evolve.
  3. *Universality of metalaws.* A set of model metalaws are studied, and there is shown to be a notion of universality of metalaws, such that one can't distinguish them from each other (Smolin, 2015b, 2008).

- *A causal theory of views* (Smolin, 2018) arises by the combination of the real ensemble formulation of quantum mechanics with the energetic causal sets. The beables are the causal views of each event.
- *The emergence of the Einstein equations from the thermodynamics of an energetic causal set.* Conditions are given to demonstrate the emergence of general relativity from the thermodynamic limit of an energetic causal set theory (Smolin, 2014, 2017a, 2017c) to which certain additional conditions have been imposed. These conditions are expressed in Smolin (2017a) as four principles, the first of which is equivalent to the statement that a quantum spacetime is described by an energetic causal set model.

  I refer the reader to the original paper (Smolin, 2017a) for a full discussion and motivation of these principles. The key point for our discussion in this paper is that I am able to show that these principles suffice to recover the Einstein equations from a certain kind of energetic causal set model (Smolin, 2017a). To do this, I follow a strategy pioneered by Jacobson (1995, 2016) and (Padmanabhan, 2004, 2010a, 2010b).

8.7 Conclusions

What is beyond spacetime?

I have described the main ideas, and main results to date, of a research program that aims to find out. We postulate that time, causation, energy, and momentum are fundamental and irreducible, but space and spacetime are emergent. Quantum physics is also emergent from a more fundamental dynamics, based on the principle of maximal variety (Barbour and Smolin, 1992). From the resulting viewpoint, locality is discovered to be a proxy for similarity of views by which a subsystem of the universe sees their position in the network of relations and interactions that defines the world.

Acknowledgments

I first of all want to thank my collaborators with whom the program described here was formulated: Roberto Mangabeira Unger and Marina Cortês. I am grateful also to Giovanni Amelino Camelia, Stephon Alexander, Henrique Gomes, Jerzy Kowalski Glickman, Andrew Liddle, Joao Magueijo, and Yigit Yargic for collaborations on related work.

A lifetime of conversations and/or collaborations with Abhay Ashtekar, Julian Barbour, Laurent Freidel, Stuart Kauffman, Renate Loll, Fotini Markopoulou, Carlo Rovelli, Rafael Sorkin, and Antony Valentini have been essential to my intellectual life.

Dialogues with philosophers have been essential to my work on these issues, among many I am especially grateful to David Albert, Harvey Brown, Jim Brown, Jennan Ismael, Simon Saunders, and Steve Weinstein. While I was still a student, my life was changed by encounters with two great philosophers of science, Abner Shimony and Paul Feyerabend. I also wish to thank Nick Huggett, Keizo Matsubara and Christian Wuthrich for the invitation to contribute to this collection.

This research was supported in part by Perimeter Institute for Theoretical Physics. Research at Perimeter Institute is supported by the Government of Canada through Industry Canada and by the Province of Ontario through the Ministry of Research and Innovation. This research was also partly supported by grants from NSERC and FQXi. I am especially thankful to the John Templeton Foundation for their generous support of this project.

References

Abraham, E., Colin, S. and Valentini, A. (2014). Long-time relaxation in pilot-wave theory. *J. Phys.*, A47, 39.

Alder, S. L. (2004). *Quantum Theory as an Emergent Phenomenon: The Statistical Mechanics of Matrix Models as the Precursor of Quantum Field Theory*. Cambridge: Cambridge University Press.

Ambjorn, J., Jordan, S., Jurkiewicz, J. and Loll, R. (2013). Quantum spacetime, from a practitioner's point of view. *AIP Conf. Proc.*, 1514, 60–66.

Amelino-Camelia, G. (2001). Testable scenario for relativity with minimum-length. *Phys. Lett. B*, 510, 255–263.

Amelino-Camelia, G. (2002). Relativity in space-times with short-distance structure governed by an observer-independent (Planckian) length scale. *Int. J. Mod. Phys. D*, 11, 35–59.

Amelino-Camelia, G., Freidel, L., Kowalski-Glikman, J. and Smolin, L. (2011). The principle of relative locality. *Phys. Rev. D*, 84, 084010.

Barbour, J. and Smolin, L. (1992). Extremal variety as the foundation of a cosmological quantum theory. arXiv:hep-th/9203041.

Bohm, D. (1952). A suggested interpretation of quantum theory in terms of hidden variables, I. *Phys. Rev.*, 85, 166–179.

Bombelli, L., Lee, J. Meyer, D. and Sorkin, R. D. (1987). Spacetime as a causal set. *Phys. Rev. Lett.*, 59, 521–524.

Colin, S. and Valentini, A. (2016). Robust predictions for the large-scale cosmological power deficit from primordial quantum nonequilibrium. *Int. J. Mod. Phys.*, D25, 6, 165–168.

Conlon, J. (2016). *Why String Theory*. Boca Raton, FL: CRC Press.

Contaldi, C. R., Magueijo, J. and Smolin, L. (2008). Anomalous CMB polarization and gravitational chirality. *Phys. Rev. Lett.*, 101, 141101.

Cortês, M., Gomes, H. and Smolin, L. (2015). Time asymmetric extensions of general relativity. *Phys. Rev. D*, 92, 043502.

Cortês, M., Liddle, A. R. and Smolin, L. (2016). Cosmological signatures of time-asymmetric gravity. *Phys. Rev. D*, 94, 12.

Cortês, M. and Smolin, L. (2014a). The universe as a process of unique events. *Phys. Rev. D*, 90, 084007.

Cortês, M. and Smolin, L. (2014b). Energetic causal sets. *Phys. Rev. D*, 90, 044035.

Cortês, M. and Smolin, L. (2016). Spin foam models as energetic causal sets. *Phys. Rev. D*, 93, 084039.

Cortês, M. and Smolin, L. (2017). Reversing the irreversible: from limit cycles to emergent time symmetry. *Phys. Rev. D*, 97, 026004.

Criscuolo, A. and Waelbroeck, H. (1999). Causal set dynamics: a toy model. *Class. Quant. Grav.*, 16, 1817–1832.

Eichhorn, A. (2018). Status of the asymptotic safety paradigm for quantum gravity and matter. *Found. Phys.*, 48, 1407–1429.

Einstein, A. (1916). Näherungsweise Integration der Feldgleichungen der Gravitation. *Sitzungsberichte der Königlich Preußischen Akademie der Wissenschaften*, 688–696.

Freidel, L. and Smolin, L. (2011). Gamma ray burst delay times probe the geometry of momentum space. arXiv:hep-th/arXiv:1103.5626.

Furey, C. (2006). Notes on algebraic causal sets, unpublished notes. *Cambridge Part III Research Essay*.

Ghirardi, G. C., Grassi, R. and Rimini, A. (1990). Continuous-spontaneous-reduction model involving gravity. *Phys. Rev. A*, 42, 1057.

Gomes, H., Gryb, S. and Koslowski, T. (2011). Einstein gravity as a 3D conformally invariant theory. *Class. Quant. Grav.*, 28, 045005.

Hossenfelder, S. (2010). Bounds on an energy-dependent and observer-independent speed of light from violations of locality. *Phys. Rev. Lett.*, 104, 140402.

Jacobson, T. (1995). Thermodynamics of space-time: the Einstein equation of state. *Phys. Rev. Lett.*, 75, 1260–1263.

Jacobson, T. (2016). Entanglement equilibrium and the Einstein equation. *Phys. Rev. Lett.*, 116, 20.

Kowalski-Glikman, J. (2002). De Sitter space as an arena for doubly special relativity. *Phys. Lett. B*, 547, 291–296.

Kuchar, K. (1981). General relativity: dynamics without symmetry. *J. Math. Phys.*, 22, 2640–2654.

Lee, J., Kim, H. and Lee, J. (2013). Gravity from quantum information. *J. Korean Phys. Soc.*, 63, 1094–1098.

Leibniz, G. W. (1714). *The Monadology.*

Leibniz, G. W. (1717). *Leibniz's letters to Samuel Clarke.*

Magueijo, J. and Smolin, L. (2002). Lorentz invariance with an invariant energy scale. *Phys. Rev. Lett.*, 88, 190403.

Markopoulou, F. (2000). Quantum causal histories. *Class. Quant. Grav.*, 17, 2059–2072.

Markopoulou, F. and Smolin, L. (2004). Quantum theory from quantum gravity. *Phys. Rev. D*, 70, 124029.

Markopoulou, F. and Smolin, L. (2007). Disordered locality in loop quantum gravity states. *Class. Quant. Grav.*, 24, 3813–3824.

Mercati, F. (2018). *Shape Dynamics: Relativity and Relationalism.* Oxford: Oxford University Press.

Nelson, E. (1966). Derivation of the Schrödinger equation from Newtonian mechanics. *Phy. Rev.*, 150, 1079–1085.

Nelson, E. (1985). *Quantum Fluctuations.* Princeton, NJ: Princeton University Press.

Oriti, D. (2017). Spacetime as a quantum many-body system. Pages 365–379 of: Angilella, G. G. N. and Amovilli, C. (eds.). *Many-Body Approaches at Different Scales: A Tribute to Norman H. March on the Occasion of His 90th birthday.* New York: Springer.

Padmanabhan, T. (2004). Entropy of static spacetimes and microscopic density of states. *Class. Quant. Grav.*, 21, 4485–4494.

Padmanabhan, T. (2010a). Thermodynamical aspects of gravity: new insights. *Rep. Prog. Phys.*, 73, 046901.

Padmanabhan, T. (2010b). Equipartition of energy in the horizon degrees of freedom and the emergence of gravity. *Mod. Phys. Lett.*, A25, 1129–1136.

Pearle, P. (1976). Reduction of the state vector by a nonlinear Schrödinger equation. *Phys. Rev. A*, 113, 857–868.

Penrose, R. (1979). Singularities and time asymmetry. *General Relativity: An Einstein Centenary Survey*, 581–638.

Rovelli, C. and Vidotto, F. (2015). *Covariant Loop Quantum Gravity: An Elementary Introduction to Quantum Gravity and Spin Foam theory.* Cambridge: Cambridge University Press.

Schutzhold, R. and Unruh, W. G. (2003). Large-scale nonlocality in 'doubly special relativity' with an energy-dependent speed of light. *JETP Lett.*, 78, 431–435.

Smolin, L. (1985). Derivation of quantum mechanics from a deterministic non-local hidden variable theory, I. The two dimensional theory. Pages 148–173 of: Isham, C. J., and Penrose, R. (eds.). *Stochastic Mechanics, Hidden Variables and Gravity in Quantum Concepts in Space and Time.* Oxford: Oxford University Press.

Smolin, L. (1992). Did the universe evolve? *Classical and Quantum Gravity*, 9, 173–191.

Smolin, L. (1997). *The Life of the Cosmos*. New York/London: Oxford University Press/Weidenfeld and Nicolson.

Smolin, L. (2002). Matrix models as non-local hidden variables theories. *AIP Conf. Proc.*, 607, 244–261.

Smolin, L. (2005). The case for background independence. arXiv: hep-th/0507235

Smolin, L. (2006). Could quantum mechanics be an approximation to another theory? arXiv:quant-ph/0609109.

Smolin, L. (2008). Matrix universality of gauge and gravitational dynamics. arXiv:0803.2926.

Smolin, L. (2012a). Precedence and freedom in quantum physics. arXiv:1205.3707.

Smolin, L. (2012b). A real ensemble interpretation of quantum mechanics. *Found. Phys.*, 42, 1239–1261.

Smolin, L. (2013a). A perspective on the landscape problem. *Found. Phys.*, 43, 21–45.

Smolin, L. (2013b). *Time Reborn*. Boston/Toronto: Houghton Mifflin Harcourt/Penguin Random House Canada.

Smolin, L. (2014). General relativity as the equation of state of spin foam. *Class. Quant. Grav.*, 31, 19.

Smolin, L. (2015a). Temporal naturalism. Stud. Hist. Philos. Sci. B. Stud. Hist. Philos. Mod. Phys. [Special Issue], 52, 86–102.

Smolin, L. (2015b). Unification of the state with the dynamical law. *Found. Phys.*, 45, 1–10.

Smolin, L. (2016a). Quantum mechanics and the principle of maximal variety. *Found. Phys.*, 46, 6, 736–758.

Smolin, L. (2016b). Dynamics of the cosmological and Newton's constant. *Class. Quant. Grav.*, 33, 2.

Smolin, L. (2017a). Four principles for quantum gravity. *Fund. Theor. Phys.*, 187, 427–450.

Smolin, L. (2017b). What are we missing in our search for quantum gravity? arXiv:1705.09208. Ithaca, NY: Cornell University.

Smolin, L. (2017c). The thermodynamics of quantum spacetime histories. *Phys. Rev. D*, 96.

Smolin, L. (2018). The dynamics of difference. *Found. Phys.*, 48, 121–134.

Sorkin, R. D. (2007). Is the cosmological "constant" a nonlocal quantum residue of discreteness of the causal set type? *AIP Conf. Proc.*, 957, 142–153.

Starodubtsev, A. (2003). A note on quantization of matrix models. *Nucl. Phys. B*, 674, 533–552.

Underwood, N. G. and Valentini, A. (2016). Anomalous spectral lines and relic quantum nonequilibrium. arXiv:1609.04576.

Unger, R. M and Smolin, L. (2014). *The Singular Universe and the Reality of Time: An Essay in Natural Philosophy*. Cambridge: Cambridge University Press.

Valentini, A. (1991). Signal-locality, uncertainty, and the subquantum H-theorem. I. *Phys. Lett.*, A 156, 5–11.

Valentini, A. (2002). Signal-locality in hidden-variables theories. *Phys. Lett.*, A 297, 273–278.

Valentini, A. and Westman, H. (2005). Dynamical origin of quantum probabilities. *Proc. Roy. Soc. Lond.*, A461, 253–272.

Wieland, W. M. (2015). New action for simplicial gravity in four dimensions. *Class. Quantum Grav.*, 32, 015016.

9

Back to Parmenides

9.1 A Summary of the Construction

Quantum mechanics arose in the 1920s. General relativity has been around since
the 1910s. But, as of 2018, we still have no quantum theory of the gravitational
field. What is taking us so long? I believe the most challenging obstacle in our way
is understanding the quantum superposition of general relativistic causal structures.
This obstacle is couched on facets of the "problem of time" (Kuchar, 2011)—an
inherent difficulty in reconciling a picture of time evolution in quantum mechanics
to a "block time" picture of general relativity.

I also believe we can overcome this obstacle only if we accept a fundamental
distinction between time and space. The distinction is timid in general relativity —
even in its ADM form (Arnowitt, Deser, and Misner, 1962)—and here I want to
push it further. In this spirit, I consider space to be fundamental and time to be a
derived concept—a concept at which we arrive from change (a loose quote from
Ernst Mach).

I here investigate the consequences of this distinction between time and space.
The distinction allows only a restricted class of fundamental physical fields—the
ones whose content is spatially relational—and it thus also restricts the sort of
fundamental theories of reality. It is consequential in that with this view we must
reassess our interpretation of quantum mechanics and its relationship to gravity.
The new interpretation is compatible with a version of timelessness that I explain
herein. With timelessness, comes the requirement of explaining history without
fundamental underlying dynamics. The role of dynamics is fulfilled by what I define
as *records*.

This paper consists mainly of two parts: one justifying the timeless approach
through problems in quantum gravity, and another describing physics within a
general timeless theory proposed here. In the following section, I introduce more
technical reasons for my interest in timeless theories. These have to do with quantum gravity. I thus start with a brief description of what would count as a theory of

quantum gravity, before moving on to the sort of problems it has, which I believe timelessness might cure.

9.2 The Problems with Quantizing Gravity

I start in Section 9.2.1 with what I believe are the main principles of quantum mechanics and gravity. I then follow in Section 9.2.2 with a brief idiosyncratic exposition of issues in quantizing gravity. Then, in Section 9.2.3 I move on to an illustration of which of these problems signal a fundamental difference between time and space. Finally, in Section 9.2.4, I posit what sorts of different fundamental symmetries—i.e., to be imposed also off-shell, or at the quantum mechanical level—would assuage the clash between dynamics and symmetry. These turn out to be the spatial relational symmetries.

9.2.1 A Tale of Two Theories

General relativity is one of the pillars of our modern understanding of the universe, deserving a certain degree of familiarity from all those who purport to study nature, whether from a philosophical or mathematical point of view. The theory has such pristine logical purity that it can be comprehensively summarized by John A. Wheeler's famous quip (Misner, Thorne, and Wheeler, 1973)

Matter tells spacetime how to curve, and spacetime tells matter how to move.

$$(9.1)$$

We should not forget however, that ensconced within Wheeler's sentence is our conception of spacetime as a dynamical geometrical arena of reality: no longer a fixed stage where physics unfolds, it is part and parcel of the play of existence.

In mathematical terms, we have:

$$\underbrace{R_{\mu\nu} - \frac{1}{2}Rg_{\mu\nu}}_{\text{spacetime curving}} \quad \propto \quad \underbrace{T_{\mu\nu}}_{\text{sources for curving}} . \qquad (9.2)$$

Given the sources, one will determine a geometry given by the spacetime metric $g_{\mu\nu}$—the "matter tells spacetime how to curve" bit. Conversely, it can be shown that very light, very small particles will roughly follow geodesics defined by the geometry of the left-hand side (LHS) of the equation—the "spacetime tells matter how to move" part.[1]

---

[1] This distinction is not entirely accurate, as the right-hand side (RHS) of (9.2) usually also contains the metric, and thus the equation should be seen as a constraint on which kind of spacetimes with which kind of matter distributions one can obtain, "simultaneously." I.e., it should be seen as a spacetime, block universe, pattern, not as a causal relation (see Lehmkuhl [2011]).

A mere decade after the birth of general relativity (GR), along came quantum mechanics. It was a framework that provided unprecedented accuracy in experimental confirmation, predictions of new physical effects and a reliable compass for the construction of new theories. And yet, it has resisted the intuitive understanding that was quickly achieved with general relativity. A much less accurate characterization than Wheeler's quip for general relativity has been borrowed from the pessimistic adage "everything that can happen, does happen."[2] The sentence is meant to raise the principle of superposition to the status of core concept of quantum mechanics (whether it expresses this clearly or not is very much debatable).

In mathematical terms, the superposition principle can be seen in the Schrödinger equation:

$$\hat{H}\psi = -i\hbar\frac{d}{dt}\psi, \qquad (9.3)$$

whose linearity implies that two solutions ψ_1 and ψ_2 add up to a solution $\psi_1 + \psi_2$. In the path integral representation, superposition is built in. The very formulation of the generating function is a sum over all possible field configurations ϕ,

$$\mathcal{Z} = \int \mathcal{D}\phi \, \exp\left[i \int \mathcal{L}(\phi)/\hbar\right], \qquad (9.4)$$

where $\mathcal{L}(\phi)$ is the Lagrangian density for the field ϕ and $\mathcal{D}\phi$ signifies a summation over all possible values of this field (the second integral is over spacetime).

Unfortunately, for the past 90 years, general relativity and quantum mechanics have not really gotten along. Quantum mechanics soon claimed a large chunk of territory in the theoretical physics landscape, leaving a small sliver of no-man's land also outside the domain of general relativity. In most regimes, the theories will stay out of each other's way—domains of physics where both effects need to be taken into account for an accurate phenomenological description of nature are hard to come by. Nonetheless, such a reconciliation might be necessary even for the self-consistency of general relativity: by predicting the formation of singularities, general relativity "predicts its own demise," to borrow again the words of John Wheeler. Unless, that is, quantum effects can be suitably incorporated to save the day at such high curvature regimes.

9.2.2 The Problems of Quantum Gravity

At an abstract level, the question we need to face when trying to quantize general relativity is: how to write down a theory that includes all possible superpositions and yet yields something like (9.2) in appropriate classical regimes? Although

---

[2] Recently made the title of a popular book on quantum mechanics (Cox and Forshaw, 2011).

the incompatibility between general relativity and quantum mechanics can be of technical character, it is widely accepted that it has more conceptual roots. In the following, I describe only two such roots.

9.2.2.1 Is Non-Renormalizability the Only Problem?

The main technical obstacle cited in the literature is the issue of perturbative renormalizabilty. Gravity is a nonlinear theory, which means that geometrical disturbances around a flat background can act as sources for the geometry itself. The problem is that unlike what is the case in other nonlinear theories, the "charges" carried by the nonlinear terms in linearized general relativity become too "heavy"—the gravitational coupling constant has negative mass dimensions—generating a cascade of ever-increasing types of interactions once one goes to high enough energies. This creates problems for treating such phenomena scientifically, for we would require an infinite amount of experiments to determine the strength of these infinite types of interactions. This problem can be called *loss of predictability*.

There are theories, such as Horava–Lifschitz gravity (Horava, 2009), which seem to be naively perturbatively renormalizable. The source of renormalizability here is the greater number of spatial derivatives as compared to that of time derivatives. This imbalance violates fundamental Lorentz invariance, breaking up spacetime into space and time. Unfortunately, the theory introduces new degrees of freedom that appear to be problematic (i.e., their influence does not disappear at observable scales).

And perhaps perturbative non-renormalizability is not the only problem. Indeed, for some time we have known that a certain theory of gravity called *conformal gravity* (or *Weyl squared*) is also perturbatively renormalizable. The problem there is that the theory is sick. Conformal gravity is not a unitary theory, which roughly means that probabilities will not be conserved in time. But, which time? And is there a way to have better control over unitarity? This can be called *the problem of unitarity*.[3]

9.2.2.2 A Dynamical Approach

In covariant general relativity, the fundamental field, $g_{\mu\nu}$, already codifies causal relations, whether or not the equations of motion (the Einstein equations) have been imposed. So a first question to ask is if we can give a formulation of quantum gravity that reflects the fundamental distinction between causal and acausal. One way of

---

[3] Another approach to quantum gravity called asymptotic safety (Reuter, 1998) also suffers from such a lack of control of unitarity. This approach also explores the possible existence of gravitational theories whose renormalization will generate dependence only on a finite number of coupling constants, thus avoiding the loss of predictibility explained earlier.

approaching this question is to first use a more dynamical account of the theory. We
don't need to reinvent such an account, it is already standard in the study of gravity,
going by the acronym of ADM (Arnowitt–Deser–Misner) (Arnowitt et al., 1962).
The main idea behind a dynamical point of view is to set up initial conditions on
a spatial manifold M and construct the spacetime geometry by evolving in a given
auxiliary definition of time.[4] Indeed most of the work in numerical general relativity
requires the use of the dynamical approach. Such formulations allow us to use the
tools of the Hamiltonian formalism of quantum mechanics to bear on the problem
of quantizing gravity. With these tools, matters regarding unitarity are much easier
to formulate, because there is a time with respect to which probabilities are to be
conserved.

However, since the slicing of spacetime is merely an auxiliary structure, the
theory comes with a constraint—called the Hamiltonian constraint —which implies
a freedom in the choice of such artificial time slicings. The metric associated to
each equal time slice, g_{ab}, and its associated momenta, π^{ab}, must be related by the
following relation at each spatial point $x \in M$:

$$H(x): = R(x) - \frac{1}{g}\left(\pi^{ab}\pi_{ab} - \frac{1}{2}\pi^2\right)(x) = 0 \qquad \forall x \in M, \qquad (9.5)$$

where g stands for the determinant of the metric.

9.2.3 Problems of Time

The constraints (9.5) are commonly thought to guarantee that observables of the
theory should not depend on the auxiliary "foliation" of spacetime. As we will see
in this section, there are multiple issues with this interpretation. Famously, (9.5) also
contains the generator of time evolution. In other words, time evolution becomes
inextricably mixed with a certain type of gauge freedom, leading some to conclude
that in GR evolution is "pure gauge." This is one facet of what people have called
the problem of time (see, e.g., Isham [1992] and Anderson [2017]). It is related
to the picture of block time—the notion that the object one deals with in general
relativity is the *entire* spacetime, for which the distinction among past, present, and
future is not fundamental. The worry is that the Hamiltonian formalism might be
freezing the bathwater with the baby still inside. What can we salvage in terms of
true evolution?

Even in the simplest example, it is not clear that the refoliation invariance
interpretation is tenable, as shown by Torre and Varadarajan (1999). For a scalar

---

[4] In this constructed spacetime, the initial surface must be Cauchy, implying that one can perform this analysis
only for spacetimes that are time orientable.

field propagating in Minkowski spacetime between two fixed hypersurfaces, different choices of interpolating foliations will have unitarily inequivalent Schrödinger time evolutions.

But what about more broadly? What can we say about the attempt to represent relativity of simultaneity in the Hamiltonian—both classical and quantum mechanical—setting? In this section I investigate this question. The conclusion will be that, at least from the quantum mechanical perspective, it might make little sense to implement refoliation invariance. I contend that this is because local time reparametrization represents an effective, but not fundamental, symmetry. I.e., I contend that *invariance under refoliation is not present at a quantum mechanical level, but should be recovered dynamically for states that are nearly classical.* This view undersigns a *functionalist* approach to spacetime (Knox, 2017).

In the following two subsections, I expand first on obstacles for a quantization of refoliation symmetry in the Hamiltonian setting, and then in the Lagrangian setting. I expound on how these two types of obstacles point to timelessness as a possible resolution.

9.2.3.1 Hamiltonian Evolution

Using the covariant symplectic formalism, one can geometrically project symmetries of the Lagrangian theory onto the Hamiltonian framework. Using this formalism, one can precisely track how Lagrangian symmetries in the covariant field-space are represented as Hamiltonian flows in phase space. As shown by Wald and Lee (Lee and Wald, 1990), for this projection to be well defined in the case of non-spatial diffeomorphism invariance (acting as a spacetime Lagrangian symmetry), one needs to restrict the Lagrangian theory to consider only those fields that satisfy the equations of motion. Once restricted, indeed there exists a projection of the nonspatial diffeomorphism symmetry to the symplectic flow of the standard scalar ADM constraint (9.5). But otherwise, there isn't such a correspondence. In other words, the Hamiltonian formalism embodies the sacred principle of relativity of simultaneity only on-shell.

Once the dynamics of the gravitational field are included in the Hamiltonian formalism (home of canonical quantum mechanics), it is impossible to enforce relativity of simultaneity without also enforcing the gravitational equations of motion. This is worth repeating: for generic off-shell spacetimes, the Hamiltonian constraint does not represent relativity of simultaneity. But symmetries should hold at the quantum level irrespectively of the classical equations of motion. Which brings forth the question: what would it even mean to naively quantize (9.5)? What property would we be trying to represent with its quantization?

If one nonetheless ignores these issues and pushes quantization, one gets the infamous Wheeler–DeWitt equation:

$$\hat{H}\psi[g] = 0, \tag{9.6}$$

where $\psi[g]$ is a wave-functional over the space of 3-geometries. Following this route, we see the classical problem of time transported into the quantum regime:[5] One could look at (9.6) as a *time-independent* Schrödinger equation, which brings us again to the notion of frozen time, from (9.3). A solution of the equation will not be subject to time evolution; it will give a frozen probability wave-function on the space of 3-geometries. To talk about its solutions, or the spectrum of the Hamiltonian, one needs to talk about observables, which are nonlocal in phase space (once one includes refoliation invariance). In other words, one needs to "solve the theory" to even begin building a meaningful physical Hilbert space. Other canonical approaches have so far similarly found insurmountable problems with the quantization of this constraint. I contend that it should simply not be quantized, as it does not represent a fundamental symmetry principle of the quantum theory.

9.2.3.2 Covariant Quantum Gravity

At a more formal level, to combine (9.2) with our principle of superposition, one should keep in mind that spacetimes define causal structures, and it is far from clear how one should think about these in a state of superposition. For instance, which causal structure should one use in an algebraic quantum field theory approach when declaring that spacelike separated operators commute? Quantum field theory is formulated in a fixed spacetime geometry, while in general relativity spacetime is dynamical. Without a fixed definition of time or an a priori distinction between past and future, it is hard to impose causality or interpret probabilities in quantum mechanics.

In quantum mechanics, we have time-evolution operators $e^{-i\hat{H}t}$ taking us from an initial physical state to a final one. In the language of path integrals, it is more convenient to express evolution in terms of a gauge-fixed propagator (the inversion of the quantum mechanical propagator requires gauge-degrees of freedom to have already been gauge-fixed) $W(\phi_1, \phi_2)$, where ϕ (e.g. $\phi = (x,t)$) is deemed to at least

---

[5] For the expert readers, I should note that a derivation of (9.6) exists from the path integral formalism (Halliwell and Hartle, 1991). But that derivation already assumes that the Hamiltonian symmetries act on all the variables in the path integral, and thus it does not resolve the problem Wald and Lee pointed to. Even ignoring all of these issues, (9.6) has some further problems of its own. It has operator ordering ambiguities, functional derivatives of the metric acting at a singular point, no suitable inner product on the respective Hilbert space with respectable invariance properties, etc. One could also attempt to interpret (9.6) as a Klein–Gordon equation with mass term proportional to the spatial Ricci scalar, but unlike Klein–Gordon, it is already supposed to be a quantized equation. Furthermore, the problem in defining suitable inner products is an obstacle in separating out a positive and negative spectrum of the Klein–Gordon operator (Anderson, 2017).

contain all the gauge-invariant information.[6] This simple characterization already raises two types of difficulties.

The first, more conceptual problem, is that such a transition amplitude would be one between full 4-dimensional spacetimes (or sets of spacetimes). It would be a transition *outside* of time as such, but the way we usually think of quantum transition amplitudes is *within* time.

The second is that there is no known local parametrization of physical (observable) Lorentzian 4-geometries, $[^{(4)}g]$ (global obstructions—e.g., the Gribov problem [Singer, 1978]—also exist, but are less concerning). To understand what this means, we need to introduce the notion of a slice. A "slice" is a split between the physically equivalent (or gauge-equivalent) field configurations and the physically distinct ones. A local slice is one that performs this split only locally in field space, and it is equivalent to a local gauge-fixing if gauge transformations form well-defined gauge orbits. Finding a slice theorem is relatively straightforward in the case of Riemannian metrics (Euclidean signature) for any dimension. To make a long story short, applying the strategy used in the Riemanninan slice theorem does not work in the Lorentzian case due to the lack of invertibility of certain second-order differential operators. Such operators are elliptic in the Riemannian case, but for Lorentzian metrics the analogous operators are hyperbolic, invalidating the construction of the slice. It is the sign of time that gets in the way—it blocks general proofs either through the hyperbolic character of gauge-fixing equations or through the noncompactness of the Lorentz group. For instance, the latter poses limitations on physical parametrizations of Lorentzian metrics through curvature invariants (Coley et al., 2009), which is unable to distinguish Kundt spacetimes, a relatively broad class. Indeed this is related to the difficulty of finding a gauge-fixing of time (or good clocks) that are valid everywhere in phase space.

A local parametrization, or slice for Lorentzian metrics has been constructed (by Isenberg and Marsden) only for those Lorentzian metrics that (1) satisfy the Einstein equations and (2) furthermore admit a particular kind of time foliation (Isenberg and Marsden, 1982). This choice—constant mean curvature (CMC)—corresponds to synchronizing clocks so that they measure the same expansion rate of space everywhere (i.e., same local Hubble parameter).

Of course, this is suboptimal, to say the least. Quantization requires that we consider metrics that are off-shell, i.e., that do not obey the classical equations of motion. This issue is similar to the previously mentioned one, also surrounding relativity of simultaneity—it has only a Hamiltonian representation for spacetimes satisfying Einstein equations.

---

[6] One can parametrize observables by families of gauge-fixings; as in e.g., partial observables, in the sense of Rovelli (2002). Their inclusion does not change the discussion.

9.2.3.3 Gluing Transition Amplitudes for Spacetime Regions

The previous argument assumes that spacetime does not possess boundaries, where the gauge symmetries can be fixed. If one would like to have a piecewise approach—that is, considering spacetime regions and then gluing them—other types of questions arise about how to glue different transition amplitudes in gauge theories, for instance, questions about the compatibility of gauge-fixing the degrees of freedom at the boundaries (Donnelly and Freidel, 2016; Donnelly and Giddings, 2017) (which is problematic for a variety of reasons, including the previous one of a lack of slice).

In a 3+1 decomposition, at least two related difficulties arise for the path integral. One could take $\dot{N} = 0 = N_i$, for lapse N and shift N_i. This is part of the standard gauge-fixing taken for transition amplitudes in GR, such as for Hartle–Hawking (Teitelboim, 1983). But these won't cover generic spacetimes; such coordinates generically form caustics. This is related to the problem mentioned earlier: there is no (known) slice theorem for Lorentzian metrics (Isenberg and Marsden, 1982).

Moreover, as mentioned in footnote 5, the proof that such transition amplitudes obey the Hamiltonian invariance equations, (9.6), assumes invariance under the actions of the constraints, i.e., it assumes phase space invariance, not spacetime (Lagrangian) invariance. Thus the argument expounded in regard to the Wheeler–DeWitt equation applies also here: it is unclear if the transition amplitude expresses relativity of simultaneity (Lee and Wald, 1990).

The second difficulty is that the spacetime corresponding to each gauge-fixed path is taken to have boundaries(at least the initial and final time slices), and, again, the fate of diffeomorphisms in the presence of boundaries is a very current matter of discussion in the community.[7] At least in the presence of degrees of freedom, which are not strictly topological in nature, that is.

Indeed, many approaches to quantum gravity, such as spin foams (see Rovelli [2007] and references therein) are inspired by a treatment of topological quantum field theory (TQFT) originated by Atyiah and Segal (Atiyah, 1988). These treatments depict a transition amplitude from (co)boundaries of a manifold. The boundaries host states and the interior of the manifold encode the transition amplitude between these states. The transition amplitude can be obtained by a path integral of all field histories between the two boundaries. And it is true that there are many examples for which we understand quantization of TQFTs with diffeomorphism symmetry, including gravity in 2+1 spacetime dimensions (see Carlip [2005]). However, being topological in nature, in none of these theories is there a discrepancy between the full field space and that subset which satisfies the equations of motion; the equations of motion are trivially satisfied by using gauge

---

[7] These concerns have resurfaced due to the study of entanglement entropy (Donnelly, 2014).

transformations; in this case, indeed there is no difference between implementing symmetries on-shell or off-shell. Therefore, in these theories, the counterargument outlined in this section is not valid; it requires a gap between what is kinematical and what is dynamical.

9.2.3.4 Many Problems. . .

Let us take stock of the many problems related to time in GR that we have mentioned: first, *dynamics*—the Einstein equations—has little to do with causality properties; even kinematically, the field $g_{\mu\nu}$ already carries causal relations.[8] Second, the Hamiltonian relates only to redefinitions of simultaneity when the Einstein field equations are met (Lee and Wald, 1990). Relatedly, in GR the lines between symmetry and evolution are completely blurred. In other words, within the Hamiltonian formalism—the most natural formalism for quantum mechanics—one cannot implement the symmetries off-shell, that is, in a truly quantum mechanical manner. Indeed, as mentioned in the beginning of the section, even in the simplest example of a field theory (with local degrees of freedom) testing invariance under refoliations in the quantum mechanical realm is problematic (Torre and Varadarajan, 1999). Third, in the covariant approach, no generic (i.e., also off-shell) gauge-fixing is known. This is again a problem of the signature of operators related to the time direction.

9.2.4 Symmetries, Relationalism and Laws of the Instant

Many of the problems in the previous section seem to point in the same direction: parametrizing physical degrees of freedom in the presence of relativity of simultaneity is a difficult task. Perhaps restricting the fundamental symmetries of the theory to disallow this mixing with evolution would cure some of these issues. As I have emphasized, we only need to recover refoliation invariance on-shell, i.e., only after the equations of motion have been imposed. This frees up the theory to accept different fundamental symmetry principles and delegate the fulfillment of refoliation invariance to on-shell properties. In this section, we uncover what this new sort of symmetries can be.

9.2.4.1 A Silver Lining

There are makeshift patches to the problems of time we have mentioned, and in them, we can find a silver lining. For instance, as I mentioned, a very weak form

---

[8] As a background, that is. One can move to the 3+1 context, in which, after gauge-fixing, one finds hyperbolic equations of motion propagating field disturbances. But then, one is back to the other issues I have mentioned: first, regarding generic gauge-fixings of the 4-diffeomorphisms. Second, regarding the inextricable relation between evolution and gauge symmetry; is it possible to satisfy one but not the other?

of a slice theorem—essentially generic gauge-fixings—exists for GR (Isenberg and Marsden, 1982). It requires that the spacetime satisfy the Einstein equations and admit a CMC foliation, i.e., synchronizing clocks so that the expansion of space is constant everywhere. Moreover, going to the 3+1 framework to study *dynamics*, one finds that most formal proofs for existence and uniqueness of solutions also require the use of CMC foliations, through the so-called York method (York, 1971, 1973). Indeed, most if not all of the numerical simulations used to model black hole mergers have worked within CMC (Pretorius, 2005), even helping interpret Laser Interferometry Gravitational Observatory (LIGO) data; and every test that has ever been passed by GR is known to be consistent with a CMC foliation.

The constraints these foliations need to satisfy generate certain dynamical (symplectic) flows: changes of spatial scale, i.e., *local conformal transformations*, as we will see shortly. Indeed, CMC foliations have special properties in GR: the evolution of the spatial conformal geometry decouples from the evolution of the pure scale degrees of freedom of the metric. Surprisingly, both the York method and the slice theorem show that, *although GR is not fundamentally concerned with spatial conformal geometries, it is deeply related to them.* As we now comment on, this is not an accident; the most general sort of symmetries that act pointwise in configuration space and locally in space are indeed conformal transformations and diffeomorphisms.

In the following subsection, I introduce these symmetries. In Section 9.2.4.3, I describe this result: these symmetries are the most general ones with an inherently "instantaneous," local, action.

9.2.4.2 Relationalism

In Hamiltonian language, the most general symmetry transformation acts through the Poisson bracket $\{ \cdot \ \ \cdot \}$, on configurations:

$$\delta_\epsilon g_{ij}(x) = \left\{ \int d^3x \, F[g, \pi; x'] \epsilon(x') \, , \, g_{ij}(x) \right\}, \tag{9.7}$$

where ϵ is the not necessarily scalar gauge parameter, which in this infinite dimensional context is a function on the closed spatial manifold, M, and we are using DeWitt's mixed functional dependence; i.e., F depends functionally on g_{ij} (not just on its value at x'), as denoted by square brackets, but it yields a function with position dependence—the "$; x'$") at the end.

Regarding the presence of gauge symmetries in configuration space, we would like to implement the most general relational principles that are applicable to space (as opposed to spacetime). At face value, the strictly relational symmetries should be:

- **Relationalism of locations.** In Newtonian particle mechanics this would imply that only relative positions and velocities of the particles, not their absolute position and motion, are relevant for dynamics. In the center of mass frame, i.e., in which the total linear momentum vanishes ($\vec{P} = 0$), the supposition holds only if the total angular momentum of the system also vanishes (see Barbour [2010] and Mercati [2014]), $\vec{L} \approx 0$. In the gravitational field theory case, relationalism of locations is represented by the (spatial) diffeomorphism group Diff(M) of the manifold M. It is generated by a constraint $F[g, \pi; x'] = \nabla_i \pi^i{}_j(x) \approx 0$, which yields on configuration space the transformation $\delta_{\vec{\epsilon}} g_{ij}(x) = \mathcal{L}_{\vec{\epsilon}} g_{ij}(x)$, which is just the infinitesimal dragging of tensors by a diffeomorphism.
- **Relationalism of scale.** In Newtonian particle mechanics this relational symmetry would imply that only the relative distance of the particles, not the absolute scale, is relevant for dynamics. It holds only if the total dilatational momentum of the system vanishes (see Barbour [2010] and Mercati [2014]). In the gravitational field theory, this symmetry is represented by the group of scale transformations (also called the Weyl group), $\mathcal{C}(M)$, which is symplectically generated by $F[g, \pi; x'] = g_{ij}\pi^{ij}(x) \approx 0$ (it yields on configuration space the scale transformation $\delta_\epsilon g_{ij}(x) = \epsilon(x) g_{ij}(x)$). In this case the infinitesimal gauge parameter ϵ is a scalar function, as opposed to a vector field $\vec{\epsilon}$ for the diffeomorphisms.

Unlike what is the case with the constraints emerging from the Hamiltonian ADM formalism of general relativity, these symmetries form a (infinite-dimensional) closed Lie algebra.

9.2.4.3 *Laws of the Instant*

Even if we disregard considerations about relationalism, local time reparametrizations, or refoliations, also don't act as a group in spatial configuration space and thus do not allow one to form a gauge-invariant quotient from its action. That is, given a particular linear combination—defined by a smearing[9] $\lambda^\perp$—of the Hamiltonian constraints given in (9.5), it generates the following transformation:

$$\delta_{\lambda^\perp} g_{ab}(x) = \frac{2\lambda^\perp(\pi_{ab} - \frac{1}{2}\pi g_{ab})}{\sqrt{g}}(x), \qquad (9.8)$$

which depends not only on the metric but also on the momenta. This dependence is clearly in contrast to the symmetries related to relationalism of scale and of position, mentioned earlier. It means that the 3-metric by itself carries no gauge-invariant information.

Indeed, for the associated symmetry to have an action on configuration space that is independent of the momenta, a given constraint $F[g, \pi, \lambda] \approx 0$

---

[9] The $\perp$ notation is standard to represent parameter acting transversally to the constant-time surfaces.

must be linear in the momenta. This already severely restricts the forms of the functional to:[10]

$$F[g, \pi, \lambda] = \int \tilde{F}(g, \lambda)_{ab}(x) \pi^{ab}(x), \qquad (9.9)$$

so that the infinitesimal gauge transformation for the gauge-parameter λ gives:

$$\delta_\lambda g_{ab}(x) = \tilde{F}(g, \lambda)_{ab}(x).$$

Thus I would like symmetries to act solely on configuration space. Only such symmetries are compatible with the demand that the $W(g^1_{ab}, g^2_{ab})$ give all the information we need about a theory, since only such symmetries allow g_{ab} to carry gauge-invariant information. I thus require that $\delta_\epsilon g^1_{ab}(x) = G[g^1_{ab}, \epsilon; x]$ for some mixed functional G, which crucially *only* depends on g^1_{ab}.

In other words, the action of the symmetry transformations of "now" depends only on the content of "now." The action of these relational symmetries on each configuration g_{ab} is self-determined, they do not depend on the history of the configuration or on configurations $\tilde{g}_{ab} \neq g_{ab}$. In Gomes (2016), I gave a proof that the relational symmetries of scale and position are indeed the only symmetries whose action in phase space projects down to an intrinsic action on configuration space.

The conclusion of this argument is that spatial relationalism is singled out by demanding that symmetries have an intrinsic action on configuration space. Just to be clear, this feature is not realized by the action of the ADM scalar constraint (9.5), since it is a symmetry generated by terms quadratic in the momenta and thus the transformation it generates on the metric requires knowledge of the conjugate momentum (and vice-versa).

Last (and also unlike what is the case with the scalar constraint (9.5)),[11] the action of these symmetries endows configuration space $\mathcal{M}$ with a well-defined, neat principal fiber bundle structure (see Fischer and Marsden [1977]), which enables their quantum treatment (Gomes, 2016).

For a theory that contains some driver of change, an absolute time of some sort, we would extend our configuration space with an independent time variable, t, making the system effectively deparametrizable. With this absolute notion of time, and an ontological deparametrization of the system, evolution from t_1 to t_2 would not require any further definition. At this point we could stop, claiming that we have expounded on what we expect a relational theory of space to look like. We would be able to define a Schrödinger equation as in the usual time-dependent framework and go about our business. Shape dynamics employing the complete

---

[10] Up to canonical transformations that don't change the metric, i.e., with generating functionals of the form $\int d^3x \, (\tilde{\pi}^{ab} g_{ab} + \sqrt{g} F[g])$, for F any functional of g and $\tilde{\pi}^{ab}$ the new momentum variable.

[11] Barring the occurrence of metrics with nontrivial isometry group.

relational symmetries is a theory of that sort.[12] However, the presence of time there is still disturbing from a relational point of view: where is this time if not in the relations between elements of the configurations? Therefore, to fully satisfy our relational fetishes, we must again tackle the question: *without a driver for change, what is the meaning of a transition amplitude?*

9.3 Timelessness, Quantum Mechanics, and Configuration Space

9.3.1 The Special Existence of the Present: Parmenides and Zeno

It could be argued that we do not "experience" spacetimes. We experience one instant at a time, so to say. We, of course, still appear to experience the passage of time, or perhaps more accurately, we (indirectly) experience changes in the spatial configuration of the world around us, through changes of the spatial configuration of our brain states.

But if present experience is somehow distinguished, how does "change" come about? This is where Parmenides has something to say that is relevant for our discussion. Parmenides was part of a group called the Eleatics, whose most prominent members were himself and Zeno, and whose central belief was that all change is illusory. The reasoning that led them to this conclusion was the following: if the future (or past) is real, and the future is not existing now, it would have both properties of existing and not existing, a contradiction (or a "turning back on itself"). Without past and future, the past cannot transmute itself into future, and thus there is also no possible change. Of course, the argument hinges on the distinction we perceive among present, past, and future, and in one form or another, is present in many subsequent formulations, such as McTaggart's (1908) influential.[13]

But perhaps no one better than Augustine captures our psychological dumbfoundness when faced with the defenestration of time's flow. He picked up the question posed by the Eleatics, concluding that change was an illusion and yet,

How can the past and future be, when the past no longer is, and the future is not yet? As for the present, if it were always present and never moved on to become the past, it would not be time, but eternity.... Nevertheless we do measure time. We cannot measure it if it is not yet into being, or if it is no longer in being, or if it has no duration, or if it has no beginning and no end. Therefore we measure neither the future nor the past nor the present nor time that is passing. Yet we do measure time.

---

[12] The absolute time used in the original version of shape dynamics (Gomes, Gryb, and Koslowski, 2011), is of the form $\langle \pi^{ab} g_{ab} \rangle$, where brackets denote the spatial average. This quantity is invariant only with respect to Weyl transformations that preserve the total volume of space and is thus not completely relational. One can extend the conformal transformation to the full group, acquiring an absolute time parametrization (Koslowski, 2015).

[13] See Price (2009) for a review of McTaggart's arguments and more recent counterarguments to a flow of time.

According to Augustine, time is a human invention: the difference between future and past is merely the one between anticipation and memory.

To the extent that future and past events are real, they are real now, i.e., they are somehow encoded in the present configuration of the universe. Apart from that, they can be argued not to exist. My memory of the donut I had for breakfast is etched into patterns of electric and chemical configurations of my brain, *right now*. We infer the past existence of dinosaurs because it is encoded in the genes of present species and in fossils in the soil. At any moment we are in the possession of a host of redundant records of the same event, and they had better be in mutual accord. It is from this consistent mosaic of records that we build models of the laws of Nature. We fit the pieces together into a larger explanatory framework we call "science."

But does the Eleatic argument then bring about a solipsism of the instant? How to connect a snapshot of the dinosaur dying with the snapshot of the paleontologist finding its remains? In a timeless universe, what we actually do is deduce from the present that there exists a continuous curve of configurations connecting "now" to some other configuration we call "the past."

A much more recent incarnation of the Parmenidean view is found in the work of Julian Barbour (see Barbour [1999]), which I use here as a nearer port of departure. Barbour observes that timeless configuration space should be seen as the realm containing every possible now, or instantaneous configuration of the universe. Barbour (1994), attempts to accommodate timelessness more intuitively into our experience:

An alternative is that our direct experience, including that of seeing motion, is correlated with only *configuration* in our brains: the correlate of the conscious instant is part of a point of configuration space ... Our seeing motion at some instant is correlated with a single configuration of our brain that contains, so to speak, several stills of a movie that we are aware of at once and interpret as motion. ... Time is not a framework in which the configurations of the world evolve. Time exists only so far as concrete configurations express it in their structure. The instant is not in time; time is in the instant.

I *almost* wholeheartedly agree with Barbour. We diverge only in the attribution of *experience* to *each* configuration alone. I believe there is no empirical access to single-field configurations; therefore, all statements about experience refer to some coarse graining, or regions of configuration space, where a given attribute is represented. Clearly, this empirical dilution of Barbour's radical "solipsism of the instant" does not conflict with my theoretical reasons for considering instantaneous configuration space to be ontologically fundamental. I still believe that the past doesn't become the present; it is only embedded in the present. Every present exists, every present is unique, and some presents may be entangled with other presents.

9.3.2 Timeless Path Integral in Quantum Mechanics

Configuration space for timeless field theories, which I will denote by $\mathcal{M}$, should be thought of as the set of all possible field configurations over a given closed manifold M. Each point of configuration space $q \in \mathcal{M}$ is a "snapshot" of the whole universe.[14] I will require symmetries to be laws of the instant precisely so that they are compatible with a theory defined at its most fundamental level by $W(q_1, q_2)$.

We start with a finite-dimensional system, whose configuration space, $\mathcal{M}$, is coordinatized by q^a, for $a = 1, \cdots, n$. An observation yields a complete set of q^a, which is called an event. Let us start by making it clear that no coordinate, or function of coordinates, need single itself out as a reference parameter of curves in $\mathcal{M}$. The systems we are considering are not necessarily deparametrizable—they do not necessarily possess a suitable notion of time variable.

Now let $\Omega = T^*\mathcal{M}$ be the cotangent bundle to configuration space, with coordinates q^a and their momenta p_a. The classical dynamics of a reparametrization invariant system is fully determined once one fixes the Hamiltonian constraint surface in Ω, given by $H = 0$. A curve $\gamma \in \mathcal{M}$ is a classical history connecting the events q_1^a and q_2^a if there exists an *unparametrized* curve $\bar{\gamma}$ in $T^*\mathcal{M}$ such that the following action is extremized:

$$S[\bar{\gamma}] = \int_{\bar{\gamma}} p_a dq^a, \tag{9.10}$$

for curves lying on the constraint surface $H(q^a, p_a) = 0$, and are such that $\bar{\gamma}$'s projection to $\mathcal{M}$ is γ, connecting q_1^a and q_2^a.

Feynman's original demonstration of the equivalence between the standard form of nonrelativistic quantum mechanics and his own path integral formulation relied on refining time slicings. The availability of time gave a straightforward manner by which to partition paths into smaller and smaller segments. Without absolute time, one must employ new tools in seeking to show the equivalence. For instance, a parametrized curve $\bar{\gamma} : [0, 1] \to \Omega$ need not be injective on its image (it may go back and forth). This requires one to use a Riemann–Stieltjes integral as opposed to a Riemann one in order to make sense of the limiting procedure to infinite subdivisions of the parametrization. In the end, a timeless transition amplitude becomes (Chiou, 2013):

$$W(q_1, q_2) = \int \mathcal{D}q^a \int \mathcal{D}p_a \, \delta[H] \exp\left[\frac{i}{\hbar} \int_{\bar{\gamma}} p_a dq^a\right], \tag{9.11}$$

---

[14] For instance, it could be the space of sections on a tensor bundle, $\mathcal{M} = C^\infty(TM \otimes \cdots TM \otimes TM^* \cdots TM^*)$. In the case of gravity, these are sections of the positive symmetric tensor bundle: $\mathcal{M} = C_+^\infty(TM^* \otimes_S TM^*)$.

where the path integral sums over paths whose projection starts at q_1 and ends at q_2, and H is a single reparametrization constraint. In the presence of gauge symmetries, if it is the case that these symmetries form a closed Lie algebra, one can in principle use a group averaging procedure, provided one uses a similarly translation invariant measure of integration (which is available for the single, or global, reparametrization group).

For a strictly deparametrizable system,[15] one obtains again:

$$W(t_1,q_1^i,t_2,q_2^i) \sim \int \mathcal{D}t\, G(t_1,q_1^i,t_2,q_2^i) \sim G(t_1,q_1^i,t_2,q_2^i),$$

up to an irrelevant overall factor. Further, if the Hamiltonian is quadratic in the momenta, one can integrate them out and obtain the configuration space path integral with the Lagrangian form of the action.

For gravity, given the symmetries acting ultralocally on configuration space (and the principal fiber bundle structure they form) studied in Section 9.2.4.3, we take the analog of (9.11), schematically projected down onto the space of conformal geometries:[16]

$$W([g_1],[g_2]) = \int \mathcal{D}[g] \int \mathcal{D}[\pi] \exp\left[\frac{i}{\hbar} \int_{\bar{\gamma}} [\pi^{ab}]\delta[g_{ab}]\right] \delta H([g],[\pi]), \quad (9.12)$$

where I have (again, schematically) used square brackets to denote the conformal-diffeo-equivalence classes of the metric and momenta, and where δH represents a single reparametrization constraint (not an infinite amount, as in the ADM scalar constraint). Schematically, this transition amplitude should play the role of (9.11) in the field theory case.

Although considerations of quantum gravity and its problems have led us to value the timeless representation of quantum theory in Section 9.2, from now on we denote the equivalence classes $[g]$ and all other equivalence classes of fields under the appropriate instantaneous symmetries, by the standard coordinate variable, q.

9.4 Records and Timelessness

Suppose that we have in our hands a $W(q_q,q_2)$ for which q's carry also the gauge-invariant degrees of freedom. Still, as stressed in previous sections; without some driver of change—which we usually call *time*—what is the meaning of this transition amplitude? Here we will see how such a meaning can arise from timelessness. The ultimate meaning $W(q_1,q_2)$ can give rise to is simple: the likelihood that

---

[15] I.e. one for which $[\hat{H}(t_1),\hat{H}(t_2)] = 0$. If this is not the case, the equality will hold only semiclassically.

[16] The full treatment of the gauge conditions requires a gauge-fixed Becchi–Rouet–Stora–Tyutin (BRST) formalism, which is a level of detail I don't need here. See Gomes (2016) for a more precise definition, equation (28), where we use $K(g_1,[g_2])$ as opposed to $W([g_1],[g_2])$.

records of q_1 will be found in q_2. In Section 9.4.1, I build the scaffolding for a static volume form in configuration space. This requires a definition of the space of beables and of an anchor to the path integral. Having done this, in Section 9.4.2 I introduce the structure that allows one to ascribe histories to properties of the static volume form—records. However, records are not enough to talk about conservation of probability, and at the end of the section I sketch how this can be done.

9.4.1 Born Rule and the Preferred Configuration

In a true spatially relational, an instantaneous state of an observer is encoded in a partial-field configuration. There are no subjective overtones attributed to an observer—it is merely a (partial) state of the fields. Of course, there are many regions of configuration space where no such thing as an observer will be represented.

Since each point is a possible now, and there is no evolution, each now has an equal claim on existing. This establishes the plane of existence, every now that can exist, does exist! We are at least partway toward the adage of quantum mechanics. If this were a discrete space, we could say that each element has the same weight. This is known as the principle of indifference, and it implies that we count each copy of a similar observer once.[17]

But configuration space is a continuous space, like $\mathbb{R}^2$ (but infinite dimensional). Unlike what is the case with discrete spaces, there is no preferred way of counting points of $\mathbb{R}^2$. We need to imprint $\mathcal{M}$ with a volume form; each volume form represents a different way of counting configurations.

Contrary to what occurs in standard time-dependent many worlds quantum mechanics, I will define a single, standard time-independent "volume element" over configuration space $\mathcal{M}$. Integrated over a given region, this volume element will simply give the volume, or the amount, of configurations in that region.[18] The theory posited here is realist, in the sense that it ascribes an actual, *true* volume form to configuration space. But there is also a role to be played by empirical guesses to what this volume form is; it is this empirical volume form that we call our theory, and it is the only thing rational beings update in this timeless picture—there is no collapse of the objective wavefunction.

---

[17] The intuition obtained for many worlds in the discrete configuration spaces can be misleading for our purposes. In that case, each branch can be counted, and one needs a further explanation to count them according to the Born rule. There are different ways of going about this, e.g., based on this principle, and on the epistemic principle of separability, Carroll et al. claim that the Born rule can be derived (Sebens and Carroll, 2014).

[18] Of course, these volume forms are divergent and technically difficult to define. Properties of locality of the volume form, discussed in Gomes (2017b) are essential to show that nonetheless their definition reduces to the usual Born rule for isolated finite-dimensional systems. Furthermore, only ratios of the volume form have any meaning.

9.4.1.1 Born Rule

The volume form $P(q)Dq$ is defined as a positive scalar function of the transition amplitude, $P(q) := F(W(q^*, q))$. We need to explain the notation in this equation. First, there is still a sign of the principle of indifference in the manner we choose the bare volume element, Dq; it is chosen as the translationally invariant measure in the field theory context. But since it is being multiplied by $P(q)$, the composition $P(q)Dq$ can still be anything. This measure, F, gives a way to "count" configurations, and it is assumed to act as a positive functional of the only nontrivial function we have defined on $\mathcal{M}$, namely, the transition amplitude $W(q^*, q)$.

Now if we restrict the positive function $F \colon \mathbb{C} \to \mathbb{R}^+$, so that it respects the multiplicative group structure,

$$F(z_1 z_2) = F(z_1) F(z_2), \tag{9.13}$$

we can recover Markowian properties (locality in time), some form of which seems likely to be a fundamental property of nature. From this multiplicative demand we also recover a notion of records from the transition amplitude (Gomes, 2017b). When we also demand that in the classical limit we obtain classical statistical mechanics, (9.13) uniquely leads to a derivation of the "Born volume" form for $F \colon F(W(q^*, q)) = |W(q^*, q)|^2$, i.e.,

$$P(q) = |W(q^*, q)|^2. \tag{9.14}$$

Last, in the definition of $P(q)$ I have sneaked in an "in" configuration, q^*, which defines once and for all the static volume form over (reduced) configuration space. I define q^* roughly as the simplest, most structureless configuration of the fields in question. Note that this can be a meaningful statement only if q carries its own physical content; i.e., for symmetries that are laws of the instant. It is in this sense distinct from a "past hypothesis" in GR, which requires some auxiliary foliation to be defined.

9.4.1.2 The Preferred Configuration, q^*

This section gives a set of natural choices for q^*, depending on the configuration space, gauge group, and manifold topology. Reduced configuration spaces may not form smooth manifolds but only what are called *stratified manifolds*. This is because the symmetry group $\mathcal{G}$ in question—whose action forms the equivalence relation by which we are quotienting—may act *qualitatively* differently on different orbits. If there are subgroups of the symmetry group—stabilizer subgroups—whose action leaves a point $\tilde{q}$ fixed, the symmetry does not act "fully" on $\tilde{q}$ (or on any other representative $\bar{q} \in [\tilde{q}]$); some subgroups of $\mathcal{G}$ simply fail to do anything to $\tilde{q}$. This implies the quotient of configuration space with respect to the full symmetry may vary in dimensionality.

Taking the quotient by such wavering actions of the symmetry group creates a patchwork of manifolds. Each patch is called a stratum and is indexed by the stabilizer subgroup of the symmetry group in question (e.g., isometries as a subgroup of Diff(M)). The larger the stabilizer group, the more it fails to act on $\tilde{q}$, the lower the dimensionality of the corresponding stratum. The union of these patches, or strata, is called a stratified manifold. It is a space that has nested "corners"—each stratum has as boundaries a lesser dimensional stratum. A useful picture to have in mind for this structure is a cube (seen as a manifold with boundaries). The interior of the cube has boundaries that decompose into faces, whose boundaries decompose into lines, whose boundaries decompose into points. The higher the dimension of the boundary component, the smaller the isometry group that its constituents have,[19] i.e., the more fully $\mathcal{G}$ acts on it. Thus the interior of the cube would have no stabilizer subgroups associated to it, and the 1-dimensional corners the highest dimensional stabilizer subgroups. Everything in between would follow this order: the face of the cube could be associated to a lower dimensional stabilizer subgroup than the edges, and the edges a lower one than the corners.

Configurations with the highest possible dimension of the stabilizer subgroup are what I define as q^*—they are the pointiest corners of this concatenated sequence of manifolds. And it is these topologically preferred singular points of configuration space that we define as an origin of the transition amplitude. Their simplicity coincides with—or is a reinterpretation of—characteristics we would expect from a low-entropy beginning of the universe.

Thus, depending on the symmetries acting of configuration space, and on the topology of M, one can have different such preferred configurations. For the case at hand—in which we have both scale and diffeomorphism symmetry and $M = S^3$—there exists two sorts of such preferred points: one connected to the rest of the quotient space and the other disconnected. The preferred q^* of $\mathcal{M}/(\text{Diff}(M) \ltimes \mathcal{C})$ that is connected to the rest of the manifold is the one corresponding to the round sphere. The disconnected point is the completely singular metric, $q^* = g_{ab} = 0$.[20]

If we look at just the spatial diffeomorphisms, then the natural choice becomes the singular metric $q^* = g_{ab} = 0$. In the Hartle–Hawking state, in minisuperspace

---

[19] E.g., let $\mathcal{M}_o$ be the set of metrics without isometries. This is a dense and open subset of $\mathcal{M}$, the space of smooth metrics over M. Let I_n be the isometry group of the metrics g_n, such that the dimension of I_n is d_n. Then the quotient space of metrics with isometry group I_n forms a manifold with boundaries, $\mathcal{M}_n/\text{Diff}(M) = \mathcal{S}_n$. The boundary of $\mathcal{S}_n$ decomposes into the union of $\mathcal{S}_{n'}$ for $n' > n$ (see Fischer [1970]).

[20] For conformal transformation, we can see it is disconnected in the quotient space, because we have no access to $g_{ab} = 0$. Choosing any given reference $\bar{g}_{ab}$, we can conformally project g_{ab}, i.e. $[g_{ab}] = \left(\frac{\bar{g}}{g}\right)^{1/3} g_{ab}$. As g_{ab} becomes degenerate, its determinant goes to zero, and any such conformal projection diverges. Any such $[g_{ab}]$ is therefore, at best, infinitely far.

(where refoliations act as a single reparametrization, as they would here), this is (equivalent to) the initial state chosen.

9.4.2 Records

9.4.2.1 Semiclassical Records

We are now in position to relinquish "a driver of change." With the notion of *records*, about to be introduced, we can recover all the appearances of change, without it having to be introduced by fiat as extraneous structure.

Having defined q^*, we can set it as q_1 and obtain a meaningful transition amplitude $W(q_1, q_2)$ to now, represented by q_2. At a fundamental level, q^*, together with a definition of F and the action, completely specify the physical content of the theory by giving the volume of configurations in a given region of $\mathcal{M}$.

It is this anchoring of the amplitude on q^* that allows probabilities to depend only on the past. It is also what permits the existence of another class of object that I call *records*.

The system one should have in mind as an example of such a structure is the Mott (1929) bubble chamber. In it, emitted particles from α-decay in a cloud chamber condense bubbles along their trajectories. A quantum mechanical treatment involving a timeless Schrödinger equation finds that the wave-function peaks on configurations for which bubbles are formed collinearly with the source of the α-decay. In this analogy, a "record holding configuration" would be any configuration with n collinear condensed bubbles, and any configuration with $n' \leq n$ condensed bubbles along the same direction would be the respective "record configuration." In other words, the $n + 1$-collinear bubbles configuration holds a record of the n-bubbles one. For example, to leading order, the probability amplitude for n bubbles along the θ direction obeys:

$$P[(n,\theta), \cdots, (1,\theta)] \simeq P[(n',\theta), \cdots, (1,\theta)]P[(n',\theta), \cdots, (1,\theta)|(n,\theta), \cdots, (1,\theta)],$$
$$(9.15)$$

where $n' < n$, and $P[A|B]$ is the conditional probability for B given A.

Let us sketch how this comes about in the present context. When semiclassical approximations may be made for the transition amplitude between q^* and a given configuration, we have:

$$W_{\mathrm{cl}}(q^*, q) = \sum_{\gamma_j} \Delta_j^{\frac{1}{2}} \exp\left((i/\hbar)S_{\mathrm{cl}}[\gamma_j]\right), \qquad (9.16)$$

where the γ_j are curves that extremize the action, which on-shell we wrote as $S_{cl}[\gamma_j]$), and Δ are certain weights for each one (called Van-Vleck determinants[21]). This formula is approximately valid when the $S_{cl}[\gamma_j]) > \hbar$.

Roughly speaking, when all of γ_j go through a configuration $q_r \neq q$, I will define q as *possessing a semiclassical record* of q_r. Note that this is a statement about q, i.e., it is q that contains the record (a more precise definition is left for Gomes [2017a]). I will call $\mathcal{M}_{(r)}$ the entire set that contains q_r as a record.

For $q \in \mathcal{M}_{(r)}$, it can be shown that the amplitude suffers a decomposition (this is shown in Gomes [2017a]):

$$W(q^*, q) \simeq W(q^*, q_r) W(q_r, q). \tag{9.17}$$

To show this in a simplified setting of a deparametrizable system—i.e., when the Hamiltonian admits a split $H(q, p) = p_o + H_o(q_i, p^i, q_o)$, with $[H_o(t), H_o(t')] = 0$—one uses the same techniques as those used to prove the semigroup properties of the semiclassical amplitude. For $q_o = t$:

$$W_{cl}((q_1^i, t_1), (q_3^i, t_3)) = \int dq_2 W_{cl}((q_1^i, t_1), (q_2^i, t_2)) W_{cl}((q_2^i, t_2), (q_3^i, t_3)). \tag{9.18}$$

Since the extremal curves all go through the single point corresponding to t_2^r, we immediately recover (9.17):

$$W_{cl}(q^*, q) = W_{cl}((q_1^i, t_1), (q_3^i, t_3)) = W_{cl}((q_1^i, t_1), (q_2^{i(r)}, t_2^r)) W_{cl}((q_2^{i(r)}, t_2^r), (q_3, t_3))$$
$$\simeq W_{cl}(q^*, q_r) W_{cl}(q_r, q).$$

Calculating the probability of q from (9.17), we get an equation of conditional probability, of q on q_r,

$$P(q_r) = P(q|q_r) P(q_r). \tag{9.19}$$

The expression (9.19) thus reproduces the Mott bubble equation, (9.15) for q_i, q_j along the same classical trajectory, and separated by more than $\hbar$.

Furthermore, it is easy to show that when q_1 is a record of q_2 and there is a unique classical path between the two configurations, then the entire path has an ordering of records. Namely, parametrizing the path, $\gamma(t)$, such that $\gamma(0) = q_1, \gamma(t^*) = q_2$, then $\gamma(t)$ is a record of $\gamma(t')$ iff $t < t'$. We call such types of objects, strings of records, and it is through them that we recover a notion of classical time.

---

[21] The weights of each extremal path are given by the Van-Vleck determinant, $\Delta_i = \frac{\delta \pi_1^i}{\delta \phi}$, where π_1^i is the initial momentum required to reach that final ϕ. Having small Van-Vleck determinant means that slight variations of the initial momentum give rise to large deviations in the final position. Let me illustrate the meaning of a Van-Vleck determinant with a well-known heuristic example: suppose that ϕ_1 contains a broken egg. If ϕ represents a configuration with that same egg unbroken (still connected to ϕ_1 by an extremal curve), small deviations in initial velocity of configuration change at ϕ_1 will result in a final configuration very much different (very far from) ϕ.

9.4.2.2 The Recovery of Classical Time

If records are present, it would make absolute sense for observers in q to attribute some of q's properties to the "previous existence" of q_r. It is as if configuration q_r had to "happen" in order for q to come into existence. If q has some notion of history, q_r participated in it.

When comparing relative amplitudes between possibly finding yourself in configurations q_1 or q_2, both possessing the same records q_r, the amplitude $W(q^*, q_r)$ factors out, becoming irrelevant. This says that we don't need to remember what the origin of the universe was, when doing experiments in the lab, the required elements for q are already encoded in q_r.[22]

I believe that indeed, it is difficult to assign meaning to some future configuration q in the timeless context. Instead, what we do, is *to compare expectations now, with retrodictions, which are embedded in our records, or memories*. We compare earlier records with more recent ones, and apply Bayesian updating of our theories accordingly.

In the classical limit, without any interference, for a coarse graining for which records are separated by on-shell actions large with respect to the Planck scale, we recover a complete notion of history.

If there is nothing to empirically distinguish between our normal view of history (i.e., as having actually happened) on one hand, and the tight correlation between the present and the embedded past on the other, why should we give more credence to the former interpretation? Bayesian analysis can pinpoint no pragmatic distinction, and I see no reasons for preferences, except psychological ones.

9.4.2.3 Records and Conservation of Probability

Now, one of the main questions that started our exploration of theories that are characterized by the timeless transition amplitude, was the difficulty in defining concepts such as conservation of probability for quantum gravity, which has no fixed causal structure. Are we in a better position now?

What we are talking about so far is volume in configuration space. How does that relate to probabilities of the sort that is conserved? In the presence of a standard time parameter, we first distinguish between the total probability P_t at one time, t, from $P_{t'}$ at another, t'. To translate this statement to one that uses only records and configuration space, we want a notion that reproduces this separation. This separation is accomplished by first restricting configurations in $\mathcal{M}_{(r)}$ to subsets, $\mathcal{S}_\alpha$, $\alpha \in \Lambda$, such that there is no pair $\phi_i^\alpha, \phi_j^\alpha \in \mathcal{S}_\alpha$ for which $\phi_i \in \mathcal{M}_{(j)}$. We call

---

[22] But note that whenever a record exists, the preferred configuration q^* is also a record. In fact, one could have defined it as *the* record, of all of configuration space. Indeed, it does have the properties of being as unstructured as possible, which we would not be amiss in taking to characterize an origin of the universe.

these sets, $\mathcal{S}_\alpha$, *screens*. In other words, in each one of these sets, no configuration is a record of any other configuration. This is taken to say that configurations belonging to a single screen are not "causally related." In relativistic terminology—which can be misleading, since here we are in configuration space and not in real space(time)—this would represent events that happen "at the same time" (for some equal time surface). But here this property also holds for many worlds type theories in configuration space, where there is no real time.

Now, each $\mathcal{S}_\alpha \subset \mathcal{M}_{(r)}$ does not contain redundant records. But there are many redundant records along each extremal trajectory, at least in the no-interference case. In that simple case, there is precisely one extremal trajectory γ_j between ϕ_r and each element ϕ_j^α of a given screen, which is thus parametrized by the set $J \ni j$. Define a screen $\mathcal{S}_1 = \{\phi_j^1 = \gamma_j(t_j^1), j \in J\}$, where t_j^1 is a given parameter along the jth extremal curve. We can then find another screen $\mathcal{S}_2 = \{\phi_j^2 = \gamma_j(t_j^2), t_j^2 > t_j^1, \forall j\}$. In these simple cases, and at least for certain types of action functionals, it can be shown that, for the translationally invariant measure and Born volume, the infinitesimal volumes respect: $V(\mathcal{S}_2) \simeq V(\mathcal{S}_1)$ (Gomes, 2017a). This is as close as we can get to a statement about conservation of probability.

9.5 What Are We Afraid Of? The Psychological Obstacles

What usually unsettles people, including me, about this view is the damage it does to the idea of a continuous conscious self. The egalitarian status of each and all instantaneous configurations of the universe—carrying on their backs our own present conscious states—raises alarms in our heads. Could it be that each instant exists only unto itself, that all our myriad instantaneous states of mind *exist separately*? This proposal appears to conflict with the construed narrative of our selves—of having a continuously evolving and self-determining conscious experience.

But perhaps, upon reflection, it shouldn't bother us as much as it does. First of all, the so-called Block Time view of the self does not leave us in much better shape in certain respects of this problem. After all, the general relativistic worldline does not imply an evolving now—it implies a collection of them, corresponding to the entire worldline. For the (idealized) worldline of a conscious being, each element of this collection will have its own, unique, instantaneous experience.

Nonetheless, in at least one respect, the worldline view still seems to have one advantage over the one presented in this paper. The view presented here appears more fractured, less linear than the worldline view, even in the classical limit—for which aspects of a one-parameter family of configurations becomes embedded in a now. The problem is that we start off with a one-parameter family of individual conscious experiences, and, like Zeno, we imagine that an inverse limiting procedure focusing on the now will eventually tear one configuration from the next, leaving us stranded in the now, separated from the rest of configuration space by an infinitesimal chasm. This is what I mean here by *solipsism of the instant*. In Section 9.5.1, I will explain how this intuitive understanding can find footing only in a particular choice of (nonmetric) topology for configuration space. That topology is not compatible with our starting point of applying differential geometry to configuration space.

9.5.1 Zeno's Paradox and Solipsism of the Instant: A Matter of Topology

It might not seem like it, but the discussion about whether we have a collection of individual instants as opposed to a continuous curve of instants hinges, albeit disguisedly, on the topology we assume for configuration space. Our modern dismissal of Zeno's paradox relies on the calculus concept of a limit. But in fact, a *limit point* in a topological space first requires the notion of topology: a limit point of a set C in a topological space X is a point $p \in X$ (not necessarily in C) that can be "approximated" by points of C in the sense that every neighborhood of p with respect to the topology on X also contains a point of C other than p itself.

In the finest topology—the discrete topology—each subset is declared to be open. On the real line, this would imply that every point is an open set. Let us call an abstract precurve in X the image of an injective mapping from $\mathbb{R}$ (endowed with the usual metric topology) to the set X. Thus no precurve on X can be continuous if X is endowed with the finest topology. Because the mapping is injective, the inverse of each point of its image (which is an open set in the topology of X) is a single point in $\mathbb{R}$, which is not an open set in the standard metric topology of $\mathbb{R}$. Likewise, with the finest topology, Zeno's argument becomes inescapable—when every point is an open set, there are no limit points and one indeed cannot hop continuously from one point to the next. We are forever stuck here, wherever here is.

In my opinion, the idea that Zeno and Parmenides were inductively aiming at was precisely that of a discrete topology, where there is a void between any two given points in the real line. If X is taken to be configuration space, this absolute "solipsism of the instant" would indeed incur on the conclusions of the Eleatics, and frozen time would necessarily follow. However, *the finest topology cannot be obtained by inductively refining metric topologies.*

With a more appropriate, e.g., metric, topology, we can only iteratively get to open neighborhoods of a point, neighborhoods that include a continuous number of other configurations. That means, for example, that smooth functions on configuration space, like $P(q)$, are too blunt an instrument—in practice its values cannot be used to distinguish individual points. No matter how accurately we measure things, there will always be open sets whose elements we cannot parse. And so it is with any subjective, empirical notion of *now*.

The point being that with an appropriate topology we can have timelessness in a brander version than the Eleatics, even assuming that reality is entirely contained in configuration space without any absolute time. With an appropriate coarser (e.g., metric) topology on configuration space, we do not have to worry about a radical solipsism of the instant: in the classical limit there are continuous curves interpolating between a record and a record-holding configuration. I can safely assume that there is a *continuous sequence* of configurations connecting me eating that donut this morning to this present moment of reminiscence.

9.5.2 *The Continuity of the Self: Locke, Hume, and Parfit*

John Locke considered personal identity (or the self) to be founded on memory, much like my own view here. He says in "Of Ideas of Identity and Diversity":

This may show us wherein personal identity consists: not in the identity of substance, but ... in the identity of consciousness. ... This personality extends itself beyond present existence to what is past, only by consciousness.

David Hume wrote in *A Treatise of Human Nature* that when we start introspecting, "we are never intimately conscious of anything but a particular perception; man is a bundle or collection of different perceptions which succeed one another with an inconceivable rapidity and are in perpetual flux and movement."

Indeed, the notion of self, and continuity of the self, are elusive upon introspection. I believe, following Locke, that our self is determined biologically by patterns in our neural connections. Like any other physical structure, under normal time evolution these patterns are subject to change. What we consider to be a self or a personality, is inextricably woven with the notion of continuity of such patterns in (what we perceive as) time. Yes, these patterns may change, but they do so continuously. It is this continuity that allows us to recognize a coherent identity.

In *Reasons and Persons*, Derek Parfit puts these intuitions to the test. He asks the reader to imagine entering a "teletransporter," a machine that puts you to sleep, then destroys you, copying the information of your molecular structure, and then relaying it to Mars at the speed of light. On Mars, another machine re-creates you, each atom in exactly the same relative position to all the other ones. Parfit poses the question of whether or not the teletransporter is a method of travel—is the person on Mars the same person as the person who entered the teletransporter on Earth? Certainly, when waking up on Mars, you would feel like being you, you would remember entering the teletransporter in order to travel to Mars, you would also remember eating that donut this morning.

Following this initial operation, the teletransporter on Earth is modified so as to leave intact the person who enters it. Each replica left on Earth would claim to be you, and also remember entering the teletransporter, and then getting out again, still on Earth. Using thought experiments such as these, Parfit argues that any criteria we attempt to use to determine sameness of personal identity will be lacking. What matters, to Parfit (1984), is simply what he calls "Relation R": psychological connectedness, including memory, personality, and so on.

This is also my view, at least intellectually if not intuitively. And it applies to configuration space and the general relativistic worldline in the same way as it does in Parfit's description. In our case there exists a past configuration, *represented* (but not contained) in configuration now in the form of a record. This past configuration has in it neural patterns that bear a strong resemblance to neural patterns contained in configuration now. Crucially, these two configurations are connected by *continuous extremal paths in configuration space*, ensuring that indeed we can act as if they are psychologically connected. We can—and should!—act as if one classically evolved from the other. Indeed, in such cases our brain states are consistent with evolved relations between all subsystems in the world we have access to. Different sets of records all agree and are compatible with arising from joint evolution. Furthermore, I would have a stronger Relation R with what I associate with future configurations of my (present) neural networks, than to other brain configurations (e.g., associated to other people). There seems to be no further reason for this conclusion to upset us, beyond those reasons that already make us uncomfortable with Parfit's thought experiment.

9.6 Summary and Conclusions

9.6.1 Crippling and Rehabilitating Time

Time does not exist. There is just the furniture of the world that we call instants of time. Something as final as this should not be seen as unexpected. I see it as the only simple and plausible outcome of the epic struggle between the basic principles of quantum mechanics and general relativity. For the one—on its standard form at least—needs a definite time, but the other denies it. How can theories with such diametrically opposed claims coexist peacefully? They are like children squabbling over a toy called time. Isn't the most effective way to resolve such squabbles to remove the toy?

Barbour, 1999

Loosely following the Eleatic view of the special ontological status of the present, here we have carved time away from spacetime, being left with timeless configuration space as a result. If time is the legs that carry space forward, we might seem to have emerged from this operation with a severely handicapped universe.

The criticism is to the point. Even if time does not exist as a separate entity in the universe, our conception of it needs to be recovered somehow. If there is no specific variable devoted to measuring time, it needs to be recovered from relational properties of configurations. This essay showed that this can be done.

9.6.2 Psychological Hang-Ups

The idea of timelessness is certainly counterintuitive.

But our own personal histories can indeed be pieced together from the static landscape of configuration space. Such histories are indiscernible from, but still somehow feel less real than, our usual picture of our pasts. Even beyond the world-line view of the self, the individual existence of *every* instant still seems to leave holes in the integrity of our life histories. I have argued this feeling is due to our faulty intuitions about the topology of configuration space.

Nonetheless, even after ensuring mathematical continuity of our notion of history, the idea of timelessness and of *all* possible states of being threatens the ingrained feeling that we are self-determining; since all these alternatives exist timelessly, how do we determine our future? But this is a hollow threat. Forget about timelessness; free will and personal identity are troublesome concepts all on their own, we should not fear infringing their territory. I like to compare these concepts to mythical animals: Nessie, Bigfoot, unicorns, and the like. They are constructs of our minds, and—apart from blurry pictures—shall always elude close enough inspection. Crypto-zoologists notwithstanding, Unicorns are not an endangered species. We need not be overly concerned about encroaching on their natural habitat.

9.6.3 What Gives, Wheeler's Quip or Superpositions?

Neither, really.

Perhaps our shortcomings in the discovery of a viable theory of quantum gravity are telling us that *spacetime* is the obstacle. Though at first sight we are indeed mutilating the beautiful unity of space and time, this split should not be seen as a step back from Einstein's insights. I believe the main insight of general relativity, contained in Wheeler's sentence (9.1), is about the dynamism of space and time themselves. There is no violence being done to this insight here.

Spatial geometry appears dynamic—it warps and bends throughout evolution whenever we are in the classical regime. Regarding the dynamism of time, the notion of "duration" *is* emergent from relational properties of space. Thus duration too is dynamic and space dependent.

Nonetheless, all relational properties are encoded in the *static* landscape of configuration space. The point is that this landscape is full of hills and valleys, dictated by the preferred volume form that sits on top of it. From the way that the volume form distributes itself on configuration space, certain classical field histories— special curves in configuration space—can give a thorough illusion of change. I have argued that this illusion is indistinguishable from how we perceive motion, history, and time.

Moreover, in regard to the quantum mechanics adage, the processes $W(q_r, q)$ straightforwardly embody "everything that can happen, does happen." The concept of superposition of causal structures (or even that of superposition of geometries), is to be replaced by interference between paths in configuration space. Those same hills and valleys in configuration space that encode classical field histories reveal the valleys and troughs of interference patterns. A very shallow valley around a point—for example representing an experimental apparatus *and* a fluorescent dot on a given point on a screen—indicates the scarcity of observers sharing that observation. By looking at the processes between records and record-holding con- figurations, we can straightforwardly make sense of interference, or lack thereof, between (coarse-grained) histories of the universe.

In all honesty, I don't know if formulating a theory in which space and time appear dynamical, and in which we can give precise meaning to superpositions of alternative histories, is enough to quantize gravity. Although the foundations seem solid, the proof is in the pudding, and we must further investigate tests for these ideas.

But neither do I believe that dropping time from the picture is abdicating hard- won knowledge about spacetime. Indeed, we can recover a notion of history, we can implement strict relationalism, we transfigure the measurement problem, and we can make sense of a union of the principles of quantum mechanics and geometrodynamics.[23]

It seems to me that there are many emotions against this resolution, but very few arguments; as I said at the beginning of these conclusions, accepting timelessness is deeply counterintuitive. But such a resolution would necessarily change only how we view reality, while still being capable of fully accounting for how we experience it. The consequences for quantum gravity still need to be unraveled. This whole approach should be seen as a framework, not as a particular theory. And indeed, in the nonrelativistic regime of quantum mechanics, we are not looking for new experiences of reality, but rather for new ways of viewing the ones we can already predict, a new framework to interpret these experiences with. This is the hallmark of a philosophical insight, albeit in the present case one heavily couched in physics. As Wittgenstein said: "Once the new way of thinking has been established, the old

---

[23] Perturbative techniques of course still need to be employed, even in the semiclassical limit, to make sense of the weights Δ in (9.16). This, and other issues to do with renormalizability are left for future study.

problems vanish; indeed they become hard to recapture. For they go with our way of expressing ourselves and, if we clothe ourselves in a new form of expression, the old problems are discarded along with the old garment."

Acknowledgments

I would like to thank Lee Smolin and Wolfgang Wieland for discussions and help with the writing, and Athmeya Jayaram for introducing me to the work of Derek Parfit. This research was supported by Perimeter Institute for Theoretical Physics. Research at Perimeter Institute is supported by the Government of Canada through Industry Canada and by the Province of Ontario through the Ministry of Research and Innovation.

References

Anderson, E. 2017. *The Problem of Time*. Cham: Springer.

Arnowitt, R., Deser, S., and Misner, C. 1962. The dynamics of general relativity. Pages 227–264 of: Witten, L. (ed.). *Gravitation: An Introduction to Current Research*. New York: Wiley.

Atiyah, M. 1988. Topological quantum field theories. *Pub. Math. IHES*, **68**(1), 175–186.

Barbour, J. 1994. The timelessness of quantum gravity: II. The appearance of dynamics in static configurations. *Classical Quant. Grav.*, **11**(12), 2875.

Barbour, J. 1999. *The End of Time: The Next Revolution in Physics*. Oxford: Oxford University Press.

Barbour, J. 2010. The definition of Mach's principle. *Found. Phys.*, **40**, 1263–1284.

Carlip, S. 2005. Quantum gravity in 2+1 dimensions: The case of a closed universe. *Living Rev. Rel.*, **8**, 1.

Chiou, D. 2013. Timeless path integral for relativistic quantum mechanics. *Class. Quant. Grav.*, **30**, 125004.

Coley, A., Hervik, S., Papadopoulos, G. O., and Pelavas, N. 2009. Kundt spacetimes. *Class. Quant. Grav.*, **26**, 105016.

Cox, B., and Forshaw, J. 2011. *The Quantum Universe: Everything That Can Happen Does Happen*. Allen Lane.

Donnelly, W. 2014. Entanglement entropy and nonabelian gauge symmetry. *Class. Quant. Grav.*, **31**(21), 214003.

Donnelly, W., and Freidel, L. 2016. Local subsystems in gauge theory and gravity. *JHEP*, **09**, 102.

Donnelly, W., and Giddings, S. B. 2017. How is quantum information localized in gravity? *Phys. Rev.*, **D96**(8), 086013.

Fischer, A. 1970. The theory of superspace. Page 303 of: Carmeli, M., Fickler, S. I., and Witten, L. (eds.). *Proceedings of the Relativity Conference held 2–6 June, 1969 in Cincinnati, OH*. New York: Plenum Press.

Fischer, A., and Marsden, J. 1977. The manifold of conformally equivalent metrics. *Can. J. Math.*, **29**, 193–209.

Gomes, H. 2016. A geodesic model in conformal superspace. *gr-qc 1603.01569*.

Gomes, H. 2017a. Quantum gravity in timeless configuration space. *Class. Quant. Grav.*, **34**(23), 235004.

Gomes, H. 2017b. Semi-classical locality for the non-relativistic path integral in configuration space. *Found. Phys.*, **47**(9), 1155–1184.

Gomes, H., Gryb, S., and Koslowski, T. 2011. Einstein gravity as a 3D conformally invariant theory. *Class. Quant. Grav.*, **28**, 045005.

Halliwell, J. J., and Hartle, J. B. 1991. Wave functions constructed from an invariant sum over histories satisfy constraints. *Phys. Rev. D*, **43**(Feb), 1170–1194.

Horava, P. 2009. Quantum gravity at a Lifshitz point. *Phys. Rev.*, **D79**, 084008.

Isenberg, J., and Marsden, J. 1982. A slice theorem for the space of solutions of Einstein's equations. *Phys. Rep.*, **89**, 179–222.

Isham, C. J. 1992. Canonical quantum gravity and the problem of time. In: *19th International Colloquium on Group Theoretical Methods in Physics (GROUP 19) Salamanca, Spain, June 29–July 5, 1992.*

Knox, E. 2017. Physical relativity from a functionalist perspective. *Stud. Hist. Philos. Sci. B. Stud. Hist. Philos. Mod. Phys.* **67**, 118–124.

Koslowski, T. 2015. The shape dynamics description of gravity. *Can. J. Phys.*, **93**(9), 956–962.

Kuchar, K. 2011. Time and interpretations of quantum gravity. *Int. J. Mod. Phys. D*, **20**(suppl 1), 3–86.

Lee, J., and Wald, R. M. 1990. Local symmetries and constraints. *J. Math. Phys.*, **31**(3), 725–743.

Lehmkuhl, D. 2011. Mass-energy-momentum: only there because of spacetime? *Briti. J. Philos. Sci.*, **62**(3), 453–488.

McTaggart, J. Ellis. 1908. The unreality of time. *Mind*, **17**(68), 457–474.

Mercati, F. 2014. A shape dynamics tutorial. *arXiv e-prints*, arXiv:1409.0105.

Misner, C. W., Thorne, K. S., and Wheeler, J. A. 1973. *Gravitation*. 2nd ed. New York: W. H. Freeman and Company.

Mott, N. F. 1929. The wave mechanics of α-ray tracks. *Proc. R. Soc. Lond. A Math. Physical Eng. Sci.*, **126**(800), 79–84.

Parfit, D. 1984. *Reasons and Persons*. Oxford: Oxford University Press.

Pretorius, F. 2005. Evolution of binary black hole spacetimes. *Phys. Rev. Lett.*, **95**, 121101.

Price, H. 2009. The flow of time. In: Callender, C. (ed.). *The Oxford Handbook of Philosophy of Time*. Oxford: Oxford University Press.

Reuter, M. 1998. Nonperturbative evolution equation for quantum gravity. *Phys. Rev.*, **D57**, 971–985.

Rovelli, C. 2007. *Quantum Gravity*. Cambridge: Cambridge University Press.

Sebens, C. T., and Carroll, S. M. 2014. Self-locating uncertainty and the origin of probability in Everettian quantum mechanics. *Brit. J. Phil. Sci.*, **69**, 25–74.

Singer, I. M. 1978. Some remarks on the Gribov ambiguity. *Commun. Math. Phys.*, **60**(1), 7–12.

Teitelboim, C. 1983. Proper-time gauge in the quantum theory of gravitation. *Phys. Rev. D*, **28**(Jul), 297–309.

Torre, C. G., and Varadarajan, M. 1999. Functional evolution of free quantum fields. *Class. Quant. Grav.*, **16**, 2651–2668.

York, J. W. 1971. Gravitational degrees of freedom and the initial-value problem. *Phys. Rev. Lett.*, **26**, 1656–1658.

York, J. W. 1973. Conformally invariant orthogonal decomposition of symmetric tensors on Riemannian manifolds and the initial value problem of general relativity. *J. Math. Phys.*, **14**, 456.

Part III

Issues of Interpretation

10

Why Black Hole Information Loss Is Paradoxical

DAVID WALLACE

10.1 Introduction

Not everyone understands Hawking's paradox the same way[.]

Samir Mathur[1]

The black hole information loss paradox has been a constant source of discussion in theoretical physics since Stephen Hawking (1976) first claimed that black hole evaporation is nonunitary and irreversible, but opinions about it differ sharply. The mainstream view in theoretical physics—especially that part of high-energy physics that has pursued string theory and string-theoretic approaches to quantum gravity—is that (1) the paradox is deeply puzzling and (2) it must ultimately be resolved so as to eliminate information loss. But prominent critics in physics (Unruh and Wald, 1995, 2017; Penrose, 2004) and in philosophy (e.g., Belot, Earman, and Ruetsche [1999] and Maudlin [2017]) seem frankly baffled that anyone could expect information *not* to be lost in black hole evaporation and often regard the apparent paradox as simply the result of confusion. Maudlin (2017, p. 2) goes so far as to suggest that 'no completely satisfactory non-sociological explanation' can be given for the paradox's persistence!)

This sharp disagreement arises, I argue, from an equivocation as to what is meant by 'the' information loss paradox. The form most widely discussed in the popular and semipopular literature, closest in form to Hawking's original discussion, and most directly engaged with by the critics, is based on a clash between apparently general features of quantum mechanics and the global structure of the spacetime describing a completely evaporating black hole. *That* version of the paradox is indeed less than compelling: information loss seems prima facie plausible, and in

Mathur (2009, p. 34).

209

any case the question seems to require a full understanding of quantum gravity to answer and so may be premature.

But there is a second version of the paradox, dating back to Page (1993), which instead rests on the conflict between a statistical-mechanical description of black holes and the exactly thermal nature of Hawking radiation as predicted in quantum field theory (QFT). *This* version of the paradox is far more compelling, inasmuch as very powerful arguments support both sides in the conflict and yet they appear to give rise to contradictory results. The (mathematical) evidence against information loss advanced by physicists is much more naturally understood in terms of the second version of the paradox, and that version has if anything been sharpened by work in recent years that strengthens both the case for and the case against information loss. It remains in the truest sense paradoxical: a compelling argument for a conclusion, a comparably compelling argument for that conclusion's negation.

The structure of the paper is as follows. In Section 10.2, I briefly summarize important background facts in the last 40 years of black hole physics: the extent to which classical black holes can be given a thermodynamical description, the discovery and significance of Hawking radiation, and the progress made in establishing a statistical-mechanical underpinning for black hole thermodynamics. In Section 10.3 I present the two forms of the information loss paradox, focusing on the second and more powerful form. In Sections 10.4 and 10.5 I review recent developments (respectively, anti–de Sitter spacetime conformal field theory [AdS/CFT] duality and the firewall paradox) that bear on this form of the information loss paradox; Section 10.6 is the conclusion.

Four notes before proceeding. First, I have written this article at about the level of mathematical rigor found in mainstream theoretical physics. I do not attempt full mathematical rigor, which seems premature in any case given the incomplete state of development of the theories being discussed.

Second, for what it's worth, my strong impression is that the version of the information-loss paradox I focus on is also the version that high-energy physicists mostly have in mind in their technical work on quantum gravity. However, I do not intend this paper as a historical account, and I leave to others the interesting task of disentangling the literature on the topic.

Third, quantum gravity is a large and varied field. The dominant research program in that field is string theory, and my discussions of quantum gravity in this paper (insofar as they go beyond low-energy regimes, which are somewhat better understood) are entirely restricted to the string theory program. This is partly because the second version of the paradox I discuss has been overwhelmingly developed and discussed in string theory, but mostly on grounds of space and of my own expertise. (I am not an expert on string theory by any means, but I understand the relevant results well enough to have reasonable confidence on how they work and what they rely on; I lack anything like that level of competence in other approaches to quantum gravity.) The reader should not infer anything about the progress, or lack of progress, in understanding black hole statistical mechanics and black hole information loss in other programs from my failure to discuss it here. Readers more sceptical than I about the prospects of string theory, in particular, are warmly encouraged to carry out analogous investigations in other programs. (For loop quantum gravity in particular—the largest quantum gravity program outside the string theory framework—Perez [2017] is a thorough recent review of black hole physics in that program.)

Last, unless otherwise noted I work in units where $k_B = c = G = \hbar = 1$.

10.2 Background

This is a rather brief overview of material I discuss, and critically assess, in much more detail in Wallace (2018, 2019). I give only the key references; readers are referred to these papers for more extensive information and references.

10.2.1 Black Hole Thermodynamics

In the 1970s it became clear that even classical black holes (that is, black holes as described by classical general relativity, along with phenomenological descriptions of matter fields) behaved in a great many respects as if they were ordinary thermal systems. I review this material in depth in Wallace (2018), and readers are referred there for details and further references, but in summary:

- Black holes have equilibrium (that is, stationary) states characterized (in their rest frame) only by their energy and by a small number of conserved quantities (charge and angular momentum); 'no-hair' theorems (Carter, 1979) establish the uniqueness of these equilibrium states, and perturbations away from equilibrium are damped down quickly (Thorne, Price, and Macdonald, 1986, chs. 6–7).

- Small-scale interactions with a black hole (such as lowering charged or rotating matter into the hole) can be divided into 'reversible' and 'irreversible' in a fashion closely analogous to infinitesimal adiabatic interactions with thermodynamical systems (Christodolou and Ruffini, 1971). In this analogy, black hole surface area plays the role of thermodynamic entropy: an infinitesimal change is reversible iff it leaves the area invariant, and no physically possible intervention can decrease the area.

- Stationary black holes of charge Q, mass M and angular momentum J satisfy the differential expression:

$$\mathrm{d}M = \frac{\kappa}{8\pi}\mathrm{d}A - \Omega \mathrm{d}J - \Phi \mathrm{d}Q, \tag{10.1}$$

where κ is the surface gravity, A the surface area, Ω the surface angular velocity, and Φ the surface electric potential (Bardeen, Carter, and Hawking, 1973). This expression can be derived either as an abstract mathematical statement about equilibrium black holes or as a statement about the small changes in Q, M, J, and A induced by allowing matter to fall into the black hole. This is exactly the expression (under either interpretation) that would encode the First Law of Thermodynamics for a self-gravitating body at thermal equilibrium with that charge, mass, and angular momentum, if it had temperature $\lambda\kappa/8\pi$ and entropy $A\lambda$ (for arbitrary positive λ). Some while later, Wald (1993) showed how to extend the First Law and how to define the appropriate generalization of (entropy $\propto$ area) to arbitrary diffeomorphism-invariant theories of gravity.

- Hawking's (1972) area theorem established that no intervention on a black hole could decrease its area, and so extended the (entropy $\propto$ area) idea from infinitesimal to finite processes.

The *membrane paradigm* (Thorne et al., 1986) that codified further advances in classical black hole physics in the 1970s and early 1980s, extended this thermodynamic interpretation of black holes to a local description, where a black hole can be regarded (from the perspective of any observer who remains outside) as a thin, viscous, charged, conducting membrane lying at the 'stretched horizon', just outside the true event horizon. Under the membrane paradigm, for instance:

- The increase of black hole area when a charge is dropped onto it can be understood as Ohmic dissipation as the charge flows through the surface;
- The return to equilibrium of the black hole under the same process can be understood as the spread of the charge from an initially localized region to a uniform charge distribution across the surface;
- An uncharged black hole rotating in an external magnetic field will develop eddy currents that slow its rotation;
- Dropping a mass into a black hole induces damped perturbation in its shape, in accordance with the Navier–Stokes equation for viscous fluids.

These results, collectively, provide an almost perfect interpretation of a black hole as a thermodynamic system *as long as heat exchanges with other systems are disregarded*. When they are considered, the analogy breaks down entirely as far as classical physics is concerned; the only consistent temperature that can be assigned to a black hole through considerations of thermal contact is zero, and no classically possible physical process can be reinterpreted as heat flow from a black hole to another thermal system.

10.2.2 Hawking Radiation

Hawking (1975) discovered that when quantum field theory (QFT) is applied to a black hole spacetime, it gives rise to field states that, as seen by distant observers, correspond to thermal radiation emitted from the black hole. Hawking radiation (which has since been rederived by many different methods; see Wallace [2018, sec. 4.2] for a review and discussion of the evidence) can be understood as a consequence of the entanglement of the QFT vacuum state and of any state locally similar to that state. Field modes on either side of the event horizon are entangled, and from the point of view of an observer who remains outside the event horizon, the effective state of a field mode just outside the horizon can be described by tracing over the physically inaccessible partner mode just inside the horizon. The resultant state is perfectly thermal as measured by an observer just outside the horizon; because of the black hole's gravitational field, most of this thermal radiation falls back in and the radiation seen by a distant observer differs from perfect black-body radiation by a so-called gray-body factor correction.

From our point of view, Hawking radiation has two key features:

1. It exactly completes the description of a black hole as a thermodynamic system. Hawking radiation from a black hole of surface gravity κ has temperature $\kappa/2\pi$, exactly in accord with the black hole temperature contained within the First Law, (with $\lambda = 4$), and it provides a method by which black holes can be put in thermal contact with each other and with other hot bodies (either through radiative transfer or by directly mining the atmosphere of thermal radiation surrounding the hole).
2. Hawking radiation carries energy (as defined via Noether's theorem applied at large distances) away from the black hole and so can be expected to reduce its mass. Exact calculations remain impossible (they would require a fully understood theory of quantum gravity) but a variety of different calculations, arguments and numerical simulations within semiclassical gravity (see Wallace [2018, sec. 4.3] for a review) all give the expected 'naive' result that the rate of decrease of the black hole's mass is equal to the Hawking radiation flux at infinity. As such, an isolated black hole will radiate away its mass theoretically until it vanishes entirely, in practice at least until its mass approaches the Planck mass and the assumptions of Hawking's calculation (that full quantum-gravity effects may be neglected) become invalid.

10.2.3 Black Hole Statistical Mechanics

So: black holes behave exactly like thermodynamic systems. All *other* thermodynamic systems we know behave that way because their thermodynamics is underpinned by a statistical-mechanical description, so it is natural to speculate that black holes also have a statistical-mechanical description that underlies their thermodynamic behavior. In particular, the thermodynamic entropy of a statistical-mechanical system is identified with the *microcanonical entropy*, i.e., the log of the number of (mutually orthogonal) states available to it, so a statistical-mechanical description for black holes implies that a black hole of given mass, charge, and angular momentum and of area A has $\sim \exp(A/4G)$ (mutually orthogonal) microstates available to it. This speculation was made even before the discovery of Hawking radiation, and by now there is a large amount of calculational evidence supporting it. I review that evidence in depth in Wallace (2019) (other reviews include Harlow [2016] and Hartman [2015]), but in brief, there are three sources: effective field theory, full quantum gravity, and AdS/CFT duality.

The effective field theory route treats general relativity as an ordinary (albeit non-renormalizable) quantum field theory, with the Einstein–Hilbert action as the leading term in an infinite series of interactions and with some unspecified high-energy physics cutting off the divergences in the field-theoretic description at around the Planck length. In this formalism (and temporarily reintroducing the gravitational constant G), statistical-mechanical entropy can be calculated using path-integral methods; as was shown by Gibbons and Hawking (1977) only a few years after the discovery of Hawking radiation, doing so recovers $S = A/4G$ as the leading-order term. Subsequent work has both clarified the conceptual basis of the calculation and extended it to include interaction terms beyond the Einstein–Hilbert action and higher-order quantum corrections to the leading-order result. The former exactly reproduces Wald's generalization of the entropy formula; the latter have exactly the required form to renormalize the gravitational constant, ensuring that the G in $S = A/4G$ is the empirically measured, renormalized Newtonian constant, not the bare constant that appears in the quantum Lagrangian. (More accurately, the divergent part of the quantum corrections renormalizes the constant; the finite part generates additional terms in the entropy formula proportional to $\log M$, which are negligible at classical scales.)

The full quantum gravity route depends on one's preferred theory of quantum gravity (and, since we have no fully worked-out theory of quantum gravity, is necessarily tentative). The most precise (and also the most influential) calculations have been done in string theory and are restricted to so-called extremal black holes; since we will anyway have to consider such black holes later, I pause to give a brief explanation.

Consider first a black hole with nonzero charge Q, mass M, but no angular momentum. The appropriate black hole solution to the field equations, the

Reissner–Nordström solution, describes a black hole only if $|Q| \leq M$ (for larger charges, there is a naked singularity). An extremal black hole satisfies $|Q| = M$ (and is best thought of as the limiting case of black holes closer and closer to extremality). The surface gravity, and thus thermodynamic temperature, of an extremal black hole, is zero; thus they do not radiate and can be thought of as ground states. (In general, a nonextremal black hole with a substantial charge will decay to an extremal black hole rather than evaporating entirely.) This generalizes to charged, rotating black holes, as well as black holes in higher dimensions and with more than one sort of charge; in each case, we can describe the black hole by its conserved charge(s) and angular momentum together with the statement that it is extremal.

Strominger and Vafa (1996) showed that for a certain class of extremal black hole in five dimensions, the statistical-mechanical entropy can be calculated in string theory; the result, to leading order, exactly matches the area formula. Subsequent calculations have both widened the class of extremal black holes whose entropy can be found in this manner and refined the calculations to include higher-order corrections and to allow for small perturbations from extremality. The match to the entropy deduced by low-energy methods is exact.

(It's tempting to conclude that [1] this is evidence that string theory is the correct theory of quantum gravity, and/or [2] that the Strominger–Vafa result is significant only *if* string theory is the correct quantum theory of gravity. Both conclusions are too quick. The fact that black hole entropy can be calculated in low-energy quantum gravity—and, indeed, in QFT on a fixed background—strongly suggests that any consistent ultraviolet completion of low-energy quantum gravity will reproduce the entropy formula. The match between Strominger and Vafa's result, Gibbons and Hawking's, and the semiclassical prediction is then evidence first that string theory is a *consistent* quantum theory of gravity, and second, that the sum-over-histories approach to low-energy quantum gravity really is describing the low-energy regime of a consistent theory. (See Wallace [2019] for further discussion.)

The AdS/CFT route relies on the conjectured duality (Gubser, Klebanov, and Polyakov, 1998; Maldacena, 1998; Witten, 1998a) between a quantum gravity theory with asymptotically $AdS_{n+1} \times K$ boundary conditions and a conformal quantum field theory on the n-dimensional boundary of AdS_{n+1}, where AdS_n denotes anti–de Sitter spacetime in n spacetime dimensions and K is some compact space. (Anti–de Sitter spacetime is most perspicuously thought of here as a sort of box, a covariant version of the 'periodic boundary condition' boxes often used to make quantum field theories calculationally more tractable.) The conjecture arose in string theory (indeed, arose in part through the extremal black hole calculations discussed earlier) but can be understood, and motivated, as a claim about general quantum-gravity theories on AdS; it cannot be proved formally at present but can be strongly motivated both by physical arguments and by a large number of examples of calculations of the same quantity on either side of the duality, where completely different methods nonetheless give the exact same answer.

The statistical mechanics of conformal field theories are conceptually under much better control than in the quantum-gravity case, and AdS/CFT duality allows us to calculate entropy and other thermodynamic quantities on the CFT side of the duality and then map them back to the AdS case. The results (which are uncontroversially statistical-mechanical in origin) reproduce the predictions of black hole thermodynamics; in general qualitatively (Witten, 1998b; Aharony et al., 2004), but quantitatively and exactly, in those cases where quantitative results can be obtained. (In particular, the extremal black hole calculations can be reproduced [Strominger, 1998] via AdS/CFT methods.)

In aggregate, these results seem to give strong support to the idea that black hole thermodynamics will be underpinned, in full quantum gravity, by a statistical mechanics of the same general form as that underpinning any other thermodynamic system, and that in particular, any classical stationary black hole has a large number of microscopic degrees of freedom, in accordance with the statistical-mechanical definition of entropy. The exact form of these degrees of freedom is left somewhat obscure but they appear to have to live in a thin skin around the event horizon, both on general physical grounds (degrees of freedom *within* the event horizon don't seem well placed to determine a system's thermodynamic properties) and because that seems to be where the detailed calculations place them in those cases where they give any answer at all. A natural way to think of this is as a quantization of the membrane paradigm: not only thermodynamically, but statistically mechanically, a black hole seems to an outside observer as a thin membrane around the black hole, which has a full unitary description as a quantum system interacting with surrounding radiation, which will be at the Planck temperature as measured by an observer suspended just above its surface, and whose thermodynamical properties are explained in terms of its microscopic degrees of freedom. In particular, in this description Hawking radiation is just ordinary thermal radiation from the surface of the membrane.

10.3 Paradoxes of Information Loss

'The' paradox of information loss encompasses a wide range of ideas, but two main versions of the paradox can be discerned. I begin with the version most commonly discussed in the foundational literature, which I review only briefly; for a more detailed review see, e.g., Wald (1994, ch. 7) or Belot et al. (1999). It turns on the violation of unitarity in complete black hole evaporation, and its relation to the causal structure of the evaporating black hole spacetime.

10.3.1 Non-Unitarity of Total Evaporation

Figure 10.1 depicts the complete process of black hole evaporation according to the semiclassical description (that is, assuming Hawking radiation via QFT on a fixed black hole background spacetime and decrease of the mass of that black hole via semiclassical back-reaction). We can distinguish three regions:

1. The *preformation region I* (the lower gray triangle), which is foliated by Cauchy surfaces like Σ_I.
2. The *evaporation region II* (the white region between the gray triangles), which can be written as the union of $II(int)$ (the region inside the horizon) and $II(ext)$ (the region outside). It is foliated by slices like Σ_{II}.
3. The *postevaporation region III* (the upper gray triangle), foliated by slices like Σ_{III}.

Each region, individually, is globally hyperbolic, as is the combined region $I + II$. The overall spacetime is *not* globally hyperbolic: the point X is a naked singularity.

If QFT on this spacetime accurately describes the black hole evaporation process, we would expect to be able to describe the physics in region $I \cup II$ by a unitary dynamics, transforming, e.g., a pure quantum state on Σ_I to one on Σ_{II}. The latter state, in general, would be entangled, with the reduced states on $\Sigma_{II} \cap II(int)$ and $\Sigma_{II} \cap II(ext)$ being mixed, but the overall evolution would remain unitary and retrodictable. Since the boundary between $II(int)$ and $II(ext)$ is a future horizon, we would also expect a future-deterministic evolution on $I \cup II(ext)$, so that the reduced state on $\Sigma_{II} \cap II(ext)$ is uniquely determined by the state on Σ_I, but this evolution will be neither unitary nor past-deterministic (many different initial states would lead to the same black hole exterior reduced state.) Strictly, the naked singularity means that there is no well-defined evolution at all from region II to region III, but if we stipulate some well-behaved local physics at the singularity, the evolution from II to III is again deterministic but non-unitary. Indeed, the quantum state on Σ_{out} that describes the spacetime immediately upon evaporation is obtained by unitarily evolving the state on Σ_I forward to just before the singularity and then tracing out over $II(int)$.

The end result is that the process of black hole formation and evaporation, as described by an observer outside the black hole, is a pure-to-mixed, irreversible transition. The same result can be seen more physically by noting that the quantum state of the exterior in the evaporation region just consists of Hawking radiation,

David Wallace

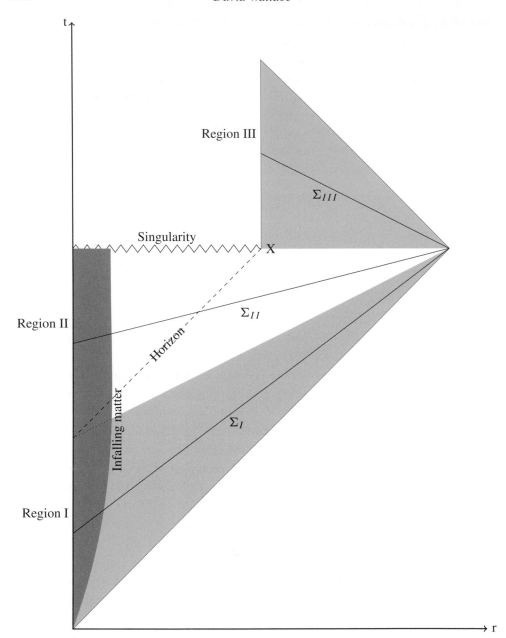

Figure 10.1 Spacetime of a completely evaporating black hole (angular coordinates suppressed).

which is (1) perfectly thermal, and hence mixed, and (2) determined only by the bulk properties of the black hole (mass, charge, angular momentum) and not by any details of its formation. On slices in region *II*, the full information remains because the formation details are encoded on the state in *II*(*int*), but that information is lost once the black hole completely evaporates. (More carefully: that information is not present in any of the slices in region *III*.)

There is a major lacuna in this argument: the final stages of black hole evaporation occur when the black hole's curvature is Planck scale and so are well into the regime where semiclassical calculations are unreliable. (This shows up formally in Figure 10.1 via the naked singularity.) But it is hard to imagine filling in the Planck-scale physics in order to save unitarity: the remaining energy does not seem sufficient to encode all the remaining information in a final burst of photons (to say nothing of the question of how it got from the black hole singularity to the evaporation point X); the persistence of the black hole as a Planck-scale 'remnant' with fantastically many internal degrees of freedom looks difficult to reconcile with other features of particle physics; the replacement of the pointlike naked singularity with a lightlike 'thunderbolt' seems unmotivated and overkill besides; and the removal of the future singularity so that the interior of the black hole forms a 'baby universe' does nothing to save unitarity from the perspective of the external observer.

This powerful argument for non-unitary evaporation is one form of the information-loss paradox, which we can usefully call the 'evaporation time paradox', yet readers might wonder why it deserves to be called a paradox and not just an argument to the interesting conclusion that information is lost. Indeed, that is how Hawking (1976) originally described it; more recently, information loss has been advocated forcefully by Unruh and Wald (1995, 2017), Penrose (2004, pp. 840–841), and Maudlin (2017) and more nuancedly by Belot et al. (1999). The stripped-down version of their arguments would be: we have a right to expect unitarity, information preservation, and retrodiction only on globally hyperbolic spacetimes; the evaporation spacetime manifestly is not globally hyperbolic; so non-unitary evolution is only to be expected.

Arguments for unitarity have been given for this version of the paradox, but frankly they are (to me) less than compelling. The most straightforward (mostly seen in informal discussion and semipopular work) is simply: unitarity is part of quantum mechanics, so non-unitary evaporation is incompatible with quantum mechanics. But this seems to equivocate on the meaning of "quantum mechanics"; of course we could just *define* quantum physics as incorporating unitary dynamics, but at least some forms of QFT seem perfectly well defined on non-globally hyperbolic spacetimes and to have dynamics that is locally unitary but globally non-unitary on those spacetimes. (See Belot et al. [1999, append.], for discussion; see also Deutsch [1991] for an example of the same phenomena in a different context.) Compare: *Hamiltonian* versions of classical electromagnetism are defined only for globally hyperbolic spacetimes, but classical electromagnetism can also be defined via its local field equations on a much more general class of spacetime.

More interesting objections come from quantum field theory. In QFT, the amplitude for a transition can generally be expressed as a sum over all ways in which the transition might come about, which for a full quantum theory of gravity ought to include processes involving formation and evaporation of Planck-scale black holes. Furthermore, high-energy processes like this are typically *not* suppressed in sum-over-histories calculations; rather, their effects can normally be absorbed into the renormalization of the empirically measured constants. Prima facie, it looks plausible that non-unitary quantum-gravity effects will make a large difference to low-energy physics and that this difference cannot be normalized away, and some calculations (Banks, Susskind, and Peskin, 1984; Srednicki, 1993) seem to support this result. But the matter is controversial (see, e.g., Hawking [1996], Unruh and Wald [1995], or Unruh [2012] for dissenting views) and at the least does not seem to provide decisive reasons to reject the argument for information loss, pending a full quantum theory of gravity in which the calculations can be done more carefully.

But while the evaporation time paradox is the form of the information loss paradox generally found in popular and foundational literature, it is not the only form. A much more compelling paradox arises when Hawking radiation is considered not just in the light of quantum mechanics in general, but in particular in the light of black hole statistical mechanics.

10.3.2 Non-Thermality of Unitary Cooling

Following Page (1993), suppose we have some ordinary thermodynamic system with Hilbert space $\mathcal{H}$, which we want to describe using the microcanonical ensemble. To do so we find some energy interval ΔE, narrow compared to typical energies of the system but wide enough so that the number of energy eigenstates between E and $E + \Delta E$ is large. Then we can define $\mathcal{H}(E)$ as the subspace spanned by eigenstates of energy with eigenvalues between E and $E + \Delta E$, and write the total Hilbert space as:

$$\mathcal{H} = \bigoplus_E \mathcal{H}(E), \tag{10.2}$$

where the sum ranges over energies $E = 0, \Delta E, 2\Delta E, \ldots N\Delta E, \ldots$. The system begins at microcanonical equilibrium at energy E_0, and for expository simplicity I assume its initial state is pure (and so is contained in $\mathcal{H}(E_0)$). It then cools through the emission of thermal radiation; that is, emission of quanta of radiation in highly mixed states. I also assume that the system is large enough that its energy remains in a narrow band as it cools, so that to high accuracy the system's state at any given time is contained within (that is, is a density operator restricted to) some $\mathcal{H}(E)$.

The original state of the system is pure, and its dynamics are unitary, so the total state of system plus radiation is pure. But if the radiation is thermal, each emitted photon will be in a mixed state and so must be entangled with some other system for the total state to be pure. In thermal radiation no two emitted quanta can be entangled, so each must be entangled with the system. More quantitatively, if the total von Neumann entropy of the radiation quanta emitted as the system cools from E_0 to some lower energy E is S, then by unitarity the von Neumann entropy $S_{VN}(E_0, E)$ of the system must also be S. But if the system has energy $E \in [E(t), E(t) + \Delta E]$ at time t, its (mixed) state at t must be contained within $\mathcal{H}(E(t))$ and so must have a von Neumann entropy less than $\log \dim \mathcal{H}(E(t))$, which is to say that the *von Neumann* entropy is bounded above by the *microcanonical* entropy $S_{MC}(E(t))$,

$$S_{VN}(E_0, E(t)) \leq S_{MC}(E(t)). \tag{10.3}$$

And the latter (assuming the system has positive temperature) is decreasing as the system cools. There will come a time—the so-called Page time, typically about halfway through the cooling process—when this inequality saturates, and after that point the radiation can no longer be exactly thermal. Instead, the late-time thermal radiation will have to be entangled with the early-time radiation. (Page provides plausibility arguments to the effect that this turnover will be fairly sharp: the radiating body will emit almost exactly thermal radiation up to the Page time, so that the von Neumann entropy of the radiating body rises initially until it equals the decreasing microcanonical entropy and then the two remain equal for the rest of the decay process.) The overall process is illustrated in Figure 10.2: the characteristic time dependence of the von Neumann entropy, initially increasing and then dropping off, is known as the *Page curve*. The initial purity of the system state plays no essential role here provided that the initial state's von Neumann entropy is much lower than its microcanonical entropy.

According to black hole thermodynamics, black holes are 'ordinary quantum-mechanical systems', and the von Neumann entropy of the matter that formed the black hole will be extremely small compared to the black hole's initial thermodynamic entropy. (And, while the stellar precursor of an astrophysical black hole is not plausibly in a pure state, the thermodynamic entropy of such a precursor is typically negligible compared to the entropy of the black hole that forms from it.) The Page time for a Schwarzschild black hole is approximately half the total evaporation time, at which point it will have radiated away about half its mass; after this time, it is not possible for the black hole's radiation to be thermal. Instead, it should be maximally entangled with the early-time radiation (albeit for any physically plausible measurement process, this entanglement will be completely undetectable; cf. Harlow and Hayden [2013]).

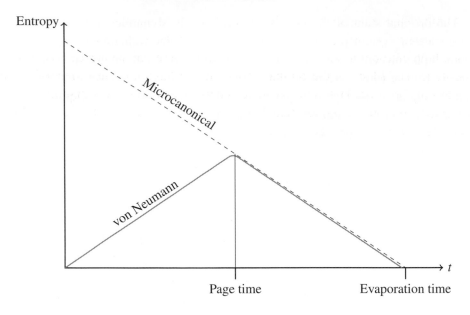

Figure 10.2 The Page curve.

The problem is that Hawking radiation—according to the quantum-field-theoretic calculations—is *exactly* thermal, displaying no entanglement whatever between early-time and late-time quanta. Indeed, if the calculation is read literally, the successively emitted quanta are sequentially redshifted down from the trans-Planckian regime at the horizon; each mode of the Hawking radiation is maximally entangled (given its overall expected energy) with a radiation mode inside the event horizon, which has no means of escaping, and so cannot also be entangled with other radiation modes (a principle sometimes called 'monogamy of entanglement').

Call this the *Page time paradox*. (This is basically the form of the paradox as presented in, e.g., Mathur [2009] and Polchinski [2016].) It is a clash between the predictions of QFT and the predictions of black hole statistical mechanics that occurs long before complete evaporation, when the black hole is still macroscopic in scale (i.e., there seems no prospect of exotic quantum-gravitational effects coming to the rescue). And while no doubt the exact results of Hawking's calculation need to be modified by various interaction terms and the like, the general form of the calculation seems robust against these modifications, and there seems little prospect that these modifications will give the large violations of thermality required to conform to statistical mechanics, at least as long as the basic QFT framework remains intact (see Mathur [2009] for a careful argument to this effect; see also the discussion in Wallace [2018, sec. 4.2]).

Remnants, or thunderbolts, or baby universes, no matter how helpful they may be in preserving unitarity, do nothing to preserve the statistical interpretation of black hole entropy or any account of black hole thermodynamics as arising from statistical mechanics in the ordinary way, and so have no role in resolving *this* version of the information loss paradox. Indeed, I will now show that this version of the information loss paradox can be stated even for situations in which the black hole does not completely evaporate.

10.3.3 Information Loss without Evaporation

For a straightforward realization of the paradox for nonevaporating black holes, consider the cooling of charged black holes. Recall from our previous discussion that charged nonrotating black holes satisfy the inequality $|Q| \leq M$. Positively charged black holes preferentially radiate positively charged particles (and vice versa), but Hawking radiation is dominated by massless particles, so typically the ratio $|Q|/M$ increases in the decay of a charged black hole, and it is possible for this inequality to saturate (i.e., reach $|Q| = M$). At this point the black hole is extremal; extremal black holes have zero temperature and do not emit Hawking radiation (a good thing for the self-consistency of black hole thermodynamics, since otherwise they would violate $|Q| \leq M$), so the black hole will not decay further.[2]

Now consider an extremal black hole that ought, according to black hole statistical mechanics, to be in a perfectly mixed state. (Such a black hole could be formed, for instance, by taking a suitably nonextremal black hole and allowing it to approach extremality by decay, such that it reaches the Page time before becoming extremal.) If some (uncharged) radiation is absorbed by the black hole, it will heat up and reradiate the absorbed energy as Hawking radiation. According to QFT, this is a pure-to-mixed transition for physics outside the stretched horizon, since the emitted radiation is perfectly thermal. But according to black hole statistical mechanics, since the quantum state of the black hole is unchanged by the process and the overall interaction is unitary, the outside-the-horizon description should likewise be unitary and the emitted radiation should be in a pure state. (This was an important test case in discussions of the information loss paradox in the 1990s; cf. Maldacena and Strominger [1997] and references therein and will be further discussed in Section 10.4.1 where we will see that it provides fairly direct evidence for unitary evaporation.)

---

[2] This assumes a black hole in a region of spacetime far from any other matter. Astrophysically realistic charged black holes would preferentially absorb oppositely charged particles (from, e.g., the interstellar medium) at a rate much faster than their evaporation by Hawking radiation, and so would tend away from extremality; thus extremal charged black holes, though a theoretically possible thought experiment, should not be understood as representing anything astrophysically realistic.

For a somewhat more complicated case (due to Maldacena [2003]), consider a mass-M black hole in a small box, in thermal equilibrium with its atmosphere and with the walls of the box. (One way to implement this is to suppose that the black hole exists in an asymptotically AdS spacetime with effective radius smaller than the black hole's Schwarzschild radius; another [York, 1986] is just to impose reflecting boundary conditions on the black hole at a radius less than $1.5\times$ the Schwarzschild radius). According to the QFT calculations, each emitted photon is in a perfectly thermal state and so is uncorrelated with any other emitted photon. The sharp statement of this in QFT (which is readily demonstrated; see Harlow [2016, sec. 6.9] and Maldacena [2003] and references therein) is that correlations between field operators at large time separations should fall off exponentially: that is,

$$C(t) \equiv \langle \widehat{O}(t)\widehat{O}(0)\rangle_\rho \equiv \mathsf{Tr}(\rho\widehat{O}(t)\widehat{O}(0)) \sim \mathrm{e}^{-c\beta t}, \qquad (10.4)$$

where ρ is the thermal state of the black hole atmosphere, $\widehat{O}(t)$ is some spatially smeared field operator localized at time t, $\beta = 8\pi M$ is the inverse temperature, and c is a dimensionless constant. Provided the operator $\widehat{O}(t)$ is chosen to project onto outward-going states, this should continue to be true, given a QFT description of Hawking radiation, even given the reflecting boundary conditions.[3]

But if black hole statistical mechanics is true, then (by the Bekenstein bound; cf. discussion in Wallace [2018, sec. 4.5]) the system of black hole plus box is an ordinary quantum system with a discrete energy spectrum $\widehat{H}\,|i\rangle = E_i\,|i\rangle$. This puts us in the realm to which the Poincaré recurrence theorem applies and so cannot be compatible with exponentially decaying correlation functions. To be more precise (here I loosely follow Harlow [2016, sec. 6.9])

$$C(t) = \sum_{i,j} \frac{\mathrm{e}^{-\beta E_i}}{Z(\beta)} |\langle i|\, O\, |j\rangle|^2 \mathrm{e}^{-it(E_i - E_j)}. \qquad (10.5)$$

If we decompose $\widehat{O}$ into diagonal and off-diagonal parts, $\widehat{O} = \widehat{O}_D + \widehat{R}$, this is:

$$C(t) = C_D + C_R(t) = \langle \widehat{O}_D^2 \rangle_\rho + \sum_{i\neq j} \frac{\mathrm{e}^{-\beta E_i}}{Z(\beta)} |\langle i|\, R\, |j\rangle|^2 \mathrm{e}^{-it(E_i - E_j)}. \qquad (10.6)$$

(To even give the correlation a chance to drop off exponentially, we should therefore choose an observable with vanishing diagonal part.) The sums here are over $\sim e^S$ states, where S is the entropy of the ensemble at temperature β, and $Z(\beta) \sim \mathrm{e}^{S-\beta E_0}$, where E_0 is the expected ensemble energy. So for this sum to be convergent at small times we can expect $|\langle i|\, R\, |j\rangle|^2$ typically to be $\sim \mathrm{e}^{-2S}$. For large times, the phase

---

[3] I am grateful to Gordon Belot for useful discussion on this point.

factors $e^{-it(E_i - E_j)}$ can effectively be treated as random, so that $C_R(t)$, very roughly, is a sum of e^{2S} terms of magnitude $\sim e^{-2S}$ and random phase. The theory of random walks predicts that this sum ought to be:

$$C_R(t) \sim e^{-2S} \times \sqrt{e^{2S}} = e^{-S}. \tag{10.7}$$

So if black hole statistical mechanics is true, the correlation factor can initially drop off exponentially, but only until it reaches a value of $\sim e^{-S}$. It will then remain at about that level for a very long time, will occasionally fluctuate back to large values, and eventually, by the quantum version of the Poincaré recurrence theorem (see Wallace [2015] for a review), will return to its original large value. All this is of course in flat contradiction with the QFT prediction of permanent exponential decay. (The sharp version of the rather heuristic assumptions about $\widehat{O}$ that I have used in this argument is the *eigenstate thermalization hypothesis* Srednicki [1994].)

10.3.4 The Strength of the Page Time Paradox

According to one definition (Quine, 1966, ch. 1), a paradox is an apparently impeccable argument to an impossible conclusion—such as a pair of apparently impeccable arguments whose conclusions contradict each other. By this definition, the Page time version of the information loss paradox is a true paradox. The arguments for black hole statistical mechanics are compelling: quite apart from the general reason to expect a statistical-mechanical underpinning for any thermodynamic system, we have the precise reproduction of the entropy formula (including subleading corrections and renormalization effects) in low-energy quantum gravity, the equally precise reproduction of that formula in a large class of extremal black holes using string-theoretic methods, and the recovery of large parts of black hole thermodynamics, qualitatively and quantitatively, from the statistical mechanics of conformal field theory and the AdS/CFT duality. It is unserious to suppose that all of this is simply coincidence. And yet, the arguments from QFT are equally compelling.

Prima facie, there are only two ways forward:

1. Accept that QFT fails as a description of the entire spacetime of an evaporating black hole, retain the statistical-mechanical underpinnings of black hole thermodynamics, and try to understand why and when the QFT description breaks down, given that the breakdown occurs in regimes that prima facie seem 'nice' and well within the applicability domain of the theory (Polchinski, 1995, Mathur, 2009).

2. Retain QFT, reject black hole statistical mechanics, and find some non-statistical-mechanical understanding of black hole thermodynamics that nonetheless makes the compatibility of black hole and ordinary thermodynamics,

and the quantitative results of various statistical-mechanical calculations of black hole entropy, non-miraculous.

Since neither is especially attractive, it's tempting to look for a middle way: to get some understanding of black hole thermodynamics that holds onto its statistical-mechanical underpinnings but permits non-unitary decay, and/or to find a way to preserve at least most of the QFT description of the evaporation process even while allowing for long-time entanglement between Hawking quanta (violating [4], for the black hole at equilibrium). But results over the last 20 years have sharpened the paradox and virtually foreclosed on the possibility of a middle way, as AdS/CFT duality has provided direct mathematical evidence for duality and as the firewall paradox has made stronger and more explicit the violation of QFT in the statistical-mechanical description.

10.4 Evidence for Unitarity from AdS/CFT

AdS/CFT duality, briefly mentioned in Section 10.2, has become a growth industry in theoretical physics since its initial discovery, and even a cursory discussion of its structure and the evidence for it would double the length of this paper. (I give a brief overview from the point of view of black hole physics in Wallace [2019]; more detailed reviews include Aharony et al. [1999], Harlow [2016], Hartman [2015], and Kaplan [2016]; for philosophical discussion see, e.g., De Haro, Mayerson, and Butterfield [2016] or Teh [2013].) Here I take the existence of the duality mostly for granted and simply discuss its implications for the unitarity of black hole decay. Specifically, in Sections 10.4.1 and 10.4.2 I reprise the two 'nonevaporating' forms of the paradox discussed in Section 10.3.3; in Section 10.4.3 I give a direct argument from AdS/CFT and from Poincaré recurrence that full evaporation is a unitary process.

10.4.1 Unitarity of Hawking Radiation from Near-Extremal Black Holes

As I noted in Section 10.2.3, the statistical-mechanical entropy of certain extremal black holes, calculated via string theory, exactly matches the predictions of black hole thermodynamics. If such a black hole absorbs a small amount of uncharged mass, it will be perturbed away from equilibrium and acquire a temperature, and the leading-order change in entropy in this process again matches the thermodynamic prediction (Horowitz and Strominger, 1996).

More strikingly, and more relevantly for our purposes, Maldacena and Strominger (1997) were able to reproduce the exact *spectrum* of the perturbed black hole's Hawking radiation. To expand: recall (Section 10.2.1) that a radiating black hole is not a black body when observed far from the horizon; radiation emitted in the near-horizon regime has some probability to be scattered back across the horizon, and that probability depends on the radiation's angular momentum and its energy. These *gray-body factors* cannot be calculated analytically for the Schwarzschild black hole, but they can be for the near-extremal black hole analyzed by Maldacena and Strominger, giving the emission rate for photons of energy ω as a—reasonably complicated—function of ω and the parameters characterizing the black hole. Maldacena and Strominger found this function and then found the equivalent function for the decay rate of the excited string state corresponding (in Strominger and Vafa's [1996] analysis) to the black hole, through string perturbation theory. They match exactly, despite the completely different calculational tools applied in the two cases (solving the wave equation on the black hole spacetime in one case, calculating emission rates for a dilute gas of string excitations in the other).

This provides pretty powerful support for the hypothesis that black hole decay is a unitary process. To quote Maldacena and Strominger (emphasis in original),

The string decay rates, extrapolated to the large black hole region, agree precisely with the semiclassical Hawking decay rates in a wide variety of circumstances. However, the string method not only supplies the decay *rates*, but it also gives a set of unitary *amplitudes* underlying the rates. We find it tempting to conclude that these extrapolated amplitudes are also correct. It is hard to imagine a mechanism which corrects the amplitudes, but somehow conspires to leave the rates unchanged.

The robust nature of the string picture is very significant because it allows us to directly confront the black hole information puzzle ... According to Hawking information is lost as a large excited black hole decays to extremality. On the other hand the string analysis ... gives a manifestly unitary answer.

In retrospect, the Maldacena–Strominger calculation can be recognized as a form of AdS/CFT correspondence, realizing the duality between the near-horizon region just outside a near-extremal black hole (which can be approximated as $AdS_3 \times K$ for compact K) and 2-dimensional conformal field theory on the boundary of that spacetime. In either case the system is coupled to degrees of freedom describing the far-from-horizon region; the CFT description of that coupling is manifestly unitary and exactly reproduces the AdS results for overall scattering levels. See Hartman (2015, pp. 114–130) for a presentation that makes the AdS/CFT aspect explicit. (And note that as such, the calculation relies only on AdS/CFT duality and not on any specific features of string theory.)

10.4.2 Unitarity of Hawking Radiation for Large Black Holes

Radiation from near-extremal black holes is one of the two no-evaporation forms of the information loss paradox I discussed in Section 10.3.3. The other—Maldacena's eternal black hole—can also be analyzed via AdS/CFT methods. Recall that for a black hole in a box, large enough to be in stable equilibrium with its atmosphere, QFT calculations of Hawking radiation contradict black hole unitarity for sufficiently large times: the former predict that long-time correlation functions decay exponentially without limit, while unitarity (in a finite box) predicts that the correlations reach a minimum value of $\sim e^{-S}$, and that they eventually undergo Poincaré recurrence back to their original values. Since a large AdS black hole is a 'black hole in a box, in equilibrium with a radiation bath', we can use AdS/CFT duality to work out the correlation functions (the operators whose correlations are calculated are well outside the horizon, in a region where the translation between AdS and CFT descriptions is fairly well understood). We get these results:

1. For comparatively short times, the exponential decay of the correlations can be recovered on the CFT side of the correspondence (Papadodimas and Raju, 2013).
2. For very long times, the discreteness of the spectrum of the CFT Hamiltonian guarantees that the exponential decay ceases, and that the correlation coefficients display the behavior predicted by unitarity, in contradiction with the predictions of QFT applied to the interior (Maldacena, 2003).

So the AdS/CFT correspondence, applied to large black holes, provides further 'compelling evidence' (Harlow, 2016, p. 92) that black hole decay is unitary.

10.4.3 Unitarity of Black Hole Decay via AdS/CFT

In the case of black holes in AdS space that evaporate completely, it is tempting to conclude immediately that that decay is unitary: after all, the AdS description is dual to a CFT description that is manifestly unitary. But this has been challenged as too quick (cf. Unruh and Wald [2017] and Maudlin [2017]), so here I give a more explicit argument.

Specifically, suppose we have a QG theory on asymptotically AdS space (with the usual reflecting boundary conditions normally assumed in AdS/CFT duality), and a CFT on the boundary of that space, with Hilbert spaces $\mathcal{H}_G$, $\mathcal{H}_{CFT}$, respectively. We choose a foliation Σ_t of AdS compatible with the time translation symmetry map on AdS (so that the boundary of that foliation defines a foliation for the CFT, and write $\widehat{H}$ for the Hamiltonian of the CFT with respect to that foliation. Then we make the following assumptions (working in the Heisenberg picture):

A. Perturbative QG sector: For any time t and any state of the interior describable semiclassically as an excitation ψ of the vacuum (by, say, gravitons or matter particles) on Σ_t, which nowhere is dense enough to form an event horizon, there is a state $|\psi; t\rangle \in \mathcal{H}_G$ which represents that state in the full quantum-gravity theory. Call the subspace spanned by such states $\mathcal{H}_{G,P} \subset \mathcal{H}_G$.

B. CFT spectrum: The spectrum of $\widehat{H}$ is discrete and bounded below, and $\widehat{H}$ is at most finitely degenerate.

C. Perturbative duality: There is a unitary map $\widehat{V}$ from $\mathcal{H}_{G,P}$ into $\mathcal{H}_{CFT}$, such that:

$$\langle \psi;t|\psi';t+\tau\rangle = \langle \psi;t|\,\widehat{V}^{\dagger}\exp(-i\tau\widehat{H})\widehat{V}\,|\psi;t\rangle, \qquad (10.8)$$

that is, the map between the interior and boundary theories commutes with the dynamics in at least the perturbative sector.

(A) is a fairly minimal requirement of representational adequacy for the quantum-gravity theory; (B) is a standard result about conformal field theories on compact spaces (see, e.g., the discussion in Harlow [2016]); and (C) is a small fragment of full AdS/CFT duality. (The map $[\widehat{V}]$ in [C] can be constructed fairly explicitly, at least to first order in perturbation theory; see Harlow and Stanford [2011].)

Now let ψ be an excitation of the AdS vacuum corresponding to a large amount of diffuse infalling matter at time t which will at a later time (with very high amplitude) form a black hole. Given (A), there is a state $|\psi,t\rangle$ that represents this excitation. Given (B), the quantum version of the Poincaré recurrence theorem applies to the boundary CFT (see Wallace [2015] for a review) and so for any perturbative state $|\psi,t\rangle$ we can find a time T such that to an arbitrarily good approximation,

$$\exp(-iT\widehat{H})\widehat{V}\,|\psi,t\rangle \simeq \widehat{V}\,|\psi',t\rangle. \qquad (10.9)$$

Given (C), $|\psi,t\rangle \simeq |\psi,t+T\rangle$; in other words, after the recurrence time, the bulk theory will describe a multiparticle state containing all the information of the pre-black-hole state, and no information has been lost. Of course, T is vastly longer than the black hole's decay time, but if the evolution is unitary all the way forward to T, in particular it is unitary during the decay process.

10.5 The Firewall Paradox

AdS/CFT duality seems to have persuaded most of the high-energy physics community (notably including Hawking [2005]) *that* black hole evaporation is unitary. But the duality remains poorly understood as far as the black hole interior is concerned, and the question remains: *how* is it unitary, in the face of the clear arguments from QFT for information loss. In the last few years this question has become more urgent, with deep problems emerging in the hitherto leading strategy for reconciling the exterior and interior descriptions. Section 10.5.1 reviews that strategy, normally called 'black hole complementarity'; Section 10.5.2 reviews the so-called firewall paradox that appears to invalidate it.

10.5.1 Black Hole Complementarity and the Black Hole Interior

At first sight, there is a conflict between black hole statistical mechanics and QFT that is much more direct than the information loss paradox. After all, QFT predicts that an observer will encounter nothing special as they fall freely across the horizon from a starting point high above the black hole, and in particular will measure radiation that deviates only weakly from empty space. This seems hard to reconcile with the description of that same infalling observer that will be given by a fiducial (i.e., hovering) observer: for that second observer, whose observations are confined to the outside-horizon region, the quantum black hole is represented (cf. discussion in Section 10.2.3) by a membrane just above the horizon, whose local temperature is order the Planck temperature, and the infalling observer will collide with that membrane and rapidly be thermalized by it.

The *semiclassical* membrane paradigm demonstrates that this is far too quick. In that paradigm (recall) a quantity of charge dropped onto a black hole will spread out uniformly over the horizon, with a known timescale, generating heat as it does so through Ohmic dissipation. This description of stretched-horizon physics is formally compatible with—indeed, is derived from—an underlying physics in which the charge drops smoothly through the horizon and continues toward the singularity. The point is simply that there is a mathematically valid description of the physics of the black hole exterior in terms of the stretched horizon; the metaphysical question of whether that description is *true* is interesting but somewhat tangential.

Susskind, Thorlacius, and Uglum (1993) proposed extending this idea—that interior information is encoded in information about surface perturbations—from semiclassical physics to the full quantum theory of the black hole. Just as in the classical case, the proposal is that the selfsame physical process can be described in terms of coordinates adapted to the interior of the black hole or in terms of the degrees of freedom of a membrane just outside the event horizon. A sketch of this idea for the infalling observer would go as follows. The fall of the observer from infinity onto the stretched horizon can be described equally well with respect to stationary observers hovering above the horizon at a fixed height and with respect to inertially falling observers (and that description is analyzable in terms of known physics; cf. Unruh and Wald [1982]). Since the inertial description tells us that the observer does not catch fire and burn up during this part of their journey, the stationary description must give the same result, so that the observer's passage through the black hole atmosphere is uneventful despite its increasingly high temperature. (The intuitive feeling that this can't happen physically can be assuaged by noting that the observer passes through the hotter part of the atmosphere at extreme relativistic speed and so interacts with the atmosphere for a very short proper time; note that an observer whose fall begins from quite close to the horizon and so takes a much longer proper time to fall in will encounter highly blue-shifted Hawking radiation right from the start of the fall, even in an inertial-frame description and so will be consistently described as being burned up in both descriptions.)

When the observer reaches the stretched horizon, according to black hole statistical mechanics, they will rapidly become thermalized (specifically, thermalization takes time $\sim M \log M$, as can be read off the thermodynamics of the membrane paradigm). But 'thermalization' is a coarse-grained notion: it means that *for any observable relevant to the exterior physics*, the expectation value of that observable is the same when calculated with the 'thermalized' state as with the true microcanonical-equilibrium state (the projector onto the energy eigensubspace of the black hole Hilbert space). This is perfectly compatible with observables relevant to the black hole *interior* having expectation values that deviate sharply from the values calculated from the microcanonical-equilibrium state, and indeed that describe the infalling observer in accordance with general relativity.

Susskind et al. (1993) called this duality of surface and interior descriptions *black hole complementarity*. The name invokes the noncommutativity of quantum-mechanical observables: just as the same physical process can be described with respect to a basis of definite position or definite momentum states and may look very different in the two descriptions, so the horizon-crossing process can be described with respect to a basis appropriate to exterior physics or one appropriate to the infalling observer's situation. (Regrettably, 'complementarity' *also* invokes Bohr's somewhat obscure philosophy of meaning—and Susskind et al. explicitly refer to that philosophy—but so far as I can see it is not essentially required in black hole statistical mechanics.)

However, the parallel with semiclassical complementarity is imperfect, precisely because of the (Page time) information loss paradox: after the Page time, QFT continues to predict exactly thermal radiation, whereas black hole statistical mechanics requires that radiation to be entangled with early-time radiation, and so if black hole statistical mechanics is correct, then QFT must actually be *wrong*, not just a redescription of the same physics, at late times. As we shall now see, the firewall paradox of Almheiri et al. [2013], based on earlier ideas by Mathur [2009], strongly suggests that this failure of QFT is not simply some kind of long-range effect but completely invalidates the semiclassical description of the black hole interior.

10.5.2 Firewalls

The firewall paradox can be stated in various more or less precise ways (see, e.g., Bousso [2013], Harlow [2016, sec. 7], Polchinski [2016, sec. 6], and Susskind [2012], as well as the original sources noted earlier), but in essence it works like this. Consider some photon mode B (more accurately, a wavepacket concentrated on some such mode) describing photons emitted well after the Page time. That mode will be in a thermal state at the appropriate time black hole temperature. We now have two apparently contradictory claims:

1. According to a QFT description, B is fully entangled with some mode $\tilde{B}$ just inside the event horizon.
2. According to black hole statistical mechanics, B is fully entangled with the early-time radiation emitted by the black hole.

No quantum state can be fully entangled with two different systems (sometimes called 'monogamy of entanglement'), so this seems close to a contradiction. We might hope to finesse this by remembering the idea of complementarity—that the same underlying physics can be described in radically different ways, so that the system has one valid description where B is entangled with an interior mode, and one where it is entangled with early-time radiation. But as Almheiri et al. point out, in principle (though not in practice; cf. Harlow and Hayden [2013]) an observer could

1. Collect all the radiation emitted from the black hole up to the Page time;
2. Carry out a complicated operation on that radiation to distil a single photon mode C that is fully entangled with B;
3. Linger close to the event horizon, and perform a joint measurement on B and C (more precisely, on many such pairs $B_1, C_1 \ldots B_N, C_N$) to verify their entanglement;
4. Jump into the black hole.

Assuming that the observer's own local physics can be consistently described by quantum mechanics , they can't consistently find B to be entangled with $\tilde{B}$; indeed, $B + \tilde{B}$ must be in a product state. This directly contradicts the QFT assumptions underpinning Hawking radiation, and so undermines the basis of at least some of the arguments for Hawking radiation that got black hole statistical mechanics going in the first place. (Anecdotally it seems to be a matter of dispute in the physics community to what extent the derivation of Hawking radiation is undermined by the firewall argument.) More dramatically, complete disentanglement of QFT modes across the event horizon corresponds to a Planck-scale wall of energy at the horizon—the 'firewall'—that seems physically inexplicable and quite at odds with the general-relativistic idea of an event horizon as a globally defined, locally inaccessible phenomenon.

The firewall paradox is only five years old at time of writing, and it would be premature to try to summarize, far less assess, the wide variety of different responses that have been offered (see Harlow [2016, sec. 8] for a partial review). But it has roiled the community, disrupted what had been a fairly solid consensus in favor of black hole complementarity, and thrown the theory of the black hole interior wide open.

10.6 Conclusion

Thus we find ourselves in the enviable situation of having an interesting problem with no really satisfying answer; if we are lucky this means that we will learn something deep.

Daniel Harlow[4]

The black hole information paradox, understood in its most powerful form, is a clash between the unitary description of Hawking radiation implied by statistical-mechanical models of the black hole horizon, and the non-unitary description given by quantum field theory. It is neither a foolish failure by physicists to appreciate the subtleties of non-globally hyperbolic spacetimes nor something harmlessly resolved by ADS/CFT duality (the latter, at most, gives us reason to expect the ultimate resolution to be unitary). It is a deep puzzle arising from enormously plausible yet apparently contradictory lines of reasoning within quantum gravity, and at present it is completely opaque how it is to be resolved.

This is a good thing. There are very few obvious empirical clues as to the nature of quantum gravity; in their place, the best we have are the highly demanding and often unexpected consistency constraints given by the internal structure of lower-energy physics. We have learned much about the form of any satisfactory quantum theory of gravity by trying to satisfy those constraints; as and when we find a satisfactory resolution to the information loss paradox, we will have learned still more.

Acknowledgments

I am grateful to Gordon Belot, Sean Carroll, Erik Curiel, Eleanor Knox, James Ladyman, and Tim Maudlin for useful discussions, and in particular to Jeremy Butterfield for detailed and careful comments on a draft version. I'm also grateful to three anonymous referees for careful and helpful comments.

---

[4] Harlow (2016, p. 117).

References

Aharony, O., Gubser, S., Maldacena, J., Ooguri, H. and Oz, Y. (1999). Large N field theories, string theory and gravity. *Physics Reports 323*, 183–286.

Aharony, O., Marsano, J., Minwalla, S., Papadodimas, K. and Van Raamsdonk, M. (2004). The Hagedorn/deconfinement phase transition in weakly coupled large N gauge theories. *Advances in Theoretical and Mathematical Physics 8*, 603–696.

Almheiri, A., Marolf, D., Polchinski, J. and Sully, J. (2013). Black holes: Complementarity or firewalls? *Journal of High Energy Physics 2013*, 62.

Banks, T., Susskind, L. and Peskin, M. E. (1984). Difficulties for the evolution of pure states into mixed states. *Nuclear Physics B 244*, 125–134.

Bardeen, J., Carter, B. and Hawking, S. (1973). The four laws of black hole mechanics. *Communications in Mathematical Physics 31*, 161–170.

Belot, G., Earman, J. and Ruetsche, L. (1999). The Hawking information loss paradox: The anatomy of a controversy. *British Journal for the Philosophy of Science 50*, 189–229.

Bousso, R. (2013). Firewalls from double purity. *Physical Review D 88*, 084035.

Carter, B. (1979). The general theory of the mechanical, electromagnetic and thermodynamic properties of black holes. Pages 294–369 of: S. Hawking and W. Israel (eds.). *General Relativity: An Einstein Centenary Survey*. Cambridge: Cambridge University Press.

Christodolou, D. and Ruffini, R. (1971). Reversible transformations of a charged black hole. *Physical Review D 4*, 3552–3555.

De Haro, S., Mayerson, D. R. and Butterfield, J. N. (2016). Conceptual aspects of gauge/gravity duality. https://arXiv.org/abs/1509.09231.

Deutsch, D. (1991). Quantum mechanics near closed timelike lines. *Physical Review D 44*(10), 3197–3217.

Gibbons, G. and Hawking, S. (1977). Action integrals and partition functions in quantum gravity. *Physical Review D 15*, 2752–2756.

Gubser, S., Klebanov, I. and Polyakov, A. (1998). Gauge theory correlators from non-critical string theory. *Physics Letters B428*, 105–114.

Harlow, D. (2016). Jerusalem lectures on black holes and quantum information. *Reviews of Modern Physics 88*, 015002.

Harlow, D. and Hayden, P. (2013). Quantum computation vs. firewalls. *Journal of High Energy Physics 2013*, 85.

Harlow, D. and Stanford, D. (2011). Operator dictionaries and wave functions in AdS/CFT and dS/CFT. https://arXiv.org/abs/1104.2621.

Hartman, T. (2015). Lectures on quantum gravity and black holes. www.hartmanhep.net/topics2015/gravity-lectures.pdf.

Hawking, S. (1972). Black holes in general relativity. *Communications in Mathematical Physics 25*, 152–166.

Hawking, S. (1975). Particle creation by black holes. *Communications in Mathematical Physics 43*, 199.

Hawking, S. (1976). Breakdown of predictability in gravitational collapse. *Physical Review D 14*, 2460–2473.

Hawking, S. W. (1996). Virtual black holes. *Physical Review D 53*, 3099–3107.

Hawking, S. (2005). Information loss in black holes. *Physical Review D 72*, 084013.

Horowitz, G. and Strominger, A. (1996). Counting states of near-extremal black holes. *Physical Review Letters 77*, 2368–2371.

Kaplan, J. (2016). Lectures on AdS/CFT from the bottom up. http://sites.krieger.jhu.edu/jared-kaplan/files/2016/05/AdSCFTCourseNotesCurrentPublic.pdf.

Maldacena, J. M. (1998). The large N limit of superconformal field theories and supergravity. *Advances in Theoretical and Mathematical Physics 2*, 231–232.

Maldacena, J. M. (2003). Eternal black holes in AdS. *Journal of High Energy Physics 0304*, 021.

Maldacena, J. and Strominger, A. (1997). Black hole greybody factors and D-brane spectroscopy. *Physical Review D 55*, 861–870.

Mathur, S. D. (2009). The information paradox: A pedagogical introduction. *Classical and Quantum Gravity 26*, 224001.

Maudlin, T. (2017). (Information) Paradox lost. https://arXiv.org/abs/1705.03541.

Page, D. M. (1993). Information in black hole radiation. *Physical Review Letters 71*, 3743–3746.

Papadodimas, K. and Raju, S. (2013). An infalling observer in AdS/CFT. *Journal of High Energy Physics 1310*, 212.

Penrose, R. (2004). *The Road to Reality: A Complete Guide to the Laws of the Universe*. London: Jonathon Cape.

Perez, A. (2017). Black holes in loop quantum gravity. *Reports on Progress in Physics 80*, 126901.

Polchinski, J. (1995). String theory and black hole complementarity. https://arXiv.org/abs/hep-th/9507094.

Polchinski, J. (2016). The black hole information problem. https://arXiv.org/abs/1609.04036.

Quine, W. (1966). *The Ways of Paradox*. Cambridge, MA: Harvard University Press.

Srednicki, M. (1993). Is purity eternal? *Nuclear Physics B 410*, 143–154.

Srednicki, M. (1994). Chaos and quantum thermalization. *Physical Review E 50*, 888.

Strominger, A. (1998). Black hole entropy from near-horizon microstates. *Journal of High Energy Physics 9802*, 009.

Strominger, A. and Vafa, C. (1996). Microscopic origin of the Bekenstein-Hawking entropy. *Physics Letters B379*, 99–104.

Susskind, L. (2012). Singularities, firewalls, and complementarity. https://arXiv.org/abs/1208.3445.

Susskind, L., Thorlacius, L. and Uglum, J. (1993). The stretched horizon and black hole complementarity. *Physical Review D 48*, 3743–3761.

Teh, N. J. (2013). Holography and emergence. *Studies in History and Philosophy of Modern Physics 44*, 300–311.

Thorne, K. S., Price, R. H. and Macdonald, D. A. (eds.). (1986). *Black Holes: The Membrane Paradigm*. New Haven, CT: Yale University Press.

Unruh, W. G. (2012). Decoherence without dissipation. *Philosophical Transactions of the Royal Society of London A 370*, 4454–4459.

Unruh, W. G. and Wald, R. M. (1982). Acceleration radiation and the generalized second law of thermodynamics. *Physical Review D 25*, 942–958.

Unruh, W. G. and Wald, R. M. (1995). On evolution laws taking pure states to mixed states in quantum field theory. *Physical Review D 52*, 2176–2182.

Unruh, W. G. and Wald, R. M. (2017). Information loss. https://arXiv.org/abs/1703.02140.

Wald, R. (1993). Black hole entropy is Noether charge. *Physical Review D 48*, 3427–3431.

Wald, R. M. (1994). *Quantum Field Theory in Curved Spacetime and Black Hole Thermodynamics*. Chicago: University of Chicago Press.

Wallace, D. (2015). Recurrence theorems: A unified account. *Journal of Mathematical Physics 56*, 022105.

Wallace, D. (2018). The case for black hole thermodynamics, part I: Phenomenological thermodynamics. *Studies in History and Philosophy of Modern Physics 64*, 52–57.

Wallace, D. (2019). The case for black hole thermodynamics, part II: Statistical mechanics. *Studies in History and Philosophy of Modern Physics 66*, 103–117.

Witten, E. (1998a). Anti de Sitter space and holography. *Advances in Theoretical and Mathematical Physics 2*, 253–291.

Witten, E. (1998b). Anti-de Sitter space, thermal phase transition, and confinement in gauge theories. *Advances in Theoretical and Mathematical Physics 2*, 505–532.

York, J. W. (1986). Black-hole thermodynamics and the Euclidean Einstein action. *Physical Review D 33*, 2092–2099.

11

Chronic Incompleteness, Final Theory Claims, and the Lack of Free Parameters in String Theory

11.1 Introduction

The present chapter aims at understanding the relation between two remarkable characteristics of string physics. On the one hand, the theory's position within the fabric of physical reasoning as well as some characteristics of the theory itself suggest that string theory, or the theory it would have morphed into once fully developed, might represent a *final theory*: no empirical data can be gathered in this world that would be at variance with the theory's predictions. On the other hand, half a century of intense work on string theory have not resulted in a fully fledged theory. At this point, the prospects that this will happen in the foreseeable future seem, if anything, less promising than at earlier stages of the theory's evolution.

Each of the described characteristics of string physics raises distinct and substantial philosophical questions. A final theory claim raises the question: what would the research process in fundamental physics look like once a universal final theory has been found? A natural question regarding the chronic incompleteness of string theory is: why is it so difficult to develop string theory into a fully fledged theory? The two questions might actually address two sides of the same medal. They could then be posed in the following more specific way. Are the seemingly insurmountable difficulties to turn string theory into a complete theory conceptually related to its character as a (potential) final theory? Should we therefore understand chronic incompleteness as a core characteristic of a final physical theory?

Dawid (2006, 2013a, 2013b) has suggested that string theory may represent a fundamentally new stage in physical theory building, where theory succession is replaced as the driving principle of conceptual development by the open-ended evolution of the conceptual understanding of a final theory that lacks a time frame for completion. If so, the advent of the final theory prospect would have changed the mode of scientific reasoning without abandoning the 'infinite' time horizon associated with the completion of fundamental physics.

This suggestion was based on two simple observations. (1) Chronic incompleteness and the final theory claim have arisen in parallel in string physics. (2) They seem to reflect generic features of quantum gravity. Without any specific reference to string theory, the current perspective on quantum gravity may be taken to suggest that the notion of an ongoing succession of theories for ever-smaller characteristic distance scales will break down at the Planck scale (see, e.g., 't Hooft [1993]). Likewise, the chronic problems of attempts to come up with a complete theory of quantum gravity are not confined to the string theory research program.

Dawid (2006, 2013a, 2013b) has not put forward any substantial claim, however, as to how the connection between the final theory status and the chronic incompleteness of string theory can be understood at a conceptual level. It is the aim of the present chapter to address this issue and consider some lines of reasoning that may throw more light on the conceptual connection between final theory status and chronic incompleteness. First, I will analyze the ways in which chronic incompleteness may affect our understanding of what a final theory actually amounts to. Second, a highly tentative idea shall be considered as to how finality may be conceptually linked to what seems to be the predicament of chronic incompleteness faced by physicists working on a universal theory.

What follows is of an explorative nature. It will not present strong conclusions but just test possible ways of thinking about the issues discussed. I do feel, however, that those issues are so crucial for understanding the current state of fundamental physics that they should be addressed also from a philosophical perspective, and be it with the highly insufficient means available to me.

Section 11.2 lists several ways in which string theory differs from previous stages in fundamental physics. The issues of finality and chronic incompleteness are then discussed specifically in Sections 11.3 and 11.4, and related to each other in Sections 11.5 and 11.6. Section 11.7 discusses the role of dualities in the given context, followed by a brief look at the case of M-theory in Section 11.8.

11.2 Five Ways in Which String Physics Differs from Previous Stages in Fundamental Physics

In the nineteenth century, when physical formalization found an ever-increasing scope of applications, a canonical understanding of the relation between the observed world and physical theory building emerged. This understanding was based on some core principles. Three of those principles that will be important to our discussion were:

I. A physical theory must be applied to a specified set of phenomena.
II. A theory can be adapted to observations by choosing values of dimensionless parameters that either are contained in the theory itself or specify the

embedding of the given theory within the physical background needed for understanding the measurements of the described set of phenomena.

III. Once one has identified a physical context that allows for a formalized account, theories (which might well be insufficient or only partly accurate, to be sure) can be developed (and most often also empirically tested) within a limited time frame.

When the basic principles of classical physics were toppled during the physical revolutions of the early decades of the twentieth century, this understanding of the role a physical theory was expected to play was complemented by two further principles:

IV. A physical theory must be expected to have a limited lifetime as a fundamental theory and to be eventually superseded by a new theory in regard to which it then plays the role of an effective theory.

V. Advanced physical theories contradict intuitions that are based on everyday world experience and contain parameters that control the deviations from those intuitions. However, theories that are at variance with intuitions make predictions in line with those intuitions in some limits: special relativity converges toward Galilean physics for velocities very small compared to the speed of light, general relativity approaches predictions of Newtonian gravity in contexts of very small spacetime curvature, and quantum mechanics is consistent with the classical dynamics of objects who interact with the environment and whose action is much larger than $\hbar$.

It seems fair to say that principles I–V shape many physicists' expectations regarding the role of physical theory building up to this day. While these principles still work very well in many fields of physics, they came under pressure in fundamental physics once fundamental physics developed a focus on contemplating theories of quantum gravity. This pressure can be felt today whichever approach toward dealing with the problem of quantum gravity is chosen. The present text focuses on string theory, where the inadequacy of the traditional view on physical theory looks particularly far reaching. In fact, string theory seems flatly incompatible with four of the five principles formulated and makes an intriguing amendment to the fifth. As opposed to principles I–V, one may make the following statements about string theory:[1]

I′. String theory is a universal theory.

II′. String theory has no fundamental dimensionless free parameters (but seems to have a huge number of groundstates).

---

[1] Here as throughout the entire paper, the term *string theory*, if not specified otherwise, denotes the overall theory that aims at describing the observed world and is identified by the present knowledge on perturbative superstring theory, duality relations, etc.

III′. String theory is chronically incomplete (and lacks a promising perspective for quantitative empirical testing in the foreseeable future).

IV′. String theory makes a final theory claim based on a minimal length scale.

 V′. String theory has several classical limits related to each other by duality relations.

Given that the five described significant shifts in regard to the role of theory building arise in the same theory, it seems natural to suspect that they are somewhat related to each other. In the following, I aim to turn that suspicion into something slightly more concrete.

11.3 Finality

A plausible starting point for finding connections between the five statements on string theory is point IV′. T-duality implies that the string scale amounts to a minimal length scale in string theory. Any statement on a distance smaller than the string scale can be formulated, in the T-dual picture, as a statement about a distance larger than the string length l. This fact may be taken to indicate that string theory, if viable[2] up to its own characteristic scale, correctly represents physics at all scales and therefore constitutes a final theory (Witten, 1996).

A final theory claim that is extracted from a specific theory may be suspected to be begging the question. If string theory were viable only as an effective theory of a more fundamental theory, smaller distance scales could carry new information based on that more fundamental theory. If so, the more fundamental theory would break the final theory claim that had been developed based on what turned out to be its effective theory. A final theory claim based on the allegedly final theory's own implications, therefore, may seem to deflate to the trivial claim: if the theory is absolutely true, it is actually absolutely true. Dawid (2013a, 2013b) argues that, despite the described issue, final theory claims can retain argumentative power in conjunction with arguments of limitations to scientific underdetermination.

In the present discussion, we will therefore assume that the final theory claim stated in point IV′, though obviously not conclusive, does carry argumentative weight. If so, points I′ and II′ and IV′ form a remarkable set of characteristics of string theory that suggest finality. Universality (point I′) removes the need for theory succession on the path to further unification and thereby frames string theory as a plausible end point of theory succession. The lack of fundamental dimensionless

---

[2] Throughout this chapter, I talk about a theory's viability rather than its truth. A theory is viable in a given regime (e.g., up to a given energy scale), if it is in agreement with all the data than can be collected in that regime (e.g., up to that energy scale). Talking about a theory's truth raises difficult philosophical questions that are not relevant to the issues addressed in this chapter. A more extensive argument in favor of the concept of viability can be found in Dawid (2019).

free parameters (point II') removes the need for theory succession on the path toward explaining specific parameter values (since all effective parameter values must result from the dynamics of the full theory, where no parameter values can be chosen at will) and thereby reinforces the former claim. Finally, the explicit final theory claim (point IV') provides a conceptual reason within the theory's theoretical framework for assuming the finality of string theory. In conjunction, the three points form a consistent and mutually reinforcing set of arguments for string theory's status as a final theory. Viewing the issue the opposite way, points I', II', and IV' are per se highly unusual properties that may be expected to arise in conjunction in a final theory.

11.4 Chronic Incompleteness

If it were just for the three discussed new aspects of string theory, one would thus face the stunning perspective of an imminent end of fundamental physics: once the final theory would have been fully formulated, no new fundamental theories should be expected to emerge. Physics would be reduced to the menial work of developing theories in more specific low-energy contexts and understanding their reductive connections to string theory.

String theory comes with another substantial shift, however, that renders this understanding inadequate. Fifty years after the birth of string theory (Veneziano, 1968) and 44 years after string theory was first proposed as a universal theory of all interactions (Scherk and Schwarz, 1974), the theory still lacks a plausible perspective for a full formulation in the foreseeable future. The theory has known periods of considerable optimism. At least twice during its evolution, in the years after the consistency of a superstring action had been shown by Green and Schwarz (1984), and in the years after the discovery of the web of dualities (Hull and Townsend, 1995; Polchinski, 1995; Witten, 1995) and anti–de Sitter spacetime conformal field theory (AdS/CFT) correspondence (Maldacena, 1998), string theorists were hopeful that a full formulation of the theory would be attainable by developing further and making full use of the concepts and tools available at the time. But each time overwhelming obstacles to achieving that goal soon surfaced, as if they were put up by magic hands defending the ultimate secrets of physics. The problems associated with completing string theory in each case turned out to transcend what string theorists at the given point had taken them to be.

Today it seems clear that, even if the final theory claim regarding string theory were true, this fact would not translate into a prospect of the imminent end of fundamental physics. Rather than bringing the time horizon for the completion of fundamental physics from virtual infinity to somewhere within our lifetime, string theory's final theory claim seems to be associated with an extension of the time horizon for the completion of this particular theory that may, once again, virtually reach toward infinity.

11.5 Chronic Incompleteness and the Lack of Free Parameters

Why is it so immensely difficult to turn string theory into a complete theory? One important reason arguably has to do with one of the three points associated with string theory's finality claim: point II′, which states the lack of dimensionless free parameters at a fundamental level, creates a new kind of problem for a theory's completion.

In order to get a better grasp of this point, one needs to start with a look at perturbation theory.

In quantum field theory (QFT), which is understood to be the theoretical framework for calculating low-energy effective theories of string theory, calculations of scattering processes are based on a perturbative expansion in the coupling constant. Small coupling constants allow for increasingly accurate results of calculations up to a few orders in perturbation theory.

It is one important aspect of quantum field theory that the strong coupling regime is directly related to the deep quantum regime (see, e.g., Polchinskin [2017]). Higher orders in the coupling constants correspond to higher orders in $\hbar$. A weak coupling limit therefore corresponds to a small $\hbar$ limit and represents a physical situation where elementary objects can be localized fairly well. Physics then is in a near-classical regime.

Quantum mechanics frames measurements in terms of confronting a quantum system with a quasi-classical measuring apparatus. The entire setup of a quantum theory is therefore developed from the perspective of a classical limit.

As long as the theory is in a regime where the coupling constant is small, perturbation theory works. Once one aims to describe a strong coupling regime, perturbation theory breaks down and one needs to rely on nonperturbative methods and auxiliary techniques such as lattice theory. Such a situation corresponds to a fully quantum regime where elementary objects are not well localized.

In the standard model of particle physics, the electroweak sector lives in a weak coupling regime and therefore allows for perturbative calculations. QCD, at low energies, to the contrary, is strongly coupled and requires nonperturbative techniques.

String theory has been developed based on perturbation theory in the same mold as perturbation theory for QFT. It describes the propagation and scattering of oscillating quantum strings on a spatiotemporal background. It is clear, however, that this is an insufficient approach for spelling out the full theory. The spatiotemporal background is itself generated by stringy dynamics. And string theoretical scenarios may be placed in either strong or weak coupling regimes. In other words, it is clear that a full understanding of string theory must reach out beyond the perturbative regime.

Duality relations constitute the most powerful tools in a string theoretical context to reach out beyond the perturbative regime. The power of dualities was first clearly understood in the mid-1990s, when Witten (1995) conjectured that the five types of superstring theory plus a sixth enigmatic 11-dimensional M-theory were related to each other by S- and T-duality relations that turned them into one different representation of the same theory.

Exact dualities relate different theories or models to each other that are empirically equivalent.[3] The isomorphy between the dual models is established based on a 'translation manual' that indicates which parameters in one theory correspond to which parameters in the dual theory and how the values of those parameters in one theory are related to the values of the dual parameter in the dual model.

Duality transformations are particularly helpful for reaching out toward the nonperturbative regime because they tend to invert parameter values and thereby relate one theory in a deep quantum regime to another one in a near classical regime. S-duality relates a theory with a strong string coupling g to a theory with a weak string coupling $1/g$. T-duality relates a theory with small compactification radius r to a theory with a large compactification radius $1/r$. In supersting theory, different types of superstring theory are connected that way. For example, S-duality relates type I superstring theory with a given string coupling g to SO(32) heterotic string theory with string coupling $1/g$, which in turn has as its T-dual $E_8 \times Es_8$ heterotic string theory with an inverted compactified dimension. A strongly coupled type I string theory can, therefor, be represented as a weakly coupled SO(32) heterotic string theory. And if that theory contains a small compact dimension, it can be represented as a $E_8 \times E_8$ heterotic string theory with a large compactification radius. The web of dualities, therefore, extends the reach of string theory by establishing that large parts of what seemed to lie beyond the reach of perturbation theory is accessible by perturbation theory in a dual representation.

But string theory faces a problem that is substantially more serious than the problem faced by a strongly coupled gauge theory. The additional problem is related to the fact that string theory is a theory without fundamental dimensionless free parameters.

In physics, it is important to distinguish between dimensionful and dimensionless parameters.[4] The absolute value of a dimensionful parameter is a matter of

---

[3] The analysis of this chapter is based on the understanding that duality relations are exact. Recently, the possibility of an inexact notion of dualities has been emphasized in a philosophical context by De Haro (2019).

[4] For an instructive analysis of the role of free parameters in physics, see Duff (2015).

choosing a physical unit. Per se, it is irrelevant for fitting theory to experiment. The value of a dimensionless parameter, to the contrary, cannot be altered by choosing different physical units. It, therefore, tells something significant about our set of observations (if fixed based on empirical data) or about a theory's empirical implications (if extracted from the theory itself). Physical theories normally involve dimensionless free parameters that can be tuned in order to fit the quantitative specifics of the empirical evidence. In many cases, for example in quantum theories that involve dimensionless coupling constants, dimensionless free parameters are elements of the theory itself. In other cases, for example in Newtonian gravity, the theory's embedding within our observed world requires a comparison of a dimensionful constant (such as the gravitational constant) with constants that characterize the empirical setup used to test that theory. Free dimensionless parameters then arise in the context of this comparison. In both cases, the free parameters involved play a crucial role for connecting the theory to observation.

String theory is a fully universal theory that gives a joint description of all fundamental phenomena. Measuring a fully universal theory cannot depend on paramater values that are not represented in the theory itself. Dimensionless free parameters thus cannot be generated by embedding a fully universal theory in a wider framework. The issue of free dimensionless parameters is reduced to internal characteristics of the theory itself.

String theory has one dimensionful free parameter, the string length l. But the theory has no dimensionless free parameter. In the context of perturbative string theory, this fact can be understood by looking at the geometric structure of Feynman diagrams. While Feynman diagrams for pointlike particles have pointlike interaction nodes where dimensionless coupling constants can be inserted, the corresponding stringy Feynman diagram is represented by a 2-dimensional surface with nontrivial topology. No coupling constants can be inserted at any point in those diagrams. The string coupling enters as the value of an oscillation mode of the string, the dilaton.

The lack of dimensionless free parameters in string theory is crucial for our analysis because dimensionless free parameters play an important role in the specification of classical limits. Describing a near-classical limit in QFT corresponds to choosing a small value of a dimensionless coupling constant. Describing a small curvature scenario corresponds to choosing a small value of the ratio between the characteristic length scale of the dynamics described and the radius of spacetime curvature. In the case of theories that involve free parameters, placing the theory close to or far from a classical limit can be controlled by choosing a value of a free dimensionless parameter. In string theory, nothing of this sort can be done. The fundamental dynamics of string theory, therefore, cannot be exemplified by choosing a near-classical limit. Whether or not the theory has solutions that lie close to a classical limit is decided by the fundamental dynamics itself.

To be sure, parameter values such as the string coupling, the radii of compact dimensions or background curvature terms can be specified in effective models that emerge from the theory's fundamental dynamics and correspond to a ground state of the theory. Those effective parameter values can then be close to or far from a classical limit. Calculating the fundamental dynamics itself, however, cannot rely on specifying a parameter value that controls whether physics is in a near classical or in a deep quantum regime. Only once effective parameter values have been extracted as a solution of the fundamental dynamics, does the categorization in terms of near classical or deep quantum make sense. Curvature terms, compactification radii, or string coupling constants can be specified only in the ground state of the theory once the fundamental dynamics of the theory has already played out.

String theory thus finds itself in the following peculiar situation. It is defined, as it stands, at a perturbative level in terms of strings moving through background space. The core parameters that characterize the theory's state are spatiotemporal and allow for calculations in a near classical limit where interaction is weak, objects are localized, and curvature is low. The propagation of individual strings only amounts to a plausible intuitive representation of the dynamics in a near classical limit. But the fundamental theory, lacking dimensionless free parameters, cannot be related to any such limit before calculation.

Framing the theory in terms of parameters that are selected in order to work well in the near classical limit, therefore, looks like an awkward choice when it comes to describing and calculating string theory's fundamental dynamics. Still, all physical parameters deployed for describing string theory, or any other physical theory, are of that kind (with one possible exception we will return to a little later). These parameters seem capable of determining some contours of the fundamental theory with the help of duality arguments and consistency requirements. But they have not proved capable of providing a workable basis for calculating the theory's fundamental dynamics.

11.6 Finality in the Face of Chronic Incompleteness

Let us, for the moment, presume that string theory is a final theory in the sense that (1) there is a string theory ground state that fully represents phenomenology at energies below the Planck scale and (2) no observation that can be made in principle would lead to the rejection of string theory. If string theory is a final theory in this sense, the pair of facts stated at the end of the previous paragraph may have at least three possible explanations.

A. Even in principle, there exists no mathematical scheme that is empirically equivalent to string theory and generates quantitative results that specify the fundamental dynamics of the theory. In that case, the fundamental theory is conceptually incomplete by its very nature. It has no fundamental dynamics and no set of solutions that can be deduced from its first principles. The fundamental theory merely serves as a conceptual shell that embeds low-energy descriptions (ground states of the theory) consistent with the principles encoded in the fundamental theory. Those low-energy descriptions contain specified parameter values and do generate quantitative results. But there is no way to establish from first principles how probable specific ground states of the system are.

B. Mathematically, the fundamental theory does have well-defined solutions. There is a fact of the matter as to whether or not a low energy state constitutes a solution of the fundamental theory and as to how probable that state is. However, there exists no algorithm that can establish those quantitative results based on calculations that are substantially less complex than the full description of the entire dynamics of the universe. Therefore, those calculations will never be carried out by physicists.

C. There is a mathematical scheme that allows for calculating the dynamics of the fundamental theory with reasonable effort. Physicists just have not found it yet.

Let us look a little closer at those three possibilities.

Alternative A would imply a fundamental break with the way science has understood the role of mathematics in representing observations. In particular, it would raise fairly intricate philosophical questions regarding the conception of a real world and its relation to its mathematical description.

If string theory were indeed viable and A were true, the dynamics of the world could not even probabilistically be extracted from a formal fundamental theory. A fundamental theory is here understood to be constituted by a formal structure that is defined based on a finite number of mathematical posits. In this scenario, quantitative treatment and the extraction of quantitative predictions could be achieved at the level string theory groundstates. That level of description could not be understood as fundamental, however, since its consistent conceptualization relies on the representation of that dynamics in terms of groundstates of a fundamental theory. The corresponding theory could be inferred in its basic characteristics but could not even in principle be expanded into a conceptual scheme that allowed for a specification of the fundamental dynamics and the extraction of quantitative predictions.

A situation of that kind would allow for various philosophical interpretations. From a strictly empiricist point of view, it could simply be characterized in terms of limitations to the predictive power of scientific theory building. The peculiar status of the fundamental theory, which is at the same time implied by the requirement of having a consistent notion of string theory groundstates but also fundamentally incapable of being turned into a predictive scheme, is not reflected by a specific philosophical conceptualization from such an empiricist perspective.

Viewed from the perspective of scientific realism, to the contrary, alternative A raises fundamental new questions regarding the relation between reality and physical theory. Those questions can once again be addressed in a number of different ways.

First, one might conclude that there is a part of reality that just cannot be represented by mathematics. But how, if not based on mathematics, should one conceive of that part of reality? Given the full dominance of mathematical conceptualization and the radical dissolution of any intuition-based concepts at the most fundamental levels of physical description, prospects for developing a convincing conceptual basis for any such understanding seem very limited. If there is any philosophical message one may consistently draw from the evolution of physics throughout the twentieth century, it hints in exactly the opposite direction. The century started with the emergence of a general consensus that human intuitions regarding material objects and their movement through space could be deployed in the context of microphysics in order to develop a deeper—atomist—understanding of the world. The further the century progressed, however, the more conspicuous it became that each further conceptual step in the development of fundamental physics amounted to jettisoning core elements of that intuitive perspective. Most ideas on how to join general relativity and quantum physics seem to imply that, as a last step in this decay process of intuitive concepts in fundamental physics, space and time themselves need to be abandoned as fundamental concepts. Having lost all guidelines that had been based on human intuitions regarding physical objects, mathematics seemed the only remaining basis for adequately expressing the fundamental physical characteristics of the world. Under those circumstances, the suggestion to aim at some nonmathematical perspective on fundamental aspects of the world seems hopelessly backwards-leaning.

In this light, it might look more promising to view alternative A from a very different perspective. Reality, on that alternative reading, does not reach beyond what can be grasped based on mathematical formalization but, quite to the contrary, should be regarded as being confined to those levels of mathematical description that allow for a quantitative analysis. If so, there would be no reality corresponding to the level of fundamental principles for the very reason that those principles don't imply a fundamental dynamics. The fundamental theory would assume the peculiar role of providing the closure of physics that is needed for consistency reasons without representing the real world.

A perspective that avoids this irritating idea would be to insist on the principle that there is a direct correspondence between the mathematical structure of a true physical theory and the real world. Structural reality could then be attributed to the fundamental level. At that level, however, it would play out in a way very different from what we expect from a physical theory. Only at the level of effective theories that allow for a quantitatively specified dynamics would reality resemble a physical system in the sense we are accustomed to.

Whatever the philosophical interpretation however, alternative A amounts to introducing a strictly nonreductionist element into the fabric of physical theory building: neither at a conceptual nor at an ontological level would it be possible to reduce the phenomenology describable by our effective theory to the principles of the fundamental theory.

Alternative B has the same practical implications as A but remains closer to known philosophical territory. Mathematics could remain in place as a complete representation of the dynamics of the world at the fundamental level. Reductionism and full consistency could be upheld as general ontic principles of physics. There would be a set of fundamental equations that fully specify the state of the world (or, at any rate, the probability of the state we find ourselves in). Alternative B would just imply that no manageable reduction of complexity regarding the fundamental dynamics of the world can be achieved that leads to the extraction of parameter values of effective theories based on any method of approximation. As briefly mentioned, physics knows contexts that are sufficiently chaotic to block any serious attempt at calculating specific outcomes of the dynamics. Just think of the exact point of landing of a leaf carried away by autumn winds.[5] In the given case, one might suspect an even more desperate situation. Given the lack of free parameters and classical initial conditions, the case we face might not even allow for choosing simple initial conditions that do allow for calculations of the fundamental dynamics in a toy model.

This outcome would retain the traditional philosophical take on the role of mathematics in physical theorizing. But it would not provide a physical perspective for understanding the fundamental basis of the dynamics we observe at low energies.

Alternative C is what physicists are hoping for. Are there any good reasons for assuming the existence of a full-fledged and calculable fundamental theory of all interactions? One reason for feeling inclined to give a positive answer to this question may lie in the history of science. The conviction that physical phenomena can be described by a fully calculable theory even if they seem to defy any such description at first glance has been the main driving force in physics ever since the time of Galilei. The striking successes of physical theory building during four centuries bear witness to the fact that this conviction has often been vindicated.

---

[5] Some philosophers of science (see, e.g., Cartwright [1983]) have drawn far-reaching antireductionist conclusions from this fact.

Two caveats should be pointed at, however. First, as mentioned, not all physical problems are assumed to have a calculable solution. In this light, the conviction described earlier is in need of qualification. Scientists first make up their minds, based on their experience, as to whether or not a given physical problem can be expected to have a calculable solution. The physics community's track record of predicting what would eventually be calculable obviously is not flawless but seems fairly good on average. In light of this track record, the physicists' expectation that a given physical problem has a calculable answer can become very strong. Such strong expectations do constitute a main driving force behind their work.

A theory of quantum gravity, or string theory for that matter, at first glance looks like the kind of theory that should be possible to develop and eventually to calculate. But, and this brings the second caveat into play, string theory and the conceptual context within which it is developed is in a number of ways substantially different from anything physicists have witnessed up to this point. Therefore, it is far from clear whether prevalent physical intuitions as to which kinds of questions can be expected to have a fully calculable theoretical answer are applicable in this case. It seems difficult to rule out that what seems to be a question that finds a fully calculable theoretical answer in fact rather resembles the case of the leaf carried by autumn winds and just defies calculation. This would bring us back to alternative B.

One specific difference between string theory and QFT brings alternative A into play as well. In QFT, the perturbative approach is a tool for extracting quantitative predictions from the theory. But QFT theory can be formulated as a full-fledged theory (albeit not with the rigor algebraic quantum field theorists would want). In the case of string theory, the perturbative approach was all there was in the early days of the theory. No formulation of the theory was known that did not rely on the perturbative approach. Though it was expected that the perturbative formulation of the theory should in a cogent way designate a coherent full theory of strings, it was by no means clear that such a theory existed.

The conspicuous abundance of string dualities is by many taken to offer a physical reason to believe in the existence of a calculable fundamental theory. The conjecture of, let us say, an exact S-duality relation, implies that deeply nonperturbative large coupling solutions exist for each of the types of string theory that are related to each other by the duality. This may be read as a stepping stone toward a general fundamental formulation of string theory.

In the late 1990s there was much confidence among string theorists that dualities could be developed into a tool for the full calculation of string theory's dynamics.[6] These hopes have turned out to have been premature. Though dualities have led to many new insights into string theory, it is now generally acknowledged that further

---

[6] A good example of the optimistic spirit at the end of the twentieth century is Greene (1999).

substantially new methods would be needed in order to fulfil their early promises. Dualities, in this light, may still be viewed as an indicator that full calculability is a serious possibility but, at this point, cannot establish that a fully calculable theory exists.

Given the power but also the limitations of duality relations in regard to finding a fully calculable string theory, how should one understand the role of dualities in this search on a more general basis?

The next section analyzes this issue by looking at the various kinds of dualities that have been discovered so far in the context of string theory.

11.7 Types of Duality in String Physics

The extent to which parameters change from a model to its dual varies depending on the type of duality and the context within which it plays out. The narrowest instantiation of a duality relates parameter values to inverted values of the same parameter in the same theory. This form of duality is called self-duality. An example in string physics is the S-self duality of Type IIB superstring theory: type IIB string theory with a string coupling g is dual to itself with a string coupling $1/g$. Another example is the T-duality of Bosonic string theory: a Bosonic string theory with compactified radius R is dual to itself with compactified radius l^2/R (where l is the string length).

In those cases, observable characteristics of individual objects may appear as different characteristics in the dual description. For example, the winding number of a string corresponds to the transversal momentum of the string in the T-dual description.

Dualities can also relate values of a given parameter to the inverted values of that parameter in a different theory. We have already encountered the example of an S-duality that relates type I superstring theory with a given string coupling g to SO(32) heterotic string theory with string coupling $1/g$, which in turn has as its T-dual $E_8 \times E_8$ heterotic string theory with an inverted compactified dimension.

Duality relations can reach out even farther, however, and relate very different parameters in very different theories to each other. An example of this kind is AdS/CFT duality. To state one specific exemplification of AdS/CFT, a type IIB superstring theory on $AdS_5 \times S_5$ space is related to a 4-dimensional $N = 4$ super-symmetric SU(n) Yang Mills theory (which is a conformal field theory). The duality relates the curvature of AdS (in units of the string length) to the 't Hooft parameter $(g_{YM}n)^{1/4}$ in the dual conformal field theory, where g_{YM} is the coupling constant of the Yang Mills theory. In other words, a superstring theory is conjectured to be dual to a theory that is conceptually so different that parameters that characterize the first theory (such as spacetime curvature) don't even show up as such in the dual. The duality relation, therefore, links values of that parameter to values of an entirely different parameter in the dual theory.

One characteristic is common to all exemplifications of duality mentioned so far: each of the parameter values linked by duality relations controls a classical or low curvature limit of the respective theory. Moving away from a classical limit in one theory corresponds to moving closer to a classical limit in its dual. We have discussed this feature already in the S- and T-duality cases. S-duality relates a coupling constant to its inverse. Moving to higher–and thereby more quantum–values of the coupling in one theory therefore means moving to smaller–and therefore more classical–values in the dual. Similarly, if one decreases the compactification radius of a theory–and thereby moves toward a situation where the classical concept of localization on that dimension becomes less meaningful– one increases the compactification radius of the T-dual, thereby moving closer to a classical intuitive understanding of that dimension.

In the AdS/CFT case, this principle still applies. String theory on AdS can be related to an effective scenario of localized massive objects moving through space only if the curvature radius is much larger than the string length. CFT, on the other hand, has the characteristics of a weakly coupled theory only if the 't Hooft parameter is small. And a small 't Hooft parameter corresponds to a small curvature radius. Therefore, each of the dual theories has a classical limit that is controlled by parameters that are linked to each other by the duality relation. Moving closer to the classical limit in one theory amounts to moving farther away from the classical limit in the dual.

As we have seen, it is exactly this typical characteristic of duality relations encountered in the context of string physics that has proved so helpful for reaching out beyond the perturbative regime. It can allow for calculations of physical situations that don't look near classical in terms of one theory but do look near-classical in its dual. However, for the reasons discussed in Section 11.5, the described characteristic is not helpful for calculating the fundamental dynamics of a theory without free parameters. The core obstacle to any such calculation is the perturbative approach whose reliability is confined to specific parameter values while the fundamental dynamics must be calculated before any such parameter values have been fixed.

A further extension of the reach of duality relations might, in principle, overcome that obstacle: a duality that relates one parameter in the initial theory that controls a classical limit to a different parameter in the dual theory that does not lead toward a classical limit at all. Obviously such a theory would encode the classical limit of its dual theory in a similar way as, to take one example, a strongly coupled type I superstring theory encodes the weak coupling limit of SO(32) heterotic string theory. But a theory of the described kind would not contain elementary objects that, for specific parameters values, can be localized and behave in a near classical way. Such a theory, if it exists, cannot restrict calculability to a near classical limit because there is no such limit close to which perturbation theory works. So either the theory is nowhere calculable (which means that there is no regime where it can make any quantitative predictions) or it must be calculable away from any classical limit. That is, it had to be calculable without perturbing around a classical solution. In that case, the theory might offer a promising framework for a full calculation of the fundamental dynamics.

If such a theory exists, it would be immensely difficult to find. New theories and models are normally found by first thinking about ways in which one can consistently move away from a classical limit. Once a new general theory has been developed, specific realizations of that theory are found by trying to reproduce empirical data, solving conceptual problems faced by specific model building, and scanning the options for consistent theory and model building within the framework provided by the general theory. To give an example, quantum field theory provided a general framework within which specific theories and models, such as the standard model and supersymmetric models were then developed.

Dualities are normally discovered by relating two theories to each other that had already been found along the lines just described but had been taken to be substantially different due to their different classical limits. When Montonen and Olive (1977) conjectured a duality between a theory with magnetic solitonic monopoles and one with electric ones, the surprising point was the duality structure that related weak to strong coupling limits rather than the point that electric solitons could be built just like magnetic ones. When S- and T-duality were understood to link the five types of superstring theory, all five theories had already been spelled out before. The same goes for the two sides of AdS/CFT duality. There is no well-established method to construct a new theory based on conjecturing a duality relation between a known and an unknown theory and constructing the unknown theory just by applying the duality.

So let us now assume that there is a duality relation that connects a theory with a classical limit to a theory that does not relate classical limits to specific values of its parameters. The latter theory cannot be found based on the traditional heuristics of theory development that are based on thinking in terms of near classical limits. It will also be very difficult to understand that the duality exists at all, given that, as pointed out above, duality relations typically are discovered based on prior knowledge about the two theories that turn out to be dual. Neither the theory nor the corresponding duality relation would be likely to be discovered in such a scenario.

One might, therefore, imagine the following scenario. There exist, in principle, two kinds of theories about fundamental physics. There are those that can provide quantitative calculations of near classical limits based on perturbation theory. Such theories are adequate for effective theories whose parameter values do sit close to a classical limit and can be discovered and developed by thinking about them in terms of a near-classical limit. But they are not adequate for calculating the dynamics of the final theory (that is, a theory without free parameters).

Then, there are those theories that are not represented by any parameters that lead toward a classical limit. Therefore they don't have parameter values that are preferable for calculations. To the extent those theories are calculable at all, they may be expected to allow for the calculation of the entire parameter space and, in this light look like adequate theories for representing the dynamics of a fundamental theory without free parameters. But they are hidden in a conceptual realm that cannot be accessed by following near-classical intuitions.

11.8 The Case of M-Theory

There is one important theory in the context of string physics that may represent an example of the described scenario: M-theory.

Type IIA string theory with a given string coupling g is conjectured to be dual to an 11-dimensional theory of membranes and 5-branes (M-theory) with a compact-ified 11th dimension. $E_8 \times E_8$ heterotic string theory is conjectured to be dual to an 11-dimensional theory of membranes and 5-branes where the 11th dimension is bounded by two 9-branes, each of them carrying one set of E_8 gauge fields. The size of the 11th dimension (and the membrane extending through that 11th dimension) in each case corresponds to the size of the string coupling. The strong coupling limits of the string theories in both cases corresponds to a dual limit of M-theory, with an infinitely extended 11th dimension.

The relation between M-theory and the two types of string theories dual to it is of a peculiar kind. In the low coupling limit, the radius of the 11th dimension of M-theory goes to zero. The theory dual to M-theory therefore does have a classical limit of the kind known from other duality relations. But M-theory with a large 11th dimension does not amount to a new classical limit. It cor-responds to full uncompactified M-theory. M-theory does have a low energy limit (covering energy scales that are small compared to the string length) where 11-dimensional supergravity serves as its effective theory. In this sense, one can reach a new classical limit that amounts to the classical limit of a quantum field theory of pointlike objects in 11 dimensions. But that limit is not taken in a genuine M-theory description. Eleven-dimensional supergravity knows nothing about 2- and 5-branes.

Therefore, the relation between M-theory and type IIA or $E_8 \times E_8$ heterotic string theory is quite different from all other known duality relations. By lacking a classical limit, it resembles the scenario that was described before as a scenario that may generate the prospect of a fully calculable theory.

The lack of a perturbative regime for M-theory also constitutes one core obstacle to formulating M-theory even to the extent this has been achieved for 10-dimensional superstring theories. The perturbative 'entry point' that is available in the case of superstring theories is not available in the case of M-theory. More than 20 years after it was conjectured, it is thus still unclear how to turn a theory of 2- and 5-branes in 11 dimensions into a consistent quantum theory. The concepts of membranes and 5-branes provide a good heuristics for understanding the nature of the duality relation. In the absence of a genuinely 'membrany' regime that allows for a perturbative description, it seems doubtful, however, whether membranes and 5-branes provide a promising framework for fully formulating and calculating the theory. A lot of energy has been spent on finding better parameters for representing the theory. Matrix theory (Banks, Fischler, and Shenker, 1997) has been an influential candidate but has faced problems of its own. In other words, understanding M-theory suffers from the fact that the heuristics of theory development cannot be guided by looking at a classical limit of the theory itself.

11.9 Conclusion

This chapter has argued that the lack of free parameters of string theory plays a crucial role in conceptually connecting two striking features of string physics: string theory's final theory claim and its chronic incompleteness. For several reasons, a lack of free parameters may be taken to be a natural feature of a final and universal theory. But the lack of free parameters removes the possibility of calculating the theory's dynamics starting from a classical limit based on a perturbative expansion. Full access to a theory without free parameters thus might be expected to require representations that don't have their own classical limit. The fact that they cannot be developed by generalizing away from a classical limit seems to impede the full formulation of a final theory even once one has found it. The resulting idea of a fundamental theory whose full formulation is hidden from the physicists' grasp because its most adequate representation lacks intuitive roots has even more radical rivals, which amount to questioning the possibility of calculating the dynamics of the fundamental theory either within the bounds of human calculational power or as a matter of principle.

It has been suggested by various exponents and observers of contemporary fundamental physics (see, e.g., Smolin [2003], Woit [2003], and Hossenfelder [2018]) that the chronic incompleteness of string theory represents a substantial failure of the research program that is indicative of a strategical problem that has afflicted fundamental physics in recent decades. This conclusion is based on the implicit understanding that the challenges faced by scientific theory building, though obviously differing in their specifics, must always remain of roughly the same general kind. A substantial deviation from expectations regarding the general 'performance' of scientific theories in a field on that view must indicate that the way the scientific process plays out in the given case is flawed.

This chapter suggests a very different understanding of the current state of physics. Considering the range and character of the very substantial differences that set the current state of fundamental physics apart from any previous stage in the history of physics, there is little reason to expect that theory building at the present stage can be judged according to criteria that seemed adequate in the past. To be sure, it might still happen that a new conceptual idea will in the near future break the gridlock and, within a moderate timespan, lead up to a complete and predictively powerful theory of quantum gravity. Obviously, none of the arguments presented rules out such a scenario. But, judging the present situation on its own merits rather than based on the experience one has gathered with respect to earlier theories that played out in a very different overall context, that may not be the most plausible scenario.

It may be necessary to substantially correct our understanding of the role of theory building once physics has entered the stage when it plays out in the context of a final theory. It might therefore make sense to significantly downscale expectations on what a theory of quantum gravity can achieve in the foreseeable future. Understanding the character of the suggested shift in detail and analyzing which achievements can be reasonably aimed at under the new circumstances will then itself become an important element of physical reasoning.

References

Banks, T., Fischler, W., Shenker, S. H. and Susskind, L. (1997). M-theory as a matrix model: A conjecture. *Phys. Rev.*, D55, 5112–5128.

Cartwright, N. (1983). *How the Laws of Physics Lie*. Oxford: Clarendon Press.

Dawid, R. (2006). Underdetermination and theory succession from the perspective of string theory. *Philos. of Sci.*, 73(3), 298–322.

Dawid, R. (2013a). *String Theory and the Scientific Method*. Cambridge: Cambridge University Press.

Dawid, R. (2013b). Theory assessment and final theory claim in string theory. *Found. Phys.*, 43(1), 81–100.

Dawid, R. (2019). The significance of non-empirical confirmation in fundamental physics. In: Dardashti, R., Dawid, R. and Thebault, K. (eds.). *Why Trust a Theory?* Cambridge: Cambridge University Press.

De Haro, S. (2019). The heuristic function of duality. *Synthese*, 196(12), 5169–5203.

Duff, M. (2015). How fundamental are fundamental constants? *Contemp. Phys.*, 56(1), 35–47.

Green, M. B. and Schwarz, J. H. (1984). Anomaly cancellation in supersymmetric D = 10 gauge theory and superstring theory. *Phys. Lett. B*, 149, 117–122.

Greene, B. (1999). *The Elegant Universe*. New York: Vintage.

Hossenfelder, S. (2018). *Lost in Math—How Beauty Leads Physics Astray*. New York: Basic Books.

Hull, C. M. and Townsend, P. K. (1995). Unity of superstring dualities. *Nucl. Phys.*, B438, 109–137. arXive:hep-th/9410167.

Maldacena, J. (1998). The large N limits of superconformal field theories and supergravity. *Adv. Theor. Math. Phys*, 2, 231–252.

Montonen, C. and Olive, D. (1977). Magnetic monopoles as gauge particles? *Phys. Lett. B*, 72, 117–120.

Polchinski, J. (1995). Dirichlet-Branes and Ramond-Ramond charges. *Phys. Rev. Lett.*, 75, 4724. arXiv:hep-th/9510017.

Polchinski, J. (2017). Dualities of fields and strings. *Stud. Hist. Philos. Mod. Phys.*, 59, 6–20.

Scherk, J. and Schwarz, J. H. (1974). Dual models for nonhadrons. *Nucl. Phys.*, B81, 118–144.

Smolin, L. (2003). *The Trouble with Physics*. Boston: Houghton Mifflin.

't Hooft, G. (1993). Dimensional reduction in quantum gravity. Pages 284–296 of: Ali, A. and Amati D. (eds.). *Salamfestschrift*. Singapore: World Scientific.

Veneziano, G. (1968). Construction of a crossing-symmetric, regge behaved amplitude for linearly rising trajectories. *Nuovo Cimento*, 57A, 190.

Witten, E. (1995). String theory dynamics in various dimensions. *Nucl. Phys.*, B443, 85–126.

Witten, E. (1996). Reflections on the fate of spacetime. Pages 125–137 of: Callender, C. and Huggett, N. (eds.). *Physics Meets Philosophy at the Planck Scale*. Cambridge: Cambridge University Press (2001).

Woit, P. (2003). *Not Even Wrong: The Failure of String Theory and the Continuing Challenge to Unify the Laws of Physics*. London: Jonathan Cape.

12

Spacetime and Physical Equivalence

SEBASTIAN DE HARO

12.1 Introduction

In their programmatic special issue on the emergence of spacetime in quantum theories of gravity, Huggett and Wüthrich (2013, p. 284) write that the program of 'interpreting a theory "from above", of explicating the empirical significance of a theory, is both "philosophical", in the sense that it requires the analysis of concepts, and crucial to every previous advance in fundamental physics... As such, it must be pursued by the study of theory fragments, toy models, and false theories capturing some promising ideas, asking how empirical spacetime relates to them.'

Huggett and Wüthrich further articulate their project around three foci of attention:

1. Does quantum gravity *eliminate* spacetime as a fundamental structure?
2. If so, how does quantum gravity explain the *appearance* of spacetime?
3. What are the broader implications of quantum gravity for metaphysical accounts of the world?

These are indeed central questions for candidate theories of quantum gravity. If the answer to the first question turns out to be affirmative, the impact on the philosophy of spacetime will be spectacular: for there will be, at the fundamental level, *no* spacetime. In this essay I consider the contribution, to the answering of questions 1 and 3, of one particular quantum gravity approach: gauge/gravity duality.[1] This is an approach developed in the context of string and M theory, but it has broader ramifications: e.g., applications to condensed matter physics and heavy-ion collisions (Ammon and Erdmenger, 2015, III)). I will not consider question 2 here.[2]

---

[1] For an expository overview, see, e.g., Ammon and Erdmenger (2015). De Haro, Mayerson, and Butterfield (2016) is a conceptual review.

[2] There is a growing literature on the contribution of gauge/gravity dualities to question 2; see e.g., De Haro (2017), Dieks, Dongen, and Haro (2015), and Rickles (2012).

Thus I lay out, in the first part of the essay (Section 12.2), the conceptual scheme for dualities that I advocate. My argument proceeds by analyzing dualities (12.2.2) in terms of four contrasts, which lead up to making the distinction between theoretical equivalence and physical equivalence.

A duality is an isomorphism between (bare) theories. Then the four contrasts are

a. *Bare theory vs. interpreted theory* (Section 12.2.1): A *bare theory* is a triple of states, quantities, and dynamics, each of which are construed as structured sets, invariant under appropriate symmetries. An *interpreted theory* has, in addition, a pair of interpretative maps to physical quantities and other entities.
b. *Extendable vs. unextendable theories* (Sections 12.2.3.3–12.2.4): theories that do, respectively do not, admit suitable extensions in their domains of application. I will also allow for a weaker conception of 'unextendable theory', according to which unextendable theories may admit an extension via, e.g., *couplings* to other theories in their domain, but are such that their interpretations are *robust*, i.e., unchanged under such extensions.
c. *External vs. internal interpretations* (Sections 12.2.1.2, 12.2.3.1, 12.2.3.2): interpretations that are obtained from outside (e.g., by coupling the theory to a second theory that has already been interpreted), respectively from inside, the theory, i.e., from the role that states, quantities, and dynamics have within the theoretical structure.
d. *Theoretical vs. physical equivalence* (Section 12.2.3): formal equivalence (i.e., agreement of the bare theories, but with possible disagreement of the interpretations) vs. full equivalence of the interpreted theories: i.e., agreement of both the bare theory and the interpretive maps.

These contrasts build upon each other: so that (a) is used in the analysis of (b), (a) and (b) are jointly used in the interpretative analysis of (c), and (a)–(c) are all needed in order to reach a verdict distinguishing theoretical vs. physical equivalence, as (d) intends.

My account provides sufficient details, so that the scheme can be readily applied to other cases, and I give several examples that will work toward applying the scheme to gauge/gravity dualities. However, a *full* account of theoretical and physical equivalence, doing full justice to the intricacies of the matter, will have to be left for the future. Further formal and conceptual development of the scheme is in De Haro and Butterfield (2018).

Of course, not all of these notions are completely new. But my construal of them is largely novel (the only exception being the contrast (c), for which I am in full agreement with, and just develop further, the position of Dieks et al. [2015] and De Haro [2017]). In particular, the way I here articulate the notions of theoretical and physical equivalence in terms of the contrasts (a)–(d), so that I can successfully analyze dualities, are novel and are intended to add to the literature on both dualities and equivalence of theories.

In the second part of the essay (Section 12.3), I apply the scheme (a)–(d) to gauge/gravity duality. This will answer the following two questions, from the perspective of this quantum gravity program: (iii) the nature of spacetime in quantum gravity (Section 12.3.1), and (iv) the broader philosophical and physical implications (Section 12.3.3). Philosophers of physics have started to address the philosophical significance of dualities in recent years: and I compare with these works in Section 12.3.2, so as to clarify my own contribution.

The organization of the second part of the essay thus responds to the questions (1) and (3) posed by Huggett and Wüthrich, taken as specifically about gauge/gravity duality.

Gauge/gravity duality is one particular approach to quantum gravity. Briefly, it is the equivalence between:

(I: Gravity) On the one hand: a theory of quantum gravity in a *volume* bounded inside a certain surface.

(II: QFT) On the other: a quantum field theory (QFT) defined on that *surface*, which is usually, in most models, at 'spatial infinity', relative to the volume.

I will argue that, despite the apparent innocence of the references to spacetime appearing in this brief summary of the duality: the physical interpretation of this duality calls for a revision of the role that most of our physical and mathematical concepts are supposed to play in a fundamental theory—most notably, the role of spacetime. And the interpretation of the duality itself requires us to carefully reconsider the philosophical concepts of theoretical equivalence and physical equivalence.

12.2 Theories, Duality, and Physical Equivalence

In this section, I briefly describe the scheme (a)–(d), which I apply in Section 12.3. In Section 12.2.1, I specify more exactly what I mean by 'theory' and related notions. In Section 12.2.2, I give my conception of a duality between such theories. Section 12.2.3 describes how, for theories 'of the whole world', duality is tantamount to physical equivalence, i.e., the theories at issue being really the same theory: this will involve two conditions (Section 12.2.3.2). In Section 12.2.4, I compare my account with Glymour's notion of equivalence.

12.2.1 The Conception of a Theory

Before we engage with the interpretation of dualities (which we will do in Sections 12.2.2–12.2.3), we need to have conceptions of theory that are sufficiently articulated that they make an analysis of physical equivalence possible. I first introduce, in Section 12.2.1.1, the notion of a *bare theory*. Then, in Section 12.2.1.2, I discuss the notion of an *interpreted theory*.

12.2.1.1 Bare Theory

I take a *bare theory* to be a triple $T = \langle \mathcal{H}, \mathcal{Q}, \mathcal{D} \rangle$ comprising: (1) a set $\mathcal{H}$ of states, endowed with appropriate structure, (2) a set of physical quantities $\mathcal{Q}$, endowed with appropriate structure, (3) a dynamics $\mathcal{D}$, consistent with the relevant structure. Such a triple will generally also be endowed with *symmetries*, which are automorphisms $s : \mathcal{H} \rightarrow \mathcal{H}$ preserving (a subset of) the valuations of the physical quantities on the states (for details, see De Haro and Butterfield [2018, sec. 3.3]), and which commute with (are suitably equivariant for) the dynamics $\mathcal{D}$.

For a *quantum* theory, which will be our main (though not our sole!) focus: we will take $\mathcal{H}$ to be a Hilbert space; $\mathcal{Q}$ will be a specific subset of operators on the Hilbert space; and $\mathcal{D}$ will be taken to be a choice of a unique (perhaps up to addition by a constant) Hamiltonian operator from the set $\mathcal{Q}$ of physical quantities. In a quantum theory, the appropriate structures are matrix elements of operators evaluated on states, and the symmetries are represented by unitary operators. But as mentioned: the present notion of a bare theory applies equally well to classical and to quantum theories, and there will be *no* requirement that a duality must relate quantum theories.

A theory may contain many more quantities, but it is only after we have singled out the ones that have a physical significance that we have a *physical*, rather than a *mathematical*, theory or model. The quantities $\mathcal{Q}$, the states $\mathcal{H}$, and the dynamics $\mathcal{D}$ have a physical significance at a possible world W, and within it a domain of application D_W (which we can think of as a subset $D_W \subseteq W$), though it has not yet been specified what this significance may be, nor what the possible world 'looks like'.[3] To determine the physical significance of the triple, a physical interpretation needs to be provided: which I do in the next subsection.

So, I will dub as the *bare theory*: just the formal triple $T = \langle \mathcal{H}, \mathcal{Q}, \mathcal{D} \rangle$, together with its structure, symmetries, and rules for inferring propositions, such as: 'the value of the operator $Q \in \mathcal{Q}$ in the state $s \in \mathcal{H}$ is such and such'. But there is no talk of empirical adequacy yet.

We normally study a theory through its models. A *model* is construed as a representation of the theory, the triple $\langle \mathcal{H}, \mathcal{Q}, \mathcal{D} \rangle$,[4] and will be denoted by M. A model M

---

[3] This framework allows of course for models that do not describe actual physical reality. While one can give various metaphysical construals to concepts such as 'possible worlds', I here need only a minimal conception of possible worlds as 'how things can be', in the context of a 'putatively fundamental' spacetime physics. Thus I do not mean to subscribe to a specific metaphysics of possible worlds.

[4] *Representation* is here meant in the mathematical sense of a homomorphism. In this paper, I consider only isomorphic representations: for the nonisomorphic case, see De Haro and Butterfield (2018, secs. 2.2.2–2.2.3).

of a theory T may include, in addition to the triple, some variables that are part of the descriptive apparatus but have no physical significance (in the sense of the last but one paragraph) from the point of view of the theory: I will call this the *specific structure* of the model M. Since a theory can be represented by any of its models, my account does not seek to eliminate the specific structure but to identify the core $\langle \mathcal{H}, \mathcal{Q}, \mathcal{D} \rangle$ of the models: as that structure that is preserved across equivalent models (see Section 12.2.2). In our duality of interest, gauge/gravity duality, I treat the two sides of the duality as two models of the theory.

Notice that the notion of model I use here, as a representation (more generally, an instantiation) of a theory, is distinct from the more common characterization of a model as a specific solution for the physical system concerned (e.g., a specific trajectory in the state space, or a possible history, according to the equations of the theory), which I indeed here reject. My notion is motivated by dualities because they prompt us to move what we mean by a theory 'one level up', in abstraction: and I accordingly take what I mean by a model one level up, keeping the relation between theory and model fixed, rather than introducing a new level between theory and model. Thus, what used to be 'two distinct theories', each of them having their own models, are now 'two models of a single theory', each model having its own set of solutions, related to each other by the duality. What dual models have in common, i.e., their *common core*, is isomorphic between the models. For more on the motivation for, and the uses of, these notions of theory and of model, see De Haro and Butterfield (2018, sec. 2).

12.2.1.2 Interpreted Theory

There is a certain minimalism to the presented conception of theory, since in scientific practice one must be able to tell, in a given experiment or physical situation to which the theory is supposed to apply, what the relevant quantities are that correspond to the empirical data. The specification of a theory as a triple makes no reference as yet to this: only the existence of some such relation, for a class of possible worlds W, is assumed. So, when interpreting a theory, one wishes to also:

0. Establish the meaning of certain theoretical entities (if one is a realist), whether directly measurable or not;
1. Establish some kind of bridge principles between the physically significant parts of the theory and the world.

These two desiderata will be fulfilled by the two interpretative maps (denoted I^0, I^1). Furthermore, one may also wish to establish *theoretical principles* that, for

---

This reference argues that a more general account of a model can be had: as an instantiation, or a realization, of a theory.

example, interconnect various experimental results (symmetry and locality being just two examples of such theoretical principles often considered in physics).

So, I take a *physical interpretation* to be a pair of partial maps, preserving appropriate structure, from the bare theory to some suitable set of physical quantities. I denote the maps as $I_T := (I_T^0, I_T^1): T \to D_W$, where T is the triple[5] and D_W is the domain within the possible world W (a pair of possible worlds, since there are two maps [more on this later; for simplicity in the notation, I often drop the subscript]). The restriction to only a domain within a possible world stems from the wish to account for theories that do not describe the whole world, but only part of it. The requirement that the map need only be partial means that not all the elements of the theory necessarily refer. The more adequate the theory, the more elements will refer and the map will satisfy additional conditions, but we will not use this here.

The first partial map, I_T^0, is from the triple, $\mathcal{H}, \mathcal{Q}$, and $\mathcal{D}$, to the quantities understood conceptually (potential energy, magnetic flux, etc.), i.e., it corresponds to the theoretical meaning of the terms. Of course, this theoretical meaning is informed by experimental procedures, and it will usually involve the quantities as realized in all possible laboratory experiments, through observations, or even through relations between experiments and/or observations, in a domain D^0 of a possible world W^0. But this possible world is not to be understood as our world in its full details of context, but as a conceptual realm that fixes the reference of the theoretical concepts.[6]

The second partial map, I_T^1, is from the triple, $\mathcal{H}, \mathcal{Q}$, and $\mathcal{D}$ (or Cartesian products thereof), to the set of values (the set of numbers formed by all experimental or observational outcomes, as well as the relations between them), i.e., the domain D^1 to which the theory applies, within a possible world W^1. That set of values is typically (minimally) structured, and such minimal structure is to be preserved by the map. Thus typically, the second map maps, e.g., individual states or quantities, e.g., by pairing them: $I_T^1: \mathcal{H} \times \mathcal{Q} \to \mathbb{R}$, where $\mathbb{R}$ is endowed with addition and multiplication (for a concrete example, see Section 12.2.3.1), and likewise for the dynamics. Notice that, unlike the first map, the second map depends on contingent facts about the possible world W^1, such as the context of a measurement.

The codomains of I_T^0 and I_T^1 as thus defined might, at first sight, seem to be again just theoretical or mathematical entities, themselves in need of interpretation—a space of functions in the first example, a set of real numbers endowed with addition

[5] Or Cartesian products thereof: e.g., in quantum mechanics, the interpretation map maps, e.g., expectation values to real numbers in the world. The expectation values themselves are maps from Cartesian products of states and quantities to real numbers.
[6] Cf. De Haro and Butterfield (2018, sec. 2), where the first map is an intension, and the second an extension.

and multiplication, in the second. But this is not what is intended. We are to think of the codomain of I_T^1 as representing the real world in a straightforward way: so, for example, the reals measuring the energy represent the value of a quantity in appropriate units, given by the position of a voltmeter's pointer on a scale (see some examples in Section 12.2.3.1). The point is to map from theories to structured sets of functions and numbers, which do *not* describe more theory, but rather are identified with a set of possible physical situations, experimental outcomes, or observational and conceptual procedures.

In what follows, the distinction between I^0 and I^1 (and between the idealized world W^0 and the concrete world W^1, to which they map) will not be important, and so I gloss over the '0 vs. 1' contrast and simply talk of a pair of interpretation maps, I, from the theory to the world, W. The distinction between the two cases is brought to bear in De Haro and Butterfield (2018, secs. 2–3).

In my discussion, I have so far assumed that the interpretative maps are from the elements of *the triples themselves*, rather than from their models, to quantities etc. in the world. We will call such an interpretation *internal*: it requires nothing but the theory. But, as I discuss in detail in Section 12.2.3.1, there are often good reasons to pursue an interpretation that is obtained by, e.g., coupling the theory T to some other theory T_{meas} that is already interpreted. T thus inherits its interpretation from T_{meas}, and the coupling of T to T_{meas} will often differ for the different models M of T. In other words, T_{meas} may be coupled to M through M's specific structure. So, we distinguish:

1. *External interpretation of T*: a pair of maps, as earlier, $I_M : M \times T_{\text{meas}} \rightarrow D_W$, from the *model M* of the theory T, coupled to the theory of measurement T_{meas} (or some other relevant theory that provides an interpretation), to the domain $D_W \subseteq W$.
2. *Internal interpretation of T*: a pair of maps, as earlier, $I_T : T \rightarrow D_W$, from the *theory*, T, to the domain, $D_W \subseteq W$.

I will call a bare theory, once it is equipped with an interpretation, the *interpreted theory*. It is the physical interpretation that enables the theory to be empirically successful and physically significant.

The interpretation maps involve, of course, philosophically laden issues. But the aim here is not to settle these issues, but rather to have a scheme in which the formal, the empirical, and the conceptual are clearly identified and—as much as possible—distinguished. The interpretational scheme should make the ontological commitments *explicit* (more on this in Section 12.2.4).

The fact that the first map, I_T^0, is conceptual, and that the two maps also respect relations between quantities (which might not be directly and independently measurable, while still being physically significant), reflect the fact that the set of quantities Q need not be restricted to measurable or observable quantities. I illustrate this with an example in Section 12.2.4.

The formulation of a bare theory as a triple is minimalist. But, with the interpretation maps added, it is strong enough—because of the complete specification of the set of physical quantities —that, under two conditions mentioned later, it will determine whether two theories are about the same subject matter. Questions concerning the identity of two such triples will be questions concerning the sameness of models (Section 12.2.3.2), rather than standard cases of underdetermination of theory by empirical data. This is because we assume that the triple $T = \langle \mathcal{H}, \mathcal{Q}, \mathcal{D} \rangle$, together with the valuations constructed from the syntax, is well defined and consistent, and that it encompasses all the empirical data (and relations between the data), in a certain domain D_W within W. The latter condition can be restated as the requirement that the interpretative map be *surjective*, so that no element of the domain D_W is left out from the model's description.[7] I call such a model *complete*. Thus, the equivalence of two triples means that the theories agree about everything they deem physical. On the other hand, in standard cases of underdetermination considered in the philosophy of science, the theories are underdetermined by what is measurable or observable, by the available evidence, or even by the actual empirical data obtained by scientists. The completeness of the theory is thus a necessary condition for duality to which I return.

12.2.2 The Conception of Duality

With the conception of a theory considered in Section 12.2.1, a *duality* is now construed as an *equivalence of theories*. More precisely, it is an isomorphism $d : M_1 \to M_2$ between two models M_1 and M_2 of a theory T: there exist bijections between the models' respective sets of states and of quantities, such that the values of the quantities on the states are preserved under the bijections. (In the case of quantum theories, these values are the set of numbers $\langle s_1 | Q | s_2 \rangle$, where $s_1, s_2 \in \mathcal{H}$, $Q \in \mathcal{Q}$.) The duality is also required to commute with (d is to be equivariant for) the two models' dynamics, and to preserve the symmetries of the theory.

The notion of duality in this subsection is motivated by both physics and mathematics. Duality in mathematics is a formal phenomenon: it does not deal with physically interpreted structures (even though, of course, several of the mathematical dualities turn out to have a physical significance). But this is also how the term is used by physicists: it is attached to the equivalence of the formal structures of the theories, regardless of their interpretations.

---

[7] The qualification, 'in a domain within a possible world', is important because the theory need not be complete at *our world*. A model may, for example, be complete within a given range of parameters not containing the relevant values for our world. This is why completeness is to be construed as relative to D_W and W.

Duality, as a formal equivalence between two triples without the requirement of identical interpretations, is thus a special case of theoretical equivalence: an isomorphism, as earlier.

Like the conception of a theory, my conception of a duality is minimalist. On this definition, for instance, position–momentum duality in quantum mechanics is indeed a duality. The duality has two models, namely the formulations of quantum mechanics based on, respectively, the x- and the p-representations of the Hilbert space: Fourier transformation being the duality map. This duality is, of course, somewhat trivial, because the two models contain *the same amounts of specific structure*, in the sense given earlier: in short, a single variable. And indeed, I regard it as a virtue that my conception of duality is general enough that both familiar, and relatively simple, dualities, as well as the more sophisticated ones in quantum field theory and quantum gravity, all qualify as dualities, under the same general conception. Indeed, I take it that:

One of the lessons of duality is that 'widely differing theories' are (surprisingly) equivalent to each other, in the same sense of equivalence in which two 'notational variants' differ from each other.

An important question, given the conceptions of theory, model, and duality I have introduced, is whether one can verify, in some well-understood examples, that the scheme leads to the correct verdicts regarding duality. Examples of equivalences between very different looking quantum field theories are provided in De Haro and Butterfield (2018), where it is shown how duality thus construed obtains. The focus of this essay—in line with the quote by Huggett and Wüthrich given in Section 12.1—is in applying this conception, and the four contrasts (a) to (d), to shed light on *duality in quantum gravity.*

The two-pronged conception of an interpreted theory as a triple plus an interpretation, together with the notion of duality as isomorphism, allow us to introduce the notion of the *physical equivalence* of models. The discussion of duality so far indeed prompts us to distinguish *theoretical equivalence* from *physical equivalence*: the latter being the complete equivalence of two models as descriptions of *physical systems*, i.e., models with identical interpretations. The difference may be cashed out as follows: theoretically equivalent models, once interpreted, 'say the same thing' about possibly *different subject matters* (different parts of the world), whereas physically equivalent models say the same thing about the *same subject matter* (the same part of the world). There will also be a weak, but interesting, form of physical equivalence, in which two dual models describe a single given world equally well, even if in other cases they may also describe different worlds. More on this in Section 12.2.3.

Duality, then, is one of the ways in which two models can be theoretically equivalent, without its automatically implying their physical equivalence. For instance, a duality can relate a real and an imagined or an auxiliary system. In such a case, duality is a useful and powerful calculational device—and nothing more. But it is, of course, those cases in which dualities do reveal something about the nature of physical reality, that prompts the philosophical interest in dualities: cases in which the interpretation of the duality promotes it to physical equivalence.

12.2.3 From Theoretical Equivalence to Physical Equivalence

Having introduced, in the previous two sections, my conception of duality, and the four contrasts (a) to (d) mentioned in Section 12.1, we now come to the central question in this section: *When does duality amount to physical equivalence?* I first discuss, in Section 12.2.3.1, the external and internal interpretations of a theory (already briefly introduced in Section 12.2.1.2). In Section 12.2.3.2, I state two conditions for physical equivalence: internal interpretation and unextendability. In Section 12.2.3.3, I discuss the physical equivalence of dual theories that are unextendable.

12.2.3.1 External and Internal Interpretations of a Theory

In this section, I further develop the external and internal interpretations, and in particular two cases: (1) cases of external interpretations, in which *physical equivalence fails to obtain*, despite the presence of a duality, and (2) cases in which an external interpretation is not consistently available (where 'consistently' will be qualified), so that one can have only an internal interpretation, and hence there is *physical equivalence*.

Let me illustrate the external interpretation with an elementary example that should make clear the difference between duality as a case of theoretical equivalence and physical equivalence. Consider classical, 1-dimensional harmonic oscillator 'duality': an automorphism, $d : \mathcal{H} \to \mathcal{H}$, defined by $d : \mathcal{H} \ni (x, p) \mapsto (\frac{p}{m\omega}, -m\omega x)$, from one harmonic oscillator state to another, leaving the dynamics $\mathcal{D}$ invariant—namely, the Hamiltonian $H = \frac{p^2}{2m} + \frac{1}{2}kx^2$ and the equations of motion that H defines. So, it is an automorphism of $T_{\mathrm{HO}} = \langle \mathcal{H}, \mathcal{Q}, \mathcal{D} \rangle$.[8] But this automorphism of $\langle \mathcal{H}, \mathcal{Q}, \mathcal{D} \rangle$ does not imply a physical equivalence of the states:[9] the two states are clearly distinct and describe different physical situations, since

---

[8] I have described the states and dynamics of the harmonic oscillator: the quantities $\mathcal{Q}$ include, e.g., any powers (and combinations of powers) of x and p.

[9] In a world consisting of a single harmonic oscillator *and nothing else*, the two situations could not be distinguished, and one might invoke Leibniz's principle to identify them. This amounts to adopting an internal interpretation, in the sense of Section 12.2.1.2.

the map d relates an oscillator in a certain state of position and momentum, to an oscillator in a *different* state.

This difference is shown in the fact that there is an independent way to measure the 'position' of the oscillator at a given time: one sets the oscillator and a standard rod side by side, observes where on the rod the oscillator is located, and so carries out a measurement of the former's position. We can picture this as coupling harmonic oscillator theory, T_{HO}, to our theory of measurement T_{meas}, and interpret the measurement as measurement of the oscillator position. I will call such an interpretation of T_{HO} an *external interpretation* (cf. Section 12.2.1.2). It is obtained by inducing the interpretation of T_{HO} from an already interpreted theory T_{meas}, or by extension to $T_{HO + meas}$. And I call a theory, that can be coupled or extended in this way an *extendable* theory.

But there are cases—such as cosmological models of the universe, and models of unification of the four forces of nature—in which these grounds for resisting the inference from duality to physical equivalence—a resistance based on the possibility of finding an external theory T_{meas}—are *lost*. For the quantum gravity theories under examination—even if they are not *final* theories of the world (whatever that might mean!)—are presented as candidate descriptions of an *entire* (possible) physical world: let us call such a theory T.[10] So, there is *no* independent theory of measurement T_{meas} to which T should, or could, be coupled, because T itself should be a closed theory (an *unextendable* theory; see Section 12.2.3.3). In the next section, I use these ideas to spell out the conditions under which two dual models are physically equivalent.

12.2.3.2 Internal Interpretation and Unextendability Allow Sameness of Reference, and so Physical Equivalence

We return to dualities as isomorphisms of models (cf. Section 12.2.2); and so, we consider, specifically, the interpretation of two models, M_1 and M_2, 'of the whole world'. In this section, I propose a condition that will, together with the internal interpretation, secure physical equivalence, in the sense that one is justified in taking duals to be physically equivalent, and I give the arguments to that effect.

The leading idea of an internal interpretation is that the interpretation has not been fixed a priori, but will be developed starting from the duality. (Or, if by some historical accident, an interpretation has already been fixed, one should now be prepared to drop large parts of it.) The requirement that, I propose, justifies the

---

[10] These are candidate descriptions of *possible* worlds, rather than the actual world. For example, the models that we will consider in Section 12.3 entail a negative cosmological constant, whereas our universe seems best described by a *positive* cosmological constant. But the interest in such models is, of course, that (1) given their rarity, *any* consistent 4-dimensional theory of quantum gravity is interesting and (2) such idealized models contain helpful lessons for the case of a positive cosmological constant.

use of the internal interpretation such that uniqueness of reference is secured, is as follows:

Unextendability: roughly, 'the interpretation cannot be changed by coupling the theory to something else or by extending its domain'. Unextendability replaces the somewhat vague phrase *of the whole world* in the previous paragraph, and I expound it in Section 12.2.3.3.

Unextendability plays a key role in inferring physical equivalence. For it ensures that there is 'no more to be described' in the physical world, and that the models cannot be distinguished, even if their domains of application were to be extended (since no such extension exists). And so, it ensures that the internal interpretation *can be trusted* as a criterion of physical equivalence (cf. Section 12.2.3.3), as I now argue.

Starting, then, from two such dual models, M_1 and M_2 of T, the duality map lays bare the invariant content $\langle \mathcal{H}, \mathcal{Q}, \mathcal{D} \rangle$, as that content which is common to M_1 and M_2, through the duality map (cf. Section 12.2.1.2). This is the starting point of the *internal interpretation*, for both the theory and the models. I now propose that an internal interpretation of a theory, satisfying the two stated conditions, is the same for the two models (in the sense of Section 12.2.2), and in particular its reference is the same:

(i) The formalisms of the two models say the same thing: for they contain the same states, physical quantities, and dynamics (i.e., the domain of the maps is identified by the isomorphism), and (ii) their physical content is also the same, for the interpretation given to the physical quantities and states is developed from the duality *and nothing else*. So, the codomains of the maps are the same, and they coincide with the entire world. I am thus here proposing that *the domains of the worlds described by two dual models, and the worlds themselves are the same*; this is because, on an *internal* interpretation, the two worlds, in all their physical facts, are 'constructed, or obtained from', the triples.

There is a way in which this inference, from dual models with internal interpretations, to identical worlds, might fail: there might be more than one internal interpretation and, therefore, more than one codomain D_W described by the theory. For in that case, despite the isomorphism of the two models, one might be tempted to think that one model could be better interpreted in one way, and the other better interpreted in another way.

But I take this objection to be misguided: the point is that, even if there is more than one internal interpretation, the reference of a given internal interpretation is the same for any two isomorphic models. Remember that, by definition, an internal interpretation cannot discern between models, because it starts from the theory, as a triple, *and nothing else*. So, there can be no reason for the interpretation to distinguish one model from the other—they both describe the world equally well and in the same way, according to that internal interpretation. In other words: *even if a single common core admitted several internal interpretations, each of them would refer to a single possible world, which would be the single reference of the corresponding internal interpretation of all the models isomorphic to the common core.*

Let me spell out in more detail this inference from the isomorphism of models, to the identity of the internal interpretations and identity of the worlds described (hence physical equivalence), under the unextendability condition. There are two ways to make this inference. The first argument, from the unextendability of the models, will be given in Section 12.2.3.3. It is a version of Leibniz's principle of the identity of indiscernibles, which applies here because the two models describe the entire world. The second argument, given in the previous paragraph, simply follows from the definition of the internal interpretation, given in Section 12.2.1.2: an internal interpretation is constructed from the triple of the theory *and nothing else* (so, the specific structure of a model is not to find a counterpart in the world, since the interpretation must be invariant under the duality map). Thus, given an internal interpretation of the theory, the codomain of that interpretation, mapped from the two models, is the same by definition: *since the internal interpretation is insensitive to the differences in specific structure between the models, its reference must be the same.*[11] Explicitly, $I_{M_1} = I_{M_2} \circ d$, where $d: M_1 \to M_2$ is the duality map. Thus, such thorough-going dualities can be taken to give *physical equivalence* between apparently very different models.

It is important to note that this second argument for the identity of the codomains follows from the *definition* of an internal interpretation, given in Section 12.2.1.2 (together with the two stated conditions). What is surprising about a duality that is a physical equivalence, then, is not so much that two very different models describe *the same world*, but rather that *there is an internal interpretation* to be constructed from such minimal data as a triple,[12] and so, what is surprising is that there is a (rich) world for such a triple to describe! Admittedly, it would be hardly surprising if the internal interpretation described something as simple as the real line. But, in the examples we are concerned with here, the internal interpretation describes far richer worlds!

---

[11] The internal interpretation I is a partial surjective map from the theory to the world. But using the forgetful map from the model to the theory (the map that strips the model of its specific structure), we can construct an internal interpretation of the model, as the pullback of the interpretation map I by the forgetful map. It is in this sense that I here speak of internal interpretations of the models as well as of the theory.

[12] I call the triple 'minimal' data because it does not contain specific structure, which we normally think of as giving a model, and its interpretation, its particular features. So, the internal interpretation constructed from a triple may be rather abstract yet the claim is that it is an entire world!

I do not claim to have established physical equivalence, under the conditions of internal interpretation and unextendability, as a matter of logical necessity, just from the notions of theory, model, and interpretation. Doing so would require a deeper analysis of the notion of reference itself, and the conditions under which it applies to scientific theories, which is beyond the scope of this paper (cf., e.g., Lewis [1984]). What I claim to have argued is that, making some natural assumptions in particular about how terms refer in ordinary language and in scientific theories, the use of the internal interpretation and unextendability does deliver physical equivalence.

12.2.3.3 Unextendability, in More Detail

I turn to consider theories, such as theories of quantum gravity, T_{QG} say, for which there is no extra physics to which T_{QG} can be coupled or extended. Being a description of the entire physical universe, or of an entire domain of physics, I will take the interpretation I_{QG} to be *internal* to T_{QG}. Thus, as a sufficient condition for being justified in the use of an internal interpretation, I have required, in Section 12.2.3.2, that T_{QG} be an unextendable theory. An interpretation map I_{QG} requires only the triple $T_{QG} = \langle \mathcal{H}, \mathcal{Q}, D \rangle$ as input, and it only involves the triple's elements and their relations—it does not involve coupling T_{QG} to other theories. In such a case, duality preserves not only the formalism but necessarily also the structure of the concepts of two complete and mathematically well-defined models. If one model is entirely self-consistent and describes all the relevant aspects of the world, then so must the other model. And so, duality becomes physical equivalence. Thus, in other words, we are really talking about different formulations of a *single theory*.

Let me spell out the (sufficient) condition, suggested by this discussion, for a theory to admit an internal interpretation, since it will be important in Section 12.2.4. A bare theory T in a domain D_W of a possible world W is *unextendable* iff:

i. T is a complete theory in the domain of applicability D_W at W;
ii. There is no other theory T'' for the possible world W (or another possible world that includes it) and domain D_W, such that for some T' isomorphic to T, $T' \subset T''$ (proper inclusion).[13]
iii. The domain of applicability D_W coincides with the world described, i.e., $D_W = W$.
    Note that the possible world W is fixed by the interpretation. Unextendability is thus a relation between bare theories and worlds and is thus a property of *interpreted theories*.

[13] T' is a fiducial theory that may well be *identical* to T. But in general, it may be the case that $T \subset T''$ is not true but $T \cong T' \subset T''$ is. In other words, $T \subset T''$ may only be true up to isomorphism.

Recall the notion of completeness of a theory, in (i), introduced in Section 12.2.1.1: as well-defined, consistent, and encompassing all the empirical data in a certain domain D_W (i.e., a partial surjective map). Condition (ii), in addition, requires that there is no extension of the theory at W, or, in other words, the theory already describes all the physical aspects of the relevant domain at W. Since the relation of isomorphism in (ii) is formal, (ii) is a sort of 'meshing' condition between (i)—or, more generally, between the idea of 'not being extendable'—and the formal relation of isomorphism between bare theories.[14] Condition (iii) is the usual condition that T is a theory 'of the whole world', i.e., its domain of application is the entire world. Notice that (i) and (iii) do not suffice for physical equivalence; one needs something like the technical requirement (ii) (see also the next section).

I have concluded that, on an internal interpretation, there is no distinction of content between two dual models. In Section 12.2.4, I address two different *purposes* for which the distinction between two dual models is irrelevant, besides the ontological purposes so far discussed: viz., the logical and empirical purposes. This is not to deny that there are other significant—metaphysical, epistemic, and pragmatic—purposes or uses of physical theories, for which the differences are significant. For instance, one of the main pragmatic virtues of gauge/gravity dualities is that one theory is tractable in a regime of values of the parameters where the other theory is intractable.

12.2.4 Comparison with Glymour's Notion of Equivalence

How does the conclusion that the two theories related by gauge/gravity duality admit internal interpretations and that under two additional conditions duality implies physical equivalence compare with the relevant philosophical literature on equivalence of physical theories? To discuss this, I recall the usual strategy by which, faced with apparently equivalent theories, physicists try to break the equivalence and relate this to an influential discussion, by Glymour. I agree with Glymour's verdicts for his examples, but I argue that this depends on the theories in the examples being *extendable*.

It is a commonplace of the philosophy of science that, confronted with theoretically inequivalent, but empirically equivalent theories, physicists naturally imagine

---

[14] I have argued that unextendability is a sufficient, though not a necessary, condition for the coherence of an internal interpretation. The condition is not necessary because one can envisage a theory (e.g., general relativity without matter) receiving an internal interpretation (e.g., points are identified under an active diffeomorphism, taken as the lesson of the hole argument). This interpretation does not change when we couple the theory to matter fields, and I will say that such an internal interpretation is *robust* against extensions. If all possible extensions of a theory preserve an internal interpretation, then such an interpretation is justified. If the extensions suggest diverging interpretations, then one needs to specify the domain of the extension before one is justified in interpreting the theory internally. In other words, (i)–(iii) can be weakened, if what we want is not sameness of reference but of descriptive abilities.

resorting to some adjacent piece of physics that will enable them to confirm or disconfirm one of the two theories as against the other. The classic case is: confronted with differing identifications of a state of rest in Newtonian mechanics, Maxwell proposes a measurement of the speed of light. There is a parallel for dualities: when two theories are both theoretically and empirically equivalent, we can still argue, by an extension or by a resort to some adjacent piece of physics, for their physical inequivalence. This is articulated in the contrast, in Section 12.2.3.3, between extendable and unextendable theories.

Glymour's (1977) discussion of equivalence of theories uses the syntactic conception of a theory as a set of sentences closed under deducibility. He introduces the notion of 'synonymy': two theories are synonymous when they are, roughly speaking, logically equivalent. That is, there is a well-defined intertranslation between them.[15] Although my use of theories as triples puts me closer to the semantic conception (the syntactic conception's traditional rival), in fact Glymour's criterion of synonymy meshes well with my notion of a duality, construed as an isomorphism of triples, equipped with rules for forming propositions about, e.g., the value of a quantity or a state. One considers the set of well-formed sentences built from two triples T_1 and T_2, e.g., statements of the type 'the value of the quantity $Q_1 \in \mathcal{Q}$ (resp. Q_2), in such and such state, is such and such'. Duality then amounts to isomorphism between two such sets of sentences. And this is a case of synonymy in Glymour's sense.

But does this immediately lead to physical equivalence? No. And the reasons provided in Section 12.2.3.3 are similar to the ones Glymour gives. He envisages theories that are synonymous, in the sense just described, yet are not physically equivalent. Recall Glymour's thought experiment (1977, p. 237):

Hans one day announces that he has an alternative theory which is absolutely as good as Newtonian theory, and there is no reason to prefer Newton's theory to his. According to his theory, there are two distinct quantities, gorce and morce; the sum of gorce and morce acts exactly as Newtonian force does.

Glymour denies that the Newtonian 'force theory' and Hans's 'gorce-and-morce theory' are physically equivalent. He argues that they are empirically equally adequate, but not equally well *tested*. His reasons for this are, partly, ontological ('I am, I admit, in the grip of a philosophical theory'; Glymour [1977, p. 237]), and his ontology leads him to prefer the Newtonian theory: the gorce-and-morce theory contains two quantities rather than one, but there is no evidence for the existence of that additional quantity. The argument is from parsimony; he prefers a sparse

---

[15] Technically, what is required is a common definitional extension. Barrett and Halvorson (2016, sec. 4, thms. 1–2) show that Glymour's 'synonymy', i.e., there being a common definitional extension, is equivalent to an amendment of Quine's 'translatability'.

ontology. And so, Glymour's reply to Hans seems to entail that the two theories are not even theoretically equivalent, because the gorce-and-morce theory has one more quantity than the force theory.

But even if the two theories *were*, or could be made, theoretically equivalent, I would still agree with Glymour's verdict of a lack of physical equivalence, in so far as one is concerned with theories that admit external interpretations, and this for two reasons:

a. His examples deal with classical theories of gravitation, i.e., *extendable* theories admitting external interpretations, which can indeed vary widely. That these theories are extendable, beyond a certain regime of energies, can be seen from both Newton's and Einstein's theories' predictions of *singularities*, which are usually taken to be unphysical.

b. The force theory and the gorce-and-morce theory are empirically equivalent on a restricted domain, but their extensions are *not*: 'To test these hypotheses, the theory must be expanded still further, and in such a way as to make the universal force term [read instead: 'gorce'] determinable' (Glymour, 1977, p. 248).

But, as I argued before, under the conditions stated in Section 12.2.3.2, in particular in cases in which the theory already contains all the physics it can and should contain—in case the theory is unextendable—such extensions are simply not given and the inequivalence does not follow. In such a case, no further relevant theory construction could tell force apart from gorce and morce. The latter phrase then surely does not refer to anything independent and distinct from what is meant by 'force', and the two theories *are* physically equivalent. In other words, on an internal interpretation of a theory, Glymour synonymy leads to physical equivalence.

This also illustrates a point that I mentioned in the last but one paragraph of Section 12.2.1.2, namely, that the set of quantities Q need not be restricted to measurable quantities. Indeed, for an unextendable theory, we can perfectly well add gorce and morce as distinct quantities to the set of quantities Q, even if no measurement could possibly distinguish them. In other words, the difference of gorce and morce can be added as a quantity, whose value no empirical data can determine, in an unextendable theory.

Of course, Glymour's argument for the distinctness of the two theories has an ontological component: while Newton postulates one quantity, Hans postulates two. We assumed that we already knew which terms in each sentence referred to *some* things in the world (perhaps without yet knowing *which* things), on an external interpretation. Hans's theory was interpreted as saying that *two* things exist instead of just one, and this implied the inequivalence of the two theories—at least, if we assume that the interpretation maps refer as indicated in Section 12.2.3.2. To explain how this is possible, given that the theories were Glymour-synonymous, one envisaged extending the theory, thus giving an independent account of what these terms refer to: an account of what the existence of these two things would imply, upon formulation of the theory on a larger range of validity within its domain.

That the issue at stake here is not only the theoretical equivalence of the triples but also their physical equivalence, can be seen as follows. Glymour's reply to Hans is that he is committed to two quantities, gorce and morce, rather than a single force, and so, one might be tempted to say that, on my account, this implies that the two theories contain different quantities Q. They are different, as triples, and so not theoretically equivalent. But I don't think this is necessary. In Hans's defence, his claim could be taken to be that gorce and morce are two distinct quantities related by a symmetry, and that this symmetry should lead us to identify his theory with a theory of a single quantity. On this charitable account, Hans is saying that a theory with two quantities and a symmetry relating them is the same as a theory with a single quantity, i.e., he construes symmetries as equivalence relations and instructs us to mod out theories by such symmetries. So, he *is* construing the two theories as theoretically equivalent. But Hans's improved argument again fails to produce physical equivalence, even if he claims that there is theoretical equivalence: the theories are extendable, and so, an extension of the theory beyond its domain may reveal that the purported symmetry is broken and the triples are distinct (see two paragraphs below).

But on an internal interpretation, we cannot assume we possess an account of what 'force' and 'gorce and morce' mean, from outside the theory. The impossibility of an extension, therefore the lack of an independent account of what those terms mean, implies that we should not make such ontological claims *independently of* the equivalence of the two theories. Because the two theories are Glymour synonymous, and there is no extension, they can be taken to be physically equivalent, and so, there are not two quantities but just one.[16] I will now explain how the failure of the unextendability condition (cf. [a] and [b] above) makes the application of an *internal* interpretation problematic, for this particular example.

Notice that considering extensions of theories is, according to the perspective of modern QFT, a basic desideratum of any serious theory. The breakdown of Newtonian mechanics at short distances should be seen as an indication of its being an *effective theory*. [17]

Besides, there is a more specific relation to the gorce-and-morce proposal. Unless the theory is defined to have an exact symmetry, the introduction of new

---

[16] In discussing the internal interpretation, I am assuming that one's formulation of the theory is sufficiently perspicuous (e.g., as a triple), that the physical quantities can be read off from it. If this is not so, it might be equally natural to say that there is no fact of the matter about whether one is committed to one or two quantities—or that such facts are underdetermined by the relevant physical quantities. Cf. the discussion that follows on effective field theory.

[17] Effective field theories are theories that are accurate for phenomena in some range of (usually low) energies, but are corrected by higher-order terms in the Hamiltonian, which are relevant at high energies.

fields will generically introduce higher-order terms that break the seemingly symmetrical way in which those fields appear in the low-energy Hamiltonian. Thus, if gorce and morce are indeed distinct fields,[18] most high-energy theories that reduce to the gorce-and-more theory at low energies, will treat gorce and morce differently. They have different interactions (unless an exact symmetry protects them). Thus the framework of effective field theories promises to satisfy Glymour's demand of parsimony, that there should in principle be a way to determine the values of distinct quantities.

Thus, the force and the gorce-and-morce theories are *generically* not physically equivalent, even though they are Glymour synonymous. For there are very many possibilities for extension to high energies. The physical equivalence with the Newtonian force theory can thus be established only if additional requirements are imposed, such as the stipulation of a particular extension of the theory, or a symmetry. This example sheds light on the concrete question of which theories are likely to be unextendable.

12.3 Gauge/Gravity Dualities

The remarks in the previous section, introducing my conceptions of theory and related notions, were necessarily brief. And as I mentioned in Section 12.2.2, further work is required to illustrate how my conception of duality, and the contrasts (a) to (d), gives the correct verdicts in cases of dualities that are well-understood (cf., De Haro and Butterfield [2018]). In this section, I use the conception of duality, and the contrasts (a) to (d), to shed light on a less-well-understood case, which is of great relevance for our topic, of quantum gravity: gauge/gravity dualities.[19]

In Section 12.3.1, I take up question (1) in the introduction: whether spacetime is eliminated; in particular, whether 'spacetime' is part of what the theory says. I argue that the common core of the duality includes some, but not all, spacetime structure. In Section 12.3.2, I discuss other works on duality. In Section 12.3.3, I discuss some metaphysical implications.

12.3.1 Does What the Theories Say Include 'Spacetime'?

My conception of duality, and the contrasts (a) to (d), give us a way of tackling the question whether 'spacetime' is part of the content of the theory, or whether

---

[18] The sum of gorce and morce is the derivative of the sum of two potentials. I envisage these two potentials as pertaining to distinct fields—since Hans declares gorce and morce to be distinct.

[19] The complete, nonperturbative mathematical theory of gauge/gravity duality is not known. Certain limits of the duality are, however, well-known: the semiclassical limit, in particular. Assuming that some gauge/gravity dualities are exact, at least for a suitable regime of parameters, the scheme illustrates the sense in which two such models are equivalent. For more details, see De Haro, Teh, and Butterfield (2017, sec. 4.2).

it is eliminated. The conceptions of theories and models in Sections 12.2.1 and 12.2.2 suggest that, what the theory says—what is physical about it—is the content that the two models share: the content is preserved under the duality map. That content is the triple of states, quantities, and dynamics: viz., the *common core* of the theory. Thus we need to ask: (1) Whether spacetime structures are common to the two models, and thus belong to the common core. (2) Whether there is an internal interpretation that interprets the terms in the same way, in both models.

Gauge/gravity dualities relate $(d + 1)$-dimensional models of quantum gravity to d-dimensional quantum field theories (QFT models) with gauge symmetries. And suppose that two such models are dual, in the sense of Section 12.2.2. Thus we have an isomorphism between the two Hilbert spaces and between the physical quantities—so we are, in fact, considering unitary equivalence. For brevity, I reduce our two questions, (1) and (2), to the following single question: whether the common core, shared by these two models, is *interpreted as spatiotemporal.*

I will next argue that the common core *is* spatiotemporal, that it includes a d-dimensional spacetime $\mathcal{M}$, whose metric is defined only up to local (spacetime-dependent) conformal transformations (De Haro et al., 2016, secs. 6.1.2–6.1.3), i.e., a conformal manifold. This works as the 'core' theory in the following sense. We examine the properties of the two models under the duality map, and the structure that is mapped by duality, once formulated in a model-independent way, is the content of the core triple. In fact, the duality map itself (especially its formulation in De Haro et al. [2017, sec. 4.2]), already makes explicit what the states and the operators of the triple are. So, let us look at the two models and extract their common structure.

To obtain the quantities that the gravity model (under its standard interpretation) takes to be physical, one evaluates the path integral over all metrics and topologies with given *boundary conditions*:[20]

i. The first boundary condition is an asymptotic condition on the form of metric, which is itself determined only up to a conformal factor; in other words, one needs to specify a conformal d-dimensional manifold $\mathcal{M}$ together with a conformal class of metrics, denoted $[g]$, at the boundary.

The asymptotic symmetry algebra associated with this model is thus the conformal d-dimensional algebra, and the representations of this algebra form the class of possible states that belong to $\mathcal{H}$.

---

[20] For simplicity, I am now considering the case of quantum gravity without matter. This restricts our considerations to a class of states and operators, with specific conformal dimensions. Adding matter can be done, and does not affect the philosophical conclusions, but would involve some technical qualifications. Also, in the rest of this section (except for one example with $d = 4$), I restrict the discussion to the case where d is odd—again, purely for technical reasons.

ii. Second, a boundary condition needs to be imposed on the asymptotic value of the canonical momentum Π_g conjugate to the metric induced on the boundary, evaluated on the possible states. This choice further constrains the class of states in $\mathcal{H}$; it determines a subset of states of the conformal algebra. The simplest choice, $\langle s | \Pi_g | s \rangle = 0$, preserves the full conformal symmetry, and the states $s \in \mathcal{H}$ are, accordingly, representations of the d-dimensional conformal algebra. Other choices break the conformal symmetry and further constrain the state space.

Thus the boundary condition (ii) fixes a choice of the *subset of states* (representations of the conformal algebra) that forms the dynamical Hilbert space of the theory, whereas the first fixes a choice of a *source* that is turned on for the canonical momentum Π_g. We will write the resulting states as $|s\rangle_{\mathcal{M},[g]} \in \mathcal{H}$, where s is the state, modified by the addition of a source $[g]$ on $\mathcal{M}$ coupling to Π_g. The basic physical quantities are the canonical momenta $\Pi_g \in \mathcal{Q}$ conjugate to g (De Haro et al., 2017, sec. 4.2.2.1).[21]

The quantum field theory model of the theory is a conformal field theory (CFT) on a d-dimensional manifold whose metric is defined, up to a local conformal factor, by the very form of the asymptotic metric that one gets from the gravity model. In fact, we can identify this, via duality, with the conformal manifold $\mathcal{M}$. The states are representations of the conformal symmetry algebra, the same algebra that we obtain in the gravity model. The canonical momentum Π_g corresponds, through the duality map, to the stress-energy tensor T_{ij} of the CFT: $\Pi_g \equiv T$. The stress-energy tensor is the operator from which the generators of the conformal symmetry algebra can be constructed (see e.g., Ammon and Erdmenger [2015, sec. 3.2.3]).

Thus, the two models share the d-dimensional conformal manifold $\mathcal{M}$ with its conformal class of metrics $[g]$, the conformal algebra, and the structure of operators, as claimed. The conformal algebra and class determine $\mathcal{H}$, and thereby the valuations of the important subset $\{T_{ij}\}$ of operators of $\mathcal{Q}$, i.e., the infinite set of correlation functions $_{\mathcal{M},[g]}\langle s | T_{i_1 j_1}(x_1) \cdots T_{i_n j_n}(x_n) | s \rangle_{\mathcal{M},[g]}$, where $x_1 \ldots, x_n \in \mathcal{M}$, for any n. This infinite set of correlation functions contains important dynamical information about the CFT.[22] For us, the important point here is that *the common core is spatiotemporal*.

Since my aim is to illustrate the fact that the bare theory contains quantities that can be spatiotemporally interpreted, it will not be necessary to develop the full internal interpretation of the theory. Rather, our next question is whether

---

[21] The metric in the interior is not a quantity. There are two ways to see this: (i) from the fact that it is not part of the common core preserved by the duality. But also (ii) it *cannot* be (on the minimalist account of quantities given in this paper), since it is not a diffeomorphism-invariant quantity.

[22] There is no claim here that my description contains *all* of the information about the CFT. Nonlocal operators, such as Wilson loops (and perhaps additional states), also need to be compared.

there is an internal interpretation that supports the spatiotemporal interpretation of these quantities.

Two salient elements of an *internal interpretation* of the bare theory can be recognized. Notice that *both* models interpret the pair $(\mathcal{M}, [g])$ and T_{ij} as representing, respectively, (i) a d-dimensional conformal manifold with a conformal class of metrics and (ii) stress, energy, and momentum. Even if there are other aspects to the internal interpretation of $[g]$ and T_{ij}, (i) and (ii) are certainly important elements of it!

We are considering here gravity models with *pure gravity*, and no matter fields. The QFT, of course, *does* have matter fields. But the stipulation of the *specific set* of matter fields is *not* part of the invariant core. The interpretation of T_{ij} as the 'stress-energy momentum *for a specific set of matter fields of the QFT*' can thus not be part of the internal interpretation, because the specific set of matter fields are not part of the common core. Thus, the qualifications of fields as being 'gravity', or 'matter', are our *descriptions* of specific models, and not parts of the theory!

To illustrate this point, let us discuss the common core and the specific structure further in an example of $d = 4$: i.e., a 4-dimensional QFT and a 5-dimensional gravity model.[23] Specifically, we look at symmetries. The gauge symmetry group of the QFT is SU(N). This symmetry is completely absent from the gravity theory. The QFT is formulated so that this symmetry is explicit—the states $\mathcal{H}$ and the observables $\mathcal{Q}$ are invariant under it. But since the common core contains only states and quantities constructed from the triple $T = \langle \mathcal{H}, \mathcal{Q}, \mathcal{D} \rangle$, which are invariant under gauge symmetry, this gauge symmetry does *not* belong to the common core!

There is thus surprisingly little that is invariant under the duality between these two theories: yet they describe all that is physical about the theories. The $(d + 1)$-dimensional manifold $\hat{\mathcal{M}}$, most of its diffeomorphism group, gauge symmetries, the dimension of spacetime. In the present context, these are apparently all part of our description, rather than parts of the common core of the theory. Similarly, also the concepts of vector fields, tensor fields, Lie groups, differential geometry: though we use them to formulate our theories, each such a concept is not part of nature, at least not part of the common core of the models of a gauge/gravity duality. For example, tensor quantities in $d + 1$ dimensions do *not* map to tensor quantities in d dimensions under duality, and so do not belong to $\mathcal{Q}$. Rather, tensors are part of the specific structure!

Thus, the answer to the question we posed in the title of this subsection requires careful articulation, as follows. I state it in terms of what is eliminated from the gravity model:

[23] The QFT in question is a 'super Yang-Mills theory', a specific supersymmetric variant of Yang-Mills theory, but the details are irrelevant here.

i. The entire interior region of the $(d + 1)$-dimensional manifold, $\hat{\mathcal{M}}$ (including its topology), is eliminated.

ii. All that remains is the *asymptotic* conformal manifold $(\mathcal{M}, [g])$, which plays the role of asymptotic boundary data for the equations of motion, and so is arbitrary but fixed, and the states $|s\rangle_{\mathcal{M},[g]} \in \mathcal{H}$ and the stress-energy tensor $T_{ij} \in \mathcal{Q}$ at spacelike infinity. This common core is endowed with an action of the conformal group. So, in particular:

iii. All local gravitational structure has been eliminated: there are no *local* dynamical gravitational degrees of freedom left.

In short, 'most' of the spacetime structure is eliminated. This agrees with the general expectation in the quantum gravity literature.

12.3.2 Comparing with Recent Work on Dualities

Let us take stock of the distinctions we have made and discuss how they relate to extant philosophical discussions of dualities in the literature. In this section, I address recent work on dualities and state its limitations, as regards clarifying the *contrast between theoretical and physical equivalence*.

Recent work on dualities engaging with this question includes Matsubara (2013, p. 485) and Dieks et al. (2015, sec. 3.3.2). And in a special issue on dualities edited by Castellani and Rickles, several authors engage with it. My leading criticism is that the extant accounts, *qua* accounts of dualities, are not sufficiently articulated to provide a clear difference maker between theoretical and physical equivalence. In defence of these authors, I should add that it was not their aim to look for a clear-cut difference maker!

Huggett's (2017) focus seems closest to the ideas developed in this essay. His analysis of T-duality is syntactic, and as such is an interesting alternative to mine, which, as mentioned, is closer to the semantic conception of theory (though I compared with Glymour's syntactic account, in Section 12.2.4). T-duality, roughly speaking, relates one kind of string theory in a space with a circle of radius R to another kind of string theory in a space with a circle of radius $1/R$. Huggett asks: 'What happens if a duality applies to a "total" theory, in the sense that it is the complete physical description of a world, so that there is nothing outside the theory?' (2017, p. 86). His position is that, for dual models that describe subsystems, the duals are really distinguishable, by looking outside the model. But theories of the universe lack such a viewpoint, and so no distinction needs to be made between dual models.

We must of course require the bare theories to have correct rules for forming propositions and to be mathematically consistent (Sections 12.2.1.1 and 12.3.3).

But the requirement that the theory be 'a theory of *everything*' could not be taken as a general condition for an internal interpretation, for we do not need the theory to describe absolutely *all* the facts, not even all the physical facts, of a world, in order for it to admit an internal interpretation. Rather, what is needed for physical equivalence are the joint requirements of internal interpretation and unextendability (Sections 12.2.3.2 and 12.2.3.3). My construal, in Section 12.2.1.2, of a physical interpretation I_T as a pair of partial maps, which can be defined for an unextendable theory on a domain D_W, makes this point precise, so that one can now distinguish external and internal interpretations.

One interesting aspect of Rickles's (2017) account is his assertion that theoretical equivalence is a *gauge-type* symmetry for *all* cases of duality. Our accounts differ in this obvious sense, that mine is stated as an isomorphism between triples, rather that as a gauge-type symmetry. But the most important difference is that Rickles does not seem to contemplate cases of sophisticated dualities in which the theories could *fail to be* physically equivalent. If one's account moves too quickly from duality to physical equivalence, it may render the external, and multifaceted, uses of dualities (cf. the third paragraph of Section 12.1, and the last line of Section 12.2.3.3) unintelligible.

Indeed, the account of dualities as just being gauge-type symmetries does not explain the reasons for two theoretically equivalent models' physical inequivalence. The explanation of that *is* possible once we develop the idea of an external interpretation.

Finally, I turn to Fraser (2017). My main point will revolve around her chosen example, of Euclidean field theory (EFT) and QFT, not even being a case of theoretical equivalence, let alone physical equivalence. Elucidating the distinction between theoretical and physical equivalence seems indeed to be the main aim of Fraser's (2017) example of equivalence between EFT and QFT. She maintains that EFT and QFT are *theoretically (formally) equivalent* but not physically equivalent. She contrasts this with dualities in string theory, which she describes as cases of *physical equivalence*. The example would seem to be well chosen indeed, and Fraser's mastery of QFT is indisputable; but, unfortunately, she does not give a detailed account of how these two cases are supposed to differ.

More importantly, the example *itself* is deceptive if taken as a difference maker for theoretical and physical equivalence, for it is in fact *not* a case of theoretical equivalence! As Fraser admits in the first three sections of her essay, EFT and QFT are *not* isomorphic; there is a map from EFT to QFT but not the other way around. She writes: 'That the relations are entailments rather than equivalences is not one of the points of comparison between the EFT-QFT case and string theory that I want to emphasize' (Fraser, 2017, sec. 3, par. 3).

One may, of course, choose to downplay the role of equivalence and focus on a one-way entailment, if one is just interested in a *generic* contrast between the

EFT-QFT case and dualities in string theory. But one-way entailment will *not* give us the difference maker we need in order to distinguish theoretical from physical equivalence, for the very reason that these are *not cases of theoretical equivalence*.[24] As I have argued, the contrast between external and internal interpretations *is* such a difference maker.

12.3.3 What Are the Broader Implications of Duality?

In this section, I discuss the broader implications of gauge/gravity dualities, in three related comments.

The first comment concerns the interpretation of quantum gravity theories. In cases in which a duality supports an internal interpretation, duality can be taken to give rise to *physical equivalence* (cf. Section 12.2.3.2). In Section 12.3.1, the internal interpretation was developed by stripping the two *external interpretations* of irrelevant aspects, and identifying their common core, which constitute (at least parts of) the internal interpretation. This means that the interpretation of a quantum gravity theory, and the articulation of the ontology that may underlie such a theory, starts from the common core of the different models and the parts of their interpretations that are invariant under duality. This common core can turn out to have surprising properties, e.g., its spacetime dimension being different from the dimensions of the spacetimes of the models.[25]

The second comment concerns physical equivalence. Duality can be found in both unextendable and extendable theories. Gauge/gravity dualities between string or M theory and QFTs are presumed to be exact dualities between unextendable theories, which in particular should have exact formulations.[26] Other dualities, on the other hand, are also exactly defined, but only with limited regimes of applicability (cf. Section 12.2.4's example of the Newtonian force theory and Hans's gorce and morce theory). Such theories are extendable, and for extendable theories we are not always justified in believing internal interpretations, because an extension might modify that interpretation (unless the interpretation is *robust*: cf. footnote 14). This is because a coherent interpretation is always dependent on how the theory is extended. In such cases of extendable theories about which we do not know whether they are robust, we are not justified in taking the theories to be physically equivalent. The effective field theory perspective in Section 12.2.4 suggested that such theories are generically *not* physically equivalent.

But physical equivalence for *unextendable* theories that are theoretically equivalent under an internal interpretation Section (12.2.3.2) holds good. Such theories

---

[24] One might argue that something might still be learned from the contrast between *one-way theoretical entailment* and *one-way physical entailment*.

[25] For a more comprehensive discussion of the interpretation of dualities in quantum gravity, see De Haro (2019).

[26] For examples in which the duality is known to hold exactly, see De Haro and Butterfield (2018).

are highly constrained, valid for all values of the parameters, and cannot be coupled to any adjacent physics while preserving their theoretical equivalence.

This distinction may be seen as a contribution of quantum gravity considerations to discussions of physical equivalence. It also underlines the importance, for philosophy, of unextendable theories. Extendable theories abound: we have discussed Newtonian mechanics and effective QFTs. There is also general relativity, whose extendability is shown by its singularities due to gravitational collapse. Unextendable theories are rare but useful, and we do have some good examples of them: conformal field theories, topological quantum field theories (Chern–Simons theories, various versions of Yang–Mills theory, Wess–Zumino–Witten [WZW] models) and topological string theories (describing subsectors of string theories). They should provide important *case studies for the philosophical topics of theoretical and physical equivalence.*

Third, concerning theory construction: note that unextendable theories need not be 'finished' theories. Some theories (especially some 2-dimensional conformal field theories) are indeed understood with rigorous mathematics; but other examples (such as Chern–Simons theory and Yang–Mills theory), though expected to be unextendable for good mathematical reasons, are still 'theory fragments', in the sense of Huggett and Wüthrich's (2013, p. 284) quotation in the introduction. So, also for the theories mentioned earlier: completely rigorous mathematical proofs are still lacking, even if the fragments are robust enough that they already contributed to a Fields medal (for E. Witten in 1990).

12.4 Envoi

Let me end by echoing an important remark: the discussion, in Section 12.3, of the philosophical significance of gauge/gravity dualities, returns us to the idea (echoed, for instance, in the quotation by Huggett and Wüthrich) that philosophical analysis goes hand in hand with theory construction.

The analysis of gauge/gravity dualities shows specific features of this two-way street. On the one hand, concepts such as *theoretical and physical equivalence* (Section 12.2) help us construct theories of quantum gravity and so brings metaphysical analysis to bear on theory construction.

But also, on the other hand, theories of quantum gravity, in particular gauge/gravity dualities, help us achieve greater clarity about those two philosophical concepts, thereby exhibiting the virtues of a 'science first' approach to metaphysics.

Acknowledgments

I thank Jeremy Butterfield, Nick Huggett, Huw Price, Bryan Roberts, and three anonymous referees for comments on the paper. I also thank several audiences: the British Society for the Philosophy of Science 2016 annual conference, the Oxford philosophy of physics group, LSE's Sigma Club, the Munich Center for Mathematical Philosophy, and DICE2016. This work was supported by the Tarner scholarship in Philosophy of Science and History of Ideas, held at Trinity College, Cambridge.

References

Ammon, M. and Erdmenger, J. (2015). *Gauge/Gravity Duality: Foundations and Applications*. Cambridge: Cambridge University Press.

Barrett, T. W. and Halvorson, H. (2016). Glymour and Quine on theoretical equivalence. *Journal of Philological Logic*, 45(5), 468–483.

De Haro, S. (2017). Dualities and emergent gravity: Gauge/gravity duality. *Studies in History and Philosophy of Modern Physics*, 59, 109–125.

De Haro, S. (2019). The heuristic function of duality. *Synthese*, 196(12), 5169–5203.

De Haro, S. and Butterfield, J. N. (2018). A schema for duality, illustrated by Bosonization. Pages 305–376 of: Kouneiher, J. (ed.). *Foundations of Mathematics and Physics One Century after Hilbert*. Cham, Switzerland: Springer.

De Haro, S., Mayerson, D. and Butterfield, J. N. (2016). Conceptual aspects of gauge/gravity duality. *Foundations of Physics*, 46, 1381.

De Haro, S., Teh, N. and Butterfield, J. N. (2017). Comparing dualities and gauge symmetries. *Studies in History and Philosophy of Modern Physics*, 59, 68–80.

Dieks, D., Dongen, J. van and Haro, S. De. (2015). Emergence in holographic scenarios for gravity. *Studies in History and Philosophy of Modern Physics*, 52(B), 203–216.

Fraser, D. (2017). Formal and physical equivalence in two cases in contemporary quantum physics. *Studies in History and Philosophy of Modern Physics*, 59, 30–43.

Glymour, C. (1977). The epistemology of geometry. *Noûs*, 11(3), 227–251.

Huggett, N. (2017). Target space ≠ space, *Studies in the History and Philosophy of Modern Physics*, 59, 81–88.

Huggett, N. and Wüthrich, C. (2013). Emergent spacetime and empirical (in)coherence. *Studies in History and Philosophy of Modern Physics*, 44(3), 276–285.

Lewis, D. (1984). Putnam's paradox. *Australasian Journal of Philosophy*, 62 (3), 221–236.

Matsubara, K. (2013). Realism, underdetermination and string theory dualitites. *Synthese*, 190, 471–489.

Rickles, D. (2012). AdS/CFT duality and the emergence of spacetime. *Studies in History and Philosophy of Science Part B: Studies in History and Philosophy of Modern Physics*, 44(3), 312–320.

Rickles, D. (2017). Dual theories: 'Same but different' or 'different but same'? *Studies in the History and Philosophy of Modern Physics*, 59, 62–67.

13

On the Empirical Consequences of the AdS/CFT Duality

RADIN DARDASHTI, RICHARD DAWID, SEAN GRYB,
AND KARIM THÉBAULT

13.1 Introduction

The mathematical, physical, and conceptual consequence of dualities has over the last decade been subject to growing philosophical literature.[1] There has not, however, hitherto been any detailed philosophical analysis of the *empirical* consequences of dualities. This paper aims to address this deficit by considering a cluster of questions relating to the empirical consequences of the anti–de Sitter spacetime conformal field theory (AdS/CFT) duality. Before we outline the goals of our discussion it will be instructive to review the details of the duality itself.[2] The AdS/CFT duality relates the physics of 'bulk' gravity theories on asymptotically anti-de Sitter spacetimes to that of 'boundary' conformal field theories. While there are many concrete examples of this duality,[3] we will be focusing on the most discussed example of the AdS/CFT duality, namely between $SU(N)$, $\mathcal{N} = 4$ Super Yang–Mills theory in 3+1 dimensions and type IIB superstring theory on $AdS_5 \times S^5$ introduced by Maldacena (1999) and further developed in Gubser, Klebanov, and Polyakov (1998), and Witten (1998).

The strong form of the duality is defined in the 't Hooft limit where the number of color charges $N \rightarrow \infty$ but the 't Hooft coupling $\lambda := g_{YM}^2 N$ is fixed. In this context, the duality is between the expectation value of the Wilson loop operator for an SU(N), $\mathcal{N} = 4$ Super Yang–Mills theory in a flat spacetime and the semi-classical partition function of a macroscopic string in $AdS_5 \times S^5$ whose worldsheet Σ ends on the path of the Wilson loop at the boundary. This is thus a gauge/gravity duality with the Super Yang–Mills theory playing the 'gauge part' and the string in $AdS_5 \times S^5$ playing the 'gravity' part. The duality is between the mathematical

---

[1] See for example Dawid (2007), Rickles (2011), Dawid (2013), Matsubara (2013), De Haro (2017), Fraser (2017), Rickles (2017), Huggett (2017), and De Haro and Butterfield (2018); see also Chapter 12.

[2] Here and later we are, for the most part, following the excellent treatment of Ammon and Erdmenger (2015). For historical details on AdS/CFT duality, see Rickles (2014, sec. 10.2).

[3] See, e.g., Horowitz and Polchinski (2009) and Kaplan (2013).

objects that represent all possible empirical observations on two sides. It is crucial to note that the existence of this duality as an exact relation is largely conjectural: it remains mathematically unproven despite being extensively verified in explicit cases. In what follows we will put the formal status of the duality to one side and simply assume that it holds. Furthermore, for ease of reference we will refer to this precise specification of the duality below as 'AdS/CFT' with the relevant specifications implicit.

To return to the goals of the paper. Given that our aim is to assess the empirical consequences of dualities an immediate question is why we have chosen to focus on the AdS/CFT duality. There are a number of obvious reasons why both sides this duality might be taken to be empirically irrelevant. First, anti–de Sitter spacetimes are characterized by a negative cosmological constant, so the gravity side of the duality is evidently not well suited to describe the geometry of our universe, which according to observation has a positive cosmological constant.[4] Second, on the field theory side, we have good reasons to expect that the number of color charges is finite. Third, none of the empirically adequate quantum field theories that make up the standard model is conformally invariant. It is clear, therefore, that according to our best current understanding our world is not AdS and none of the known interactions are described by a conformal field theory. Thus the duality is applicable neither to the world as a whole nor for a description of nuclear interactions.

Why then consider the empirical consequences of AdS/CFT? Our two main reasons are related to two very different applications. First, we hope to learn something about our actual world by grasping the precise duality. This may happen either due to a so far undiscovered general gauge/gravity duality or, more modestly, if some aspects of AdS/CFT can be applied to other contexts as well.[5] In the latter case, one hopes for *conceptual* similarities between the AdS/CFT case and the so far insufficiently understood theory describing actual physics at the Planck scale. Second, in recent years AdS/CFT has been applied to contexts, like a quark-gluon plasma near criticality, which are not described by a CFT but that turn out to be quantitatively approximately describable by a CFT (which, in turn is calculable in the dual gravity theory). In that case, there *are* empirical similarities between predictions of the AdS/CFT theory and the real world. Thus there are good reasons to be interested in the empirical consequences of the AdS/CFT duality, despite the apparent strangeness of the topic.

The natural starting point for a philosophical investigation into the empirical consequences of an area of science is to consider questions relating to confirmation.

---

[4] Here we are assuming the duals to represent the entire universe rather than a subsystem with the relevant boundary conditions.

[5] For example, see Harlow (2016) for an application to the study of the unitarity of black hole physics.

The following strike us as particularly pertinent: (1) Must we reunderstand the role of inductive evidence once dualities come into play? (2) Are both sides of a duality equivalently confirmed by collection of supporting empirical evidence? (3) What are the implications of our view on confirmation for scientific ontology? (4) What are the empirical consequences of using dualities to model quark-gluon plasma? Although its use in scientific practice is rather varied, what we mean here and later by *confirmation* is the specific term of art as used in contemporary philosophy of science. That is, confirmation as a relation of inductive support between some evidence and some hypothesis. Significantly, we will, for the most part, later always be making reference to notions of confirmation that that are not putatively conclusive: we will almost always be talking about whether or not there is *partial* inductive support of a conclusion by evidence. A precise framework for analyzing such relations of partial inductive support in terms of 'personal probabilities' (or credences) is Bayesian confirmation theory, and we will apply this framework extensively in this chapter.[6]

Our analysis will be conducted with reference to three distinct contexts. The first context follows from the observation that, although the AdS/CFT duality is itself extremely unlikely to feature in a fundamental or effective description of our universe, it is a well-studied exemplar of the general type of gauge/gravity duality that could be expected to be physically salient. As such, we adopt the epistemic standpoint of an agent situated in a universe where AdS/CFT is a feature of the *fundamental* theoretical framework representing the world. Although in this sense rather wildly hypothetical, we expect our analysis be broadly applicable to the empirical and ontological implications of any fundamental gauge/gravity duality. The second line of inquiry considers a distinct hypothetical scenario where the duality relation is a characteristic of an *effective* description of the world that does not hold at the fundamental level. Although this context is importantly different from the first, we find that many aspects of our analysis carry through. The third line of inquiry addresses the way the AdS/CFT duality has been put to empirical use in the description of quark-gluon plasmas. In such contexts the duality is combined with a limiting relation between conformal field theories and certain parameter regimes of quantum field theories like QCD. We find that in this context the empirical consequences of the AdS/CFT duality must be understood in a very different way. In particular, we conclude that one should not take the partially empirically successful application the AdS/CFT duality to the study of quark-gluon plasmas to have told us anything regarding the empirical status of string theory.

---

[6] For models of confirmation in terms of the Bayesian framework, see Hartmann and Sprenger (2010) and Bovens and Hartmann (2004).

13.2 Fundamental Theory Context

In this section we concern ourselves with the empirical consequences of AdS/CFT seen as part of a fundamental theory. We adopt the epistemic standpoint of an agent situated in a universe where the AdS/CFT correspondence is part of a theory that is believed to provide a *fundamental predictive framework*. A fundamental theory is a framework for producing predictions concerning phenomena in a given domain that does not provide an effective theory for, or approximation to, an underlying theory that alters at least some of its core posits. Note that, while a theory is factually either fundamental or not, fundamentality can only be claimed based on an agent's epistemic standpoint, which may change in time. Newtonian mechanics was taken to be fundamental in the given sense by most physicists in the nineteenth century. It is not taken to be fundamental today.

Our notion of 'fundamental theory' is a narrow one in two senses. First, by *fundamental* we do not mean 'universal'. While fundamentality does imply that the theory remains viable to arbitrarily small distance scales, there is no assumption that a fundamental theory provides a complete description of all physical facts. For example, in the nineteenth century Newtonian mechanics could be understood as a fundamental theory but not applicable to the domain of electromagnetic phenomena. It would thus be nonuniversal. Our line of inquiry here thus differs substantially from one pursued by De Haro and Butterfield (2018) who consider the implications of dualities in the context of 'a physical theory as a putative theory of the whole universe, i.e. as a putative cosmology, so that according to the theory there are no physical facts beyond those about the system (viz. universe) it describes' (pp. 37–38). Theories that are universal in this sense are necessarily fundamental in our sense, but fundamental theories are not necessarily universal.[7]

The second aspect of narrowness in our definition of fundamental is particularly important. Although fundamentality, in our sense, does relate to nonempirical aspects of a theory it does not, on our view, render a theory entirely closed to conceptual revision. The precise definition of fundamentality is particularly sensitive in the context of string dualities. It is sometimes suggested that M-theory, which is conjectured to stand in an exact duality relation (and therefore to be exactly empirically equivalent) to certain types of superstring theory, is more fundamental than the superstring formulations because it offers more direct access to the degrees of freedom that govern string physics beyond the perturbative regime. On our definition of fundamentality, M-theory is not more fundamental than, let's say, Type IIA superstring theory for the very reason that the two theories are (by conjecture) empirically equivalent to each other.

---

[7] The question of whether or not universal theories are possible or in principle would involve wading into the deep and murky waters of the reduction/emergence debate. Something we are at pains to avoid.

Thus, what we call a fundamental theory context can be indicated more precisely in two steps. First, a theory is delimited in terms of a set of core theoretical posits and a target domain of empirical phenomena. Second, the theory is a fundamental theory if and only if there is no further theory that is more empirically adequate in the target domain and does not share the same core theoretical posits. A theory that is 'superseded' by a successor theory with the very same empirical predictions but a different set of fundamental posits will still count as a fundamental theory.

Now that we have some clear sense of what we mean when we talk about fundamental theories, it will be useful to attempt to achieve some clarity regarding what we mean by *empirically equivalent* theories. The most well known and discussed notion of equivalence in the context of scientific theories is empirical equivalence. A minimal schema for describing such form of equivalence can be specified as follows. Take a pair of *uninterpreted* theories as being each given by some pair of linguistic specifications (models, sentences, categories, equations, words,...) that we denote as T1 and T2, respectively. A minimally instrumentalist interpretation, $\mathcal{II}$, of such theories would be a mapping that connects linguistic items (or groups of items) to observables (pointer readings, scattering amplitudes, physical quantities,...) that are either introduced in terms of a further language or extra-linguistically. An ontological interpretation, $\mathcal{OI}$ of such theories would be a richer mapping that includes connections between linguistic items (or groups of items) and nonobservables (entities, objects, structures,...) such that nonobservables are understood (in some sense) to be existent things represented by the linguistic items. Empirical equivalence is then given by a situation where $\mathcal{II}(\text{T1}) = \mathcal{II}(\text{T2})$.

A sufficient, but not necessary, condition for empirical equivalence is the existence of a mapping *at the linguistic level* that is such that all linguistic items that are suitable for interpretation as observables are appropriately connected. Such a 'translation manual' between theories is precisely what is found in the case of dualities in general and the AdS/CFT duality in particular. It is easy to see why AdS/CFT implies empirical equivalence as per earlier: given the reasonable assumption that the expectation value of the Wilson loop operator on the gauge side and full partition function on the gravity side are the *only* things that minimally instrumentalist interpretation connects to observables, we immediately have that $\mathcal{II}(\text{AdS}) = \mathcal{II}(\text{CFT})$ in the respective semiclassical and 't Hooft limit. The AdS/CFT duality gives two different groups of linguistic items that describe *the same observables*.[8]

---

[8] Much more could be said, of course, toward a fuller characterization of both empirical equivalence in general and its relation to a formal characterization of dualities. Giving such an account is explicitly not our purpose here. This notwithstanding, we do take our minimal schema to be sufficient to our current purpose of making clear the empirical significance of dualities in our three contexts. For a much less minimal account of duality and equivalence, see De Haro and Butterfield (2018) which builds upon De Haro (Chapter 12). For further recent work on a similar theme, see Fraser (2017) and Rickles (2017).

These definitions and qualifications now made, we can proceed to our principal line of analysis. We would like to adopt the epistemic standpoint of an agent situated in a universe where the AdS/CFT correspondence is a core posit of a fundamental theory. To what extent should we take confirmation of the theory by evidence to be suitably dual? That is, could an agent take only one side of the duality as supported by the empirical success of the theory or must they consider empirical evidence for one side of the duality as necessarily evidence for the other? The inferential standpoint of our hypothetical agent can be modeled in terms of Bayesian confirmation theory as follows. We denote by H_1 the proposition:

H_1: Type IIB supersting theory on $AdS_5 \times S^5$ provides an empirically adequate description of the target system $\mathcal{T}$ within a certain domain of conditions D_{AdS} that include the semiclassical limit.

Similarly, we denote by H_2:

H_2: $SU(N), \mathcal{N} = 4$ SYM in $3+1$ provides an empirically adequate description of the target system $\mathcal{T}$ within a certain domain of conditions D_{CFT} that include the 't Hooft limit.

The negation of these propositions are defined accordingly. Finally, we can encode the evidence available to our agent by the propositions E, the empirical evidence obtains, and ¬E, the empirical evidence does not obtain.

Let us assume that the agent has derived the evidence from the AdS side of the duality. Since this is the weakly coupled side this will invariably be technically more straightforward. We thus have that $H_1 \to$ E which means (almost trivially) that $P(H_1|E) > P(H_1)$, given that $0 < P(H_1) < 1$ and $0 < P(E) < 1$, and so we have confirmation. It is instructive to break this very standard inference down a little. By Bayes's theorem:

$$P(H_1|E) = \frac{P(E|H_1)P(H_1)}{P(E)}. \tag{13.1}$$

The marginals, $P(H_1)$ and $P(E)$, we take to be $x \in \mathbb{R}$ s.t. $0 < x < 1$, but otherwise rationally unconstrained. Given that we have assumed that $H_1 \to$ E it follows necessarily that the likelihood is equal to one, i.e., $P(E|H_1) = 1$. It then immediately follows that the confirmation measure $\triangle_1 = P(H_1|E) - P(H_1)$ is greater than zero, and we have Bayesian confirmation of H_1 by E.

Consider then what we can say about H_2 given E in the fundamental theory context and with AdS/CFT assumed. Since we have by $\mathcal{II}(AdS) = \mathcal{II}(CFT)$ the empirical equivalence of both sides, it follows that $H_2 \to$ E as well. So again we have that the likelihood should be set to one, i.e., $P(E|H_2) = 1$. Things are a little trickier regarding the marginal $P(H_2)$. In principle, even despite the fundamental theory context and assumption of AdS/CFT one might still set $P(H_1) \neq P(H_2)$ since these are subjective degrees of prior belief regarding the truth of logically distinct statements. We think in the fundamental theory context it seems intuitively clear that one *should* set $P(H_1) = P(H_2)$ since there is no clear physical reason for assigning different priors in the fundamental case.[9] However, clearly the sign of the confirmation measure will be positive irrespective of the prior: so long as $0 < P(H_2) < 1$, we already have $\Delta_2 > 0$ and thus we have confirmation of H_2 by E. Moreover, if we define the relative confirmation measure as:

$$\bar{\Delta}_i = \frac{P(H_i|E) - P(H_i)}{P(H_i)}, \tag{13.2}$$

then just by the equation of likelihoods we immediately have that $\bar{\Delta}_1 = \bar{\Delta}_2$, irrespective of the priors for H_1 and H_2. This point will be significant for the discussion of the effective theory context later.

In addition to questions regarding empirical adequacy, we might also consider the inferences regarding further ontological interpretation that our hypothetical agent might make based upon the evidence obtained. There are three possible options corresponding to the popular philosophical positions on the realist–antirealist spectrum. Most straightforwardly, an ontic structural realist (Ladyman, 2016) would plausibly argue for an ontological interpretation such that the common structure between the two sides of the duality is reified. A shared structural vocabulary common between the two sides of the duality that would be taken to do the work in representing the 'actual structure of the world'. This would imply directly that $\mathcal{OI}(\text{AdS}) = \mathcal{OI}(\text{CFT})$. Also straightforwardly, an instrumentalist (Stein, 1989) or empiricist (Van Fraassen, 1980) would simply refuse to license further beliefs about extra-observables and would reject any commitment to entities, objects, or structures *actually in the world* as faithfully being represented or not.

The situation of a standard realist is more subtle. Such realists typically justify belief in the faithful representation of extra-observables by theoretical terms by reliance on empirical evidence combined with inference to the best explanation. Depending upon what one thinks about the explanations on offer, this might seem to allow for an ontological interpretation such that $\mathcal{OI}(\text{AdS}) \neq \mathcal{OI}(\text{CFT})$. In particular, our hypothetical realist agent might reason that the gravity side of the duality is more fundamental and that their universe *really is* described by a string theory in anti–de Sitter spacetime and not in any way by a Super Yang–Mills theory. This would be to take the CFT seriously only as an uninterpreted framework and ignore its putative claims for representational capacity. Plausibly,

---

[9] Here *should* is understood in some normatively weaker sense than 'an agent is rationally compelled to'.

such a position would be unstable since in interpreting the AdS side one is necessarily implicitly also interpreting at least the empirical sector of the CFT side. Whatever explanation of empirical regularities is appealed to for justifying belief in an ontology based upon the AdS side, a parallel justification is available for belief in an ontology of the CFT side. And since we are, by assumption, dealing with fundamental theories the virtue of such explanations for future generalizations cannot be used as a tie-breaker between them. Essentially, the joint claim that we have a fundamental predictive framework and that the ontological interpretation of such a framework can be based upon inference to the best explanation, force our hypothetical realist to accept a dual ontology at the pain of otherwise making an arbitrary, noninferential ontological commitment. In this vein, it has been argued by a wide range of philosophers of science (Dawid, 2007; Rickles, 2011, 2017; Matsubara, 2013; Huggett, 2017) that an exact duality renders the duals different formulations of one and the same theory. One way to avoid this perspective would be to take dualities as approximate rather than exact (De Haro, 2017). An alternative is to stick to the understanding that dualities are exact but move to a context in which the dual theories are understood to be effective. In the following section we consider the implications of the AdS/CFT duality in this effective theory context.

13.3 Effective Theory Context

Recall from earlier that we defined the fundamental theory context as a situation where there is no further theory that is more empirically adequate in the given target domain and does not share the same core theoretical posits. The contrast case that we consider in this section is the 'effective theory context' where there *is* a further theory that is more empirically adequate in the given target domain and does not share the same core theoretical posits. There are two reasons why this is a particularly interesting context to consider dualities within. First, it addresses a possible status of exact dualities that, to our knowledge, has not previously been discussed in the literature. Second, it brings us one step closer to the application of dualities to quark-gluon plasma calculations, which is discussed in the final section.

What makes the effective context relevant to our investigation is its potential to break the principle of equal confirmation of dual theories that seems unassailable in the context of fundamental dualities. To be sure, effective theories, even when known not to be fundamental, can be strongly confirmed by empirical data within the theories' intended domains. Most empirical confirmation plays out in such contexts. However, situations can arise where the knowledge about a theory's limited realm of applicability may have a detrimental effect on the confirmation value of data that are in agreement with that theory's predictions.

To understand this point, let us first think about the case of two conceptually distinct theories that are not fully equivalent to each other but share some empirical implications. An experimentum cruxis, that is strong empirical confirmation of one theory and strong empirical refutation of the other in a regime where the theories' predictions differ from each other, would obviously imply the rejection of the refuted theory also as a description of those contexts where the two theories make the same predictions. Now, let us assume that the given theory has not been *empirically disconfirmed* by data beyond the regime where the two theories are empirically indistinguishable but there are strong reasons to doubt that the theory can be made consistent or can be embedded in a consistent theory that covers a wider range of phenomena. Such considerations would clearly have an impact on assessments of the status of the given theory also in the constrained empirical regime where it is applicable and agrees with the available data. In other words, data that are consistent with that theory would not be taken to amount to strong confirmation of the theory because of conceptual problems related to the theory in a wider context. In Bayesian terms, this situation can be modeled by attributing a very low prior to the given theory due to the expected problems of embedding it in a wider conceptual context. A good example of such a situation is the case of modified Newtonian dynamics (MOND) (Milgrom, 1983). MOND is particularly well adapted to reproduce observed galaxy rotation curves (Sofue and Rubin, 2001). However, this is not considered by scientists to amount to substantial confirmation of MOND precisely because, despite several proposals, there is no completely satisfactory way to obtain the MOND phenomenology from a complete cosmological model of the universe.[10]

The case of dualities between effective theories can be understood as a natural generalization of this scenario. For the sake of clarity, we will now consider a more general abstract duality between two effective theories and return to specific considerations regarding the AdS/CFT afterwards. Let us first assume that we have an exact duality between two theories, T1 and T2. As discussed, this duality guarantees empirical equivalence but allows, in principle, for the possibility of distinct ontological interpretations. Using the terminology of the previous section, we would have that $\mathcal{II}(\text{T1}) = \mathcal{II}(\text{T2})$ but $\mathcal{OI}(\text{T1}) \neq \mathcal{OI}(\text{T2})$. Let us consider again the position of a strong realist who believes that the gravity side of the duality is more fundamental and that, notwithstanding the duality, their universe *really is* described by one side. In the fundamental theory context

---

[10] Although there has been significant work on relativistic extensions of MOND, e.g., Bekenstein (2004), as well as attempts to derive it from a more fundamental theory, e.g., Verlinde's (2017) 'emergent gravity' approach (Milgrom and Sanders, 2016) or Mannheim's (2012) conformal theory, even the originator of the approach accepts that such generalizations 'do not yet provide a satisfactory description of cosmology and structure formation' (Milgrom, 2014).

we suggested that inferentialist justification that most realists give for their view is in conflict with this form of distinct ontological interpretation of dual pairs. Prima facie, this need not be the case in the effective theory context. Recall that, a nonfundamental theory is one that could be replaced by a theory that is more empirically adequate in a given domain and does not share the same core theoretical posits. Now imagine that one takes the pair T1 and T2 to be effective theories to some more fundamental theory X in this sense. The case of particular interest is where the mismatch of core theoretical posits includes the dual structure of the effective description. In this situation it might seem reasonable for our strong realist to assert $\mathcal{OI}(\text{T1}) \neq \mathcal{OI}(\text{T2})$ on the basis that $\mathcal{OI}(\text{T1}) \approx \mathcal{OI}(\text{X})$ and $\mathcal{OI}(\text{T2}) \not\approx \mathcal{OI}(\text{X})$. Plausibly then, this relative closeness of ontology combined with the empirical superiority of X would provide an inferentialist basis for realist commitment to T1 over T2, despite the fact that they are dual theories. The claim is that this realist commitment to T1 over T2 can translate into the understanding that T1 gets strong confirmation by data that agree with its predictions while the empirically equivalent T2 does not. This happens if, in analogy to the MOND case, a much higher prior can be attributed to T1 based on its better alignment with the broader physical picture represented by X. In the given case, that would mean that T1 gets a higher prior than T2 because there is a fundamental theory X that retains the ontology of T1 while it seems unlikely that there is a fundamental theory that retains the ontology of T2.

A natural line of objection is to question whether matters of ontology can ever become relevant for matters of confirmation. There is at least one context in physics, however, where ontology indeed becomes relevant for the assessment of a physical hypothesis. Bohmian nonlocal hidden variable models have been established as a possible interpretation of nonrelativistic quantum mechanics (QM).[11] Making the Bohmian approach compatible with the Lorentz invariance of relativistic physics is an outstanding problem. If we assume, as most physicists do, that Lorentz invariance is a fundamental feature of relativistic physics then one could plausibly argue that a relativistic formulation of Bohmian quantum mechanics might in principal be ruled out. For our purposes it is not important whether or not those problems can be overcome in a way that upholds the Bohmian approach as an attractive interpretation of relativistic quantum mechanics. The crucial question for us is: *if* a relativistic Bohmian quantum mechanics proved impossible, *would* that then discredit nonrelativistic Bohmian quantum mechanics as well, even though the latter works in its own regime? Most observers think it would do so. The failure of a set of ontological claims at a fundamental level is, in this case, taken to disfavor those claims also at the effective level.

---

[11] See Goldstein (2017) for an up-to-date discussion of Bohmian mechanics.

One might object that Bohmian quantum mechanics is an interpretation rather than a theory. But does this disqualify the example? If Bohmian quantum mechanics is understood as an interpretation rather than a theory because it is empirically equivalent to other 'interpretations' of QM, then one might call dual theories 'interpretations' as well as soon as one takes their ontic commitments seriously.

One may also have the understanding that Bohmian quantum mechanics is an interpretation of quantum mechanics because it is essentially bound to a contingent ontic view on quantum mechanics. By its very nature, the Bohmian approach is predicated on a realist perspective. Its truth would imply the falsity of the ontological outlook of alternative interpretations of quantum mechanics. A failure of relativistic Bohmian mechanics would discredit the entire Bohmian approach to quantum mechanics for the very reason that the approach implies a position of ontological scientific realism: if the objects of Bohmian quantum mechanics are shown not to exist relativistically, then they are shown not to exist at all. In general terms, if one adopts a strong ontic realist view, then the degree of confidence one has in the adequacy of an effective theory is extremely sensitive to that theory's relationship with more fundamental theories.

Applying a similar rationale to dualities in the effective theory context thus hinges on accepting an ontic commitment as an essential part of endorsing a physical theory. Clearly for the defender of a strict distinction between a scientific theory and its interpretation, commitment to a theory need not involve endorsing an associated ontology. Such nonrealism would already mean that empirical confirmation is understood to be confined to theories and thus we should reject the possibility of different degrees of confirmation for different dual effective theories. Thus the question is whether string theory is consistent with any ontic commitment (that is, any positive position in the scientific realism debate) that can provide a basis for distinguishing between an ontic commitment to T1 and to T2. As already noted, in the fundamental theory context there are good reasons to assert that string dualities are at variance with such a strong form of ontic scientific realism (Dawid, 2007; Rickles, 2011, 2017; Matsubara, 2013). In particular, the possibility to transform a theory into its dual makes the choice between dual theories reminiscent of an (unphysical) gauge degree of freedom, which seems to block (or at least make less plausible) any attempt to single out the ontology of one of the duals as the true ontology (Rickles, 2017). The understanding that all dual representations are equally valid is further strengthened by the fact that typically all dual perspectives are needed in order to acquire a satisfactory understanding of the physical content of string theory (Dawid, 2013). These lines of reasoning tacitly assume, however, that string dualities play out at a fundamental level. It thus seems possible for a defender of the strong form of ontic scientific realism about dualities, in which the realist commitment to T1 differs from that to T2, to insist that their position is made more plausible by the effective theory context.

Whether or not a different degree of commitment to T1 than to T2 can be defended in that context thus depends on the specifics of the realist's commitment to X. From a minimal structural realist type position (Ladyman and Ross, 2007), we might give an explanation of a theory's 'real' content based on its role as an effective theory. If we apply such a view to X then it is very difficult for us not to apply it equally to T1 and T2, notwithstanding the facts about perceived 'closeness'. Equally, we can consider weak forms of scientific realism, such as perspectival realism (Massimi, 2018) or consistent structure realism (Dawid, 2013), wherein we accept as 'representations of reality' distinct empirically equivalent scientific representations. Clearly, such weak realism offers us no basis for using X to retrospectively differentiate the 'approximate truth' of the effective theories T1 and T2. An exponent of one of the weaker forms of realism would deny that approximate truth should be understood in terms of 'ontic continuity' in the given context and therefore would assert that both effective theories must be equally confirmed.

Finally, we can consider something more like a 'common core' perspective on scientific realism at the fundamental level. Such diachronic commonalities might be understood to trump synchronic 'common core' considerations at the effective level, and therefore distinguish the ontological significance of one of the effective theories. Choosing a 'divide-and-conquer' selective realism about X (Psillos, 2005), we might as well argue for a realist commitment to T1 over T2 due to the greater retention of theoretical terms key for explaining predictive success at the level of X. Let us imagine that there were a gauge/gravity duality at an effective level. A more fundamental theory X, however, is found to be strictly a gravity theory with a fundamental ontology similar to its effective string theory and did not have a dual field theory. In that case, an exponent of divide-and-conquer-style brand of scientific realism could argue that the truth of the fundamental theory X would render the effective gravity theory approximately true and the effective field theory false. Knowledge of the existence of X could then result in attributing a low prior to the effective field theory, in analogy to our examples of MOND and Bohmian mechanics. Data in agreement with both effective theories could then amount to strong confirmation of the gravity theory without being taken to be strong confirmation of the gauge theory. Adapting our Bayesian notation in an obvious way from earlier, it would then be plausible, seen from this strong realist perspective in the effective theory context, to set $P(\text{T1}) \gg P(\text{T2})$, which means that $\triangle_{T1} \gg \triangle_{T2}$. Of course, we would still have that the relative confirmation was equivalent, so $\bar{\triangle}_{T1} = \bar{\triangle}_{T2}$.

To conclude, differentiating degrees of confirmation based on the continuity of ontic commitments from the effective to the fundamental level is an option in principle given certain forms of strong scientific realism. However, such a view is in tension with a variety of weaker forms of scientific realism and, unsurprisingly, any kind of nonrealism. Moreover, such a view is also rather against the spirit of string physics, based on our current understanding of the theory. Nevertheless, we hope that in considering this option may be helpful for getting a better grasp of the range of possibilities one faces when thinking about the empirical consequences of duality relations.

13.4 Instrumental Context

The most surprising applications of the AdS/CFT duality in contemporary physics is in the context of quantum chromodynamics (QCD), the quantum field theoretic description of the strong nuclear force. That a duality which originated from string theoretic considerations has found application in the description of the phenomenology of hadronic physics is a rather beautiful irony, given string theory's own origin in attempts to find a phenomenological model of the strong interaction (Cappelli et al., 2012; Rickles, 2014). These fascinating connections notwithstanding, one of the main messages of this paper is that one should be at pains not to overinterpret the strength of the AdS/QCD theory connection. Rather, as could be expected from the title of this section, we argue that as things stand neither the ontological nor empirical implications of this application of the duality are compelling. Rather, as shall be detailed in the following, we think there good reasons to take the application of the AdS/CFT correspondence to hadronic physics as a *purely instrumental* one. That is, we do not take partially empirically successful applications of AdS/QCD to have told us anything regarding the empirical status of string theory.[12]

To understand how one might possibly use the AdS/CFT duality to model a special class of systems described by QCD, it will be instructive to consider the three fundamental senses in which QCD is *unlike* a CFT.[13] Recall that the particular example of AdS/CFT that we are considering is between Type IIB superstring theory on $AdS_5 \times S^5$ and $SU(N)$, $\mathcal{N} = 4$ Super Yang–Mills theory in 3+1. First, and most obviously, QCD is unlike $\mathcal{N} = 4$ $SU(N)$ Super Yang–Mills since it is not supersymmetric. Second, QCD is not conformally invariant. Third, QCD has a finite color number $N = 3$. Getting around the first two differences will depend on finding a limit where Super Yang–Mills is effectively nonsupersymmetric and where QCD is effectively conformal. Given what we know from condensed matter

---

[12] This is at least in tension with remarks of Rickles (2014, p. viii) who remarks, en passant, that string theory proving to be an 'empirical dud' is 'highly unlikely as a general claim' since duality symmetries present (or originating) in string theory have to led to several 'results that have experimental ramifications', including in the context of quark-gluon plasmas and superconductivity.

[13] Here we mostly follow the excellent treatment of Ammon and Erdmenger (2015).

physics about the emergence and disappearance of symmetries in different parameter regimes, the fact that such a limit can be found is not perhaps as surprising as
it might seem. In particular, if we fix the CFT to be at finite temperature, but near a
critical phase transition, then supersymmetry will be broken and if we choose QCD
to be near the deconfined phase transition then we find that the theory becomes
quasi-conformal. As we shall see later, the supersymmetry breaking is achieved
by choosing a black hole in asymptotically AdS spacetime with finite but near-
critical temperature on the gravity side of the duality. A physical basis to justify
ignoring the final difference is rather less obvious, but can again be well-illustrated
via the analogous approximation made in condensed matter physics. The CFT is
presumed to be defined via a large N expansion corresponding to the 't Hooft limit.
That is, we take the color number to infinity in the CFT. Now, clearly $3 \not\approx \infty$, so
how can we justify treating QCD as approximated by a theory in such a limit? An
immediate analogy is to the thermodynamic limit in statistical mechanics, where
one assumes that the particle number, n, is infinite although we of course know full
and well that it is of the order of 10^{23}. Making sense of the approximations and
idealizations of these kind of infinite limits has been the subject of a considerable
philosophical literature of the last decade.[14] In fact, Bouatta and Butterfield (2015)
make a direct comparison between the 't Hooft limit and the thermodynamic limit.
The important point for our purposes is that in the context of a $1/N$ expansion,
the difference between 3 and ∞ at first order is small enough that subsisting one
expansion with the other is a valid approximation provided one is only interested in
order of magnitude estimates.[15]

One context where such estimates are of interest to physicists is in the context of quark-gluon plasmas. There is experimental evidence that such forms of
matter should be understood as strongly coupled fluids. Strong coupling means
that conventional perturbative techniques are unreliable. The standard alternative is
then a lattice gauge field theory approach. However, for time-dependent processes,
lattice methods become intractable, and it is easier to appeal to the AdS/QCD correspondence to calculate physical values of the parameters of quark-gluon plasmas
using string theory. A particularly vivid example is the *jet-quenching parameter*,
$\hat{q}$.[16] This is the mean transverse momentum acquired by a 'hard probe' (i.e., a
high momentum, narrow beam of hadronic matter) per unit distance traveled. A
central field theory result is that $\hat{q}$ can be expressed in terms of the correlator
of Wilson lines along a particular contour. The gravity dual description is given
in terms of a string action in a AdS–Schwarzschild *black brane* spacetime with a

---

[14] Three particularly significant contributions are Batterman (2002), Butterfield (2011), and Norton (2012).
[15] See Coleman (1988, ch. 8) for an elementary introduction to the $1/N$ expansion in field theory.
[16] An earlier application of the AdS/QCD correspondence can be found in Kovtun, Son, and Starinets (2005).

finite but near-critical Hawking temperature.[17] In a landmark paper (Liu, Rajagopal, and Wiedemann, 2006), these techniques lead to a calculated value for $\hat{q}$ of 4.5 GeV^2/fm, which is close to the experimental range from Brookhaven Relativistic Heavy Ion Collider of 5–15 GeV^2/fm. At the time this was the best theoretical estimate of the parameter. How should we understand the implications of this result for AdS, CFT or, in fact, QCD? This question turns out to be extremely subtle, and the remainder of this paper is devoted to developing a line of analysis of this new 'instrumental context' of dualities that is consistent with the conclusions made regarding the previous two contexts.

A reasonable first step is to formulate our inferential standpoint in terms of Bayesian confirmation theory as per earlier. We denote by H$_3$ the proposition

H$_3$: QCD provides an empirically adequate description of the target system $\mathcal{T}'$ within a certain domain of conditions D_{QCD} that include the assumption of near-criticality.

We can encode the evidence available to our agent by introducing a variable E' corresponding to the two values, E$'$, the empirical evidence regarding the target system $\mathcal{T}'$ obtains, and ¬E$'$, the empirical evidence does not obtain. In the case of the jet-quenching parameter the target system $\mathcal{T}'$ would be a quark-gluon plasma and the empirical evidence would be the *measured value* of $\hat{q}$.

In order to consider the relevance of this empirical evidence to an agent's beliefs regarding the empirical adequacy of CFT and string theory as descriptions of $\mathcal{T}'$ one can introduce two propositions, H$_4$ and H$_5$:

H$_4$: $SU(N)$, $\mathcal{N} = 4$ SYM in $3 + 1$ provides an empirically adequate description of the target system $\mathcal{T}'$ within a certain domain of conditions D'_{CFT} that include the finite but near-critical temperature assumption.

H$_5$: Type IIB superstring theory on AdS$_5 \times S^5$ provides an empirically adequate description of the target system $\mathcal{T}'$ within a certain domain of conditions D'_{AdS} that include the finite but near-critical temperature assumption.

It is of course significant to note that in these definitions we are considering the empirical adequacy of AdS/CFT as a description of a quark-gluon plasma. There are a variety of obvious senses in which both sides of the duality immediately fail to provide such descriptions. This notwithstanding, in regard to the jet-quenching parameter taken in isolation, empirical adequacy of the dual theories is worth considering. In particular, we can expect that an agent can derive empirical predictions regarding $\mathcal{T}'$ using the calculational techniques available from the AdS side of the

---

[17] As Hawking and Page (1983) illustrated, black holes in AdS space undergo a thermodynamic phase transition when the black hole size is of order of the characteristic radius of the AdS space.

duality. Since this is the weakly coupled side this will invariably be technically more straightforward. In fact, as noted, at least in 2006, this was the best available technique.

We can represent this prediction of a value of the jet-quenching parameter that is derived via the AdS model in terms of a further binary variable F. We take this variable to have the value F when the *predicted value* obtains in the actual quark-gluon plasma. Thus we have that $H_4 \rightarrow$ F. The inferential structure of our current context is unlike that of the fundamental or effective theory context. In particular, there is no empirical regime in the physics of the quark-gluon plasma where the predictions from AdS actually match the data: F $\neq$ E′ in D_{QCD} since the predicted value for $\hat{q}$ was 4.5 GeV^2/fm, which is outside the experimental range of 5–15 GeV^2/fm. This means that E' actually disconfirms AdS: $P(H_5|E') < P(H_5)$. Due to the (assumed) exact duality between AdS and CFT, the latter gets disconfirmed by E' as well. So we have $P(H_4|E') < P(H_4)$.

As regards QCD as things have been set up, the relevance of E' for H_3 is not clear. In particular, since there is only a crude approximating relation with the dual theories it is not clear, as we have formulated things, whether we should take the new evidence as positive, negative, or neutral in regard to the empirically adequacy of QCD. The limiting relationship between QCD and AdS gives us a means to derive an order of magnitude approximation regarding the phenomenology of the former, using calculational techniques based upon the latter. Yet, AdS is not an effective theory of QCD phenomenology and cannot be confirmed even in the weak sense of increasing the degree of belief in the theory's empirical adequacy in a specific regime of testing. In fact, as already noted, there are simple and obvious senses in which the AdS/CFT duals fail as empirically adequate descriptions of a quark-gluon plasma: not least with regard to boundary conditions on the plasma and the finiteness of the color number. Thus there is a strong sense in which *any* empirical data collected from measurements on an actual quark-gluon plasma will straightforwardly contradict the empirical adequacy, considered in general terms, of AdS or CFT as a model of this system.[18]

One might, however, still plausibly expect that the data E′ to give us confidence in the reliability of the package of approximation techniques and theories that are being used. This is because, as noted in the original paper (Liu et al., 2006, p. 4), there are good theoretical reasons to expect that the value calculated using AdS would be a slight underestimate. There were good reasons to expect the right order

---

[18] This is the direct counterpart to incisive remarks made by Butterfield (2011, p. 1072) in the context of the thermodynamical limit. In that context, Butterfield notes that taking the particle number to be extremely large corresponds to cosmic lengths and thus immediately makes the model 'utterly unrealistic', and thus fundamentally empirically inadequate when considered in general. Norton (2012) makes a similar point in the context of his distinction between approximation and idealization.

of magnitude but also good reasons to expect some discrepancy. To be an accurate formalization of the inferential standpoint of the relevant scientists we surely must incorporate this kind of background contextual knowledge regarding the models that are being employed into our Bayesian framework.

Let us assume that there is some background contextual knowledge that when combined with the actual AdS calculation implies that the predicted value of the jet-quenching parameter should roughly approximate the measured data but not precisely coincide with it. Formally, we can introduce a proposition:

C: Type IIB superstring theory on $AdS_5 \times S^5$ gives correct to order of magnitude empirical approximation to QCD as a description of the target system $\mathcal{T}'$ within certain domains of conditions D'_{AdS} and D_{QCD} respectively.

Second, denote by δE a proposition:

δE: The difference between the predicted value derived via Type IIB superstring theory on $AdS_5 \times S^5$ and the experimental result is correct to order of magnitude. That is $F \approx E'$ in D_{QCD}.

We then have that $(H_3 \wedge C) \rightarrow \delta E$ and thus that $P(H_3 \wedge C | \delta E) > P(H_3 \wedge C)$. This is confirmation of the *conjunction* of the empirical adequacy of QCD and the approximation relation between Type IIB superstring theory on $AdS_5 \times S^5$ and QCD. Ideally one would then like to make a definite statement regarding the change in probabilities of each of the conjuncts. In particular, one would ideally like to claim in these circumstances that QCD is individually confirmed. However, it can be shown that the confirmation of a conjunction by some evidence is compatible with the disconfirmation of either (or both) the conjuncts considered individually (Atkinson, Peijnenburg, and Kuipers, 2009). Thus, the formal model allows us only to establish confirmation in the package of the empirical adequacy of QCD and the empirical approximation of QCD by string theory when taken together.[19]

As a final remark we should note the relevance of recent theoretical work in calculating the jet-quenching parameter using lattice QCD techniques (Panero, Rummukainen, and Schäfer, 2014). This approach leads to a calculated value of around 6 GeV^2/fm that is of course more consistent with the experimental range. Furthermore, that the value calculated via a more realistic model is indeed slightly above that calculated via the string theory model provides retrospective support to the original theoretical arguments for the string theory calculation being a *slight* underestimate.

---

[19] Informally, one plausibly *would* still expect to have at least limited confirmation of H_3 by δE. Establishing the additional conditions necessary to show this formally would take us outside the remit of the current work however.

13.5 Conclusion

Dualities are typically viewed in terms of exact correspondences between fundamental theories of the world. Dual models can and do occur, however, outside such fundamental theory contexts. If so, both the ontic implications and the specifics of the empirical relevance of dualities change. We have argued in this paper that there is a fairly rigid principle of equal confirmation in the case of dualities between fundamental theories. Although this principle could be broken by assuming different priors for dual theories, one finds no physical argument for doing so. To the contrary, a number of physical arguments suggest treating duals as different perspectives on the same theory, which would rule out the attribution of different priors. This situation changes once one considers a duality relation at the level of effective theories that has no correspondence at the fundamental level. Such a scenario could, given a strong form of ontic realism, be viewed as providing a physical basis for viewing the duals as different theories that might even have different plausibility as viable effective theories about the world. Though such a point of view has its problems, it should not be discarded out of hand. Once one looks at the instrumental application of dualities in contexts where the empirically viable theory is not understood to have a dual, such as the case of quark-gluon plasma calculations, the role of theory confirmation does not depend on distinguishing between confirmation values for the individual duals. Rather, neither of the two duals gets confirmed by the data. Rough agreement of string theory calculations with data can only generate confirmation for the approximation together with the theory it is taken to approximate, which is QCD.

Acknowledgments

We are very grateful to Nick Huggett and an anonymous referee for helpful comments on a draft manuscript and to audience members in Exeter for valuable feedback. KT and SG were supported by the Arts and Humanities Research Council (Grant Ref. AH/P004415/1).

References

Ammon, M., and Erdmenger, J. 2015. *Gauge/Gravity Duality: Foundations and Applications*. Cambridge: Cambridge University Press.

Atkinson, D., Peijnenburg, J., and Kuipers, T. 2009. How to confirm the conjunction of disconfirmed hypotheses. *Philosophy of Science*, **76**(1), 1–21.

Batterman, R. W. 2002. *The Devil in the Details: Asymptotic Reasoning in Explanation, Reduction, and Emergence*. Oxford: Oxford University Press.

Bekenstein, J. D. 2004. Relativistic gravitation theory for the modified Newtonian dynamics paradigm. *Physical Review D*, **70**(8), 083509.

Bouatta, N., and Butterfield, J. 2015. On emergence in gauge theories at the 't Hooft limit. *European Journal for Philosophy of Science*, **5**(1), 55–87.

Bovens, L., and Hartmann, S. 2004. *Bayesian Epistemology*. New York: Oxford University Press.

Butterfield, J. 2011. Less is different: Emergence and reduction reconciled. *Foundations of Physics*, **41**(6), 1065–1135.

Cappelli, A., Castellani, E., Colomo, F., and Di Vecchia, P. 2012. *The Birth of String Theory*. Cambridge University Press.

Coleman, S. 1988. *Aspects of Symmetry: Selected Erice Lectures*. Cambridge: Cambridge University Press.

Dawid, R. 2007. Scientific realism in the age of string theory. *Physics and Philosophy*, **501**, 011.

Dawid, R. 2013. *String Theory and the Scientific Method*. Cambridge: Cambridge University Press.

De Haro, S. 2017. Dualities and emergent gravity: Gauge/gravity duality. *Studies in History and Philosophy of Science Part B: Studies in History and Philosophy of Modern Physics* 59, 109–125.

De Haro, S., and Butterfield, J. 2017. A schema for duality, illustrated by Bosonization. Pages 305–376 of: Kouneiher, J. (ed.). *Foundations of Mathematics and Physics One Century after Hilbert*. Cham, Switzerland: Springer.

Fraser, D. 2017. Formal and physical equivalence in two cases in contemporary quantum physics. *Studies in History and Philosophy of Science Part B: Studies in History and Philosophy of Modern Physics*, **59**, 30–43.

Goldstein, S. 2017. Bohmian mechanics. In: Zalta, E. N. (ed.). *The Stanford Encyclopedia of Philosophy*, https://plato.stanford.edu/archives/sum2017/entries/qm-bohm

Gubser, S. S., Klebanov, I. R., and Polyakov, A. M. 1998. Gauge theory correlators from noncritical string theory. *Physics Letters*, **B428**, 105–114.

Harlow, D. 2016. Jerusalem lectures on black holes and quantum information. *Reviews of Modern Physics*, **88**, 015002.

Hartmann, S., and Sprenger, J. 2010. Bayesian epistemology. Pages 609–620 of: Bernecker, S., and Pritchard, D. (eds.). *Routledge Companion to Epistemology*. New York: Routledge.

Hawking, S. W., and Page, D. N. 1983. Thermodynamics of black holes in anti-De Sitter space. *Communications in Mathematical Physics*, **87**, 577.

Horowitz, G. T., and Polchinski, J. 2009. Gauge/gravity duality. In: Oriti, D. (ed.). *Approaches to Quantum Gravity*. Cambridge: Cambridge University Press.

Huggett, N. 2017. Target space $\neq$ space. *Studies in History and Philosophy of Science Part B: Studies in History and Philosophy of Modern Physics*, **59**, 81–88.

Kaplan, J. 2013. *Lectures on AdS/CFT from the Bottom Up*, https://sites.krieger.jhu.edu/jared-kaplan/files/2016/05/AdSCFTCourseNotesCurrentPublic.pdf.

Kovtun, P., Son, D. T., and Starinets, A. O. 2005. Viscosity in strongly interacting quantum field theories from black hole physics. *Physical Review Letters*, **94**, 111601.

Ladyman, J. 2016. Structural realism. In: Zalta, Edward N. (ed.). *The Stanford Encyclopedia of Philosophy*. Stanford, CA: Metaphysics Research Lab, Stanford University.

Ladyman, J., and Ross, D. 2007. *Every Thing Must Go: Metaphysics Naturalized*. Oxford: Oxford University Press on Demand.

Liu, H., Rajagopal, K., and Wiedemann, U. A. 2006. Calculating the jet quenching parameter. *Physical Review Letters*, **97**(18), 182301.

Maldacena, J. 1999. The large-N limit of superconformal field theories and supergravity. *International Journal of Theoretical Physics*, **38**(4), 1113–1133.

Mannheim, P. D. 2012. Making the case for conformal gravity. *Foundations of Physics*, **42**, 388–420.

Massimi, M. 2018. Four kinds of perspectival truth. *Philosophy and Phenomenological Research*, **96**(2), 342–359.

Matsubara, K. 2013. Realism, underdetermination and string theory dualities. *Synthese*, **190**(3), 471–489.

Milgrom, M. 1983. A modification of the Newtonian dynamics as a possible alternative to the hidden mass hypothesis. *The Astrophysical Journal*, **270**, 365–370.

Milgrom, M. 2014. The MOND paradigm of modified dynamics. *Scholarpedia*, **9**(6), 31410.

Milgrom, M., and Sanders, R. H. 2016. Perspective on MOND emergence from Verlinde's "emergent gravity" and its recent test by weak lensing. arxiv.org/abs/1612.09582.

Norton, J. D. 2012. Approximation and idealization: Why the difference matters. *Philosophy of Science*, **79**(2), 207–232.

Panero, M., Rummukainen, K., and Schäfer, A. 2014. Lattice study of the jet quenching parameter. *Physical Review Letters*, **112**(16), 162001.

Psillos, S. 2005. *Scientific Realism: How Science Tracks Truth*. London: Routledge.

Rickles, D. 2011. A philosopher looks at string dualities. *Studies in History and Philosophy of Science Part B: Studies in History and Philosophy of Modern Physics*, **42**(1), 54–67.

Rickles, D. 2014. *A Brief History of String Theory: From Dual Models to M-Theory*. Heidelberg: Springer.

Rickles, D. 2017. Dual theories: 'Same but different' or 'different but same'? *Studies in History and Philosophy of Science Part B: Studies in History and Philosophy of Modern Physics*, **59**, 62–67.

Sofue, Y., and Rubin, V. 2001. Rotation curves of spiral galaxies. *Annual Review of Astronomy and Astrophysics*, **39**(1), 137–174.

Stein, H. 1989. Yes, but ... some skeptical remarks on realism and anti-realism. *Dialectica*, **43**(1–2), 47–65.

Van Fraassen, B. C. 1980. *The Scientific Image*. Oxford: Oxford University Press.

Verlinde, E. P. 2017. Emergent gravity and the dark universe. *SciPost Physics*, **2**(3), 016.

Witten, E. 1998. Anti-de Sitter space and holography. *Advances in Theoretical and Mathematical Physics*, **2**, 253–291.

14

Extending Lewisian Modal Metaphysics from a Specific Quantum Gravity Perspective

TIZIANA VISTARINI

14.1 Possible Worlds and Accessibility Relations in Lewisian Formal Modal Semantics

Old disputes give way to new. Instead of asking the baffling question whether whatever is actual is necessarily possible, we could try asking: is the relation R symmetric?

David Lewis, *The Plurality of Worlds*, 1986, p. 19

Simplifying modal logic disputes is a main formal payoff of Lewis's possible worlds semantics. The latter is not the focus of the essay. Yet, since this essay is about extending some Lewisian metaphysical applications of this formal framework, it is convenient to sketch here some of the basic formal features of Lewisian possible world semantics. In this section, possible worlds and accessibility relations are purely formal devices since they remain uninterpreted.[1]

Lewisian possible worlds semantics unifies the plethora of different modal logics that have come into existence since Aristotle. This unification is achieved by translating heterogeneous modal logic disputes on controversial modal axioms in disputes on more homogeneous sets of formal properties, namely, those of the accessibility relations. Before the advent of the possible worlds semantics there was no unified understanding of fundamental axioms like (Lewis, 1987):

a. if necessarily p, then p;
b. if necessarily p, then necessarily necessarily p;
c. if p, then necessarily possibly p;
d. if possibly p, then necessarily possibly p.

Lewis possible worlds semantics equipped with formal devices like possible worlds and binary accessibility relations R between pairs of them (formal devices

---

[1] Others before Lewis pioneered the formalism of possible worlds semantics, but here I will not discuss non-Lewisian uses of the formal framework because that would bring us out of the essay's topics.

as long as they remain uninterpreted) completely changed the traditional formal debate. Indeed, the Lewisian framework gives an extensional formal semantics unifying the preexisting heterogeneous uses of these fundamental axioms. A simplified description of how the basics of this modal machinery formally works might be the following: a basic technique is that of quantifying over possible worlds "connecting" to the "actual" one by some accessibility relation R. Supposing w denotes the actual world, the machinery mainly works with two schemas for translating traditional axioms about necessity and possibility into formal properties of the accessibility relation R. These schemas are the following (Lewis, 1987):

1. necessarily p if and only if p is true at *every* possible world u such that $R(w,u)$;
2. possibly p if and only if p is true at *some* possible world u such that $R(w,u)$.

Then, an inquiry on whether any of the axioms—(a),(b),(c),(d)—holds can be translated into an inquiry about specific formal properties of the accessibility relation R. More precisely, axiom (a) holds just in case the accessibility relation involved is reflexive, namely, just in case for all worlds u, we have $R(u,u)$. The majority of interpreted accessibility relations are reflexive, since all possible worlds are accessible from themselves. Axiom (b) instead holds if and only if the accessibility relation is transitive, that is, (b) will hold just in the case $R(w_1,w_2)$ and $R(w_2,w_3)$ both entail $R(w_1,w_3)$. Moreover, the inquiry about whether axiom (c) holds is translated into the probe about whether the accessibility relation has the formal property of being symmetric, namely, if it is the case that $R(w_1,w_2)$ entails $R(w_2,w_1)$. Finally, axiom (d) holds in all those cases in which the accessibility relation has the formal property of being Euclidean, that is, all those cases in which $R(w_1,w_2)$ and $R(w_1,w_3)$ both entail $R(w_2,w_3)$.

Although not detailed, this description shows the bare bones of Lewisian translation of heterogenous modal discourses about necessity and possibility into homogeneous sets of modal claims about formal properties of accessibility relations. The philosophical payoff of such formal semantics branches off from its applications to specific problems in metaphysics, in epistemology and in many other contexts. In this essay I am focused on some Lewisian applications to modal claims in metaphysics. More precisely, the application that produces the relation of nomological accessibility among worlds, as well as the application that produces the relation of similarity accessibility.

Now, since it is relevant to the content of this essay, one might wonder whether Lewisian formal semantics is grounded on some clear-cut and unique way of distinguishing between the actual world and the possible ones: how does anyone single a Lewisian world out for the label of *actual*? In other words, what is the nature of actuality for Lewis? Answering requires that one moves forward with the analysis. As we will see in what follows, especially in Section 14.3.2, according to Lewis all possible worlds are equally real in virtue of their "concreteness." Understanding what the abstract-concrete distinction comes to in the Lewisian context will prove to be a nontrivial task. For now let's say only that, according to Lewis, all possibilities are realized in some world, independently of whether somebody thinks of them or not. Then, if all possible worlds are real, how should anyone pick one as actual? Lewis' notion of actuality appears to be an indexical one. Its reference indeed is described by him as varying with the speaker in the same way in which the reference of deictics like "here" and "now" do. The utterance "actual world" pronounced by me refers to the world where I am located. All possible worlds in which I am not located are actual for their inhabitants. The world I regard as actual is nonactual for individuals residing in those worlds I regard as nonactual (Lewis, 1973).

Lewis seems to hold a "contextual" view about actuality, one in which the context for regarding a world as actual is given mainly by ordinary linguistic games. The actual world in the metaphysical model that will be extending the Lewisian one is any spatiotemporal structure, within the sets of quantum string theory solutions, which is arbitrarily chosen as the central structure to which apply deformations functions producing other worlds. As we will see, any family of possible worlds has a central structure that is also called the "central fiber" of the family of worlds. Also different families can share the same central fiber. Any family of *possibilia* can be constructed without any prejudice on the choice of the central spatiotemporal structure from the set of quantum string theory solutions. Much more on this in Section 14.3.

14.2 Fleshing Out Accessibility Relations

One of the Lewisian applications of the possible worlds semantics to metaphysics considered here involves physical science. This application is characterized by a specific interpretation of possible worlds in terms of mereological spacetimes wholes as well as by a specific interpretation of the accessibility relation in terms of nomological accessibility. Then, saying that "world u is accessible from world w" means saying that "u obeys the physical laws of w."

This metaphysical application produces a notion of nomological necessity. Indeed the interpreted accessibility relation transforms the general Lewisian translation schema—(1) + (2)—into the following specific schema:

N. if p is nomologically necessary, then p is true at all possible worlds that are nomologically accessible from the actual world, namely, p is true at all possible worlds that obey the physical laws of the actual world.

Now, one might ask what axiom among those mentioned in Section 14.1—(a), (b), (c), and (d)—is satisfied by this definition of nomological necessity. An answer based on testing directly the axioms directly would be convoluted. By using a pre-Lewis language, it would be unclear, for example, whether nomological necessity satisfies axiom (c). Indeed that would require answering the convoluted question "is something that is nomologically possible also nomologically necessarily possible?"

Sometimes looking for simple questions might be the best strategy for dealing with simple answers. But a translation of some intricate question into a simpler one not always produces a question equivalent to the original. Instead, in Lewisian formal semantics, the translation via the scheme (N) accomplishes such equivalence. The scheme (N) translates the question, about whether axiom (c) is satisfied into an equivalent and more manageable one about whether the relation of nomological accessibility is symmetrical. Then, the translated question is "is it always the case that whenever world w_1 obeys the physical laws of w_2, then also w_2 obeys the physical laws of w_1?"

Now, according to the Lewisian notion of physical laws the answer is negative. The opposite might or might not hold true. Being symmetrical is not a definitional property of nomological accessibility. Broadly speaking, for Lewis there is nothing to reality except spatiotemporal distributions of local natural properties, and physical laws supervene on these distributions—this is the main core of the Lewisian thesis of Humean supervenience. Then, an instantiation of lack of symmetry might be that of a world w_1 having a spatiotemporal structure that strictly contains that of w_2, hence having an additional spatiotemporal distribution of local facts not shared with w_2. In this case world w_1 might obey more laws than those obeyed by w_2. Other examples of lack of symmetry might be constructed by using pairs of nomologically possible worlds that are one a spatiotemporal deformation of the other in such a way that the physical laws obeyed by one are strictly contained in the set of physical laws obeyed by the other.[2]

Now, the same physical law might supervene two different spatiotemporal distributions of local properties, but it also might not. Indeed, in general a deformation of a Humean mosaic would produce deformations of the supervening physical laws. This point about Humean mosaics, physical laws, and deformations will be analyzed in Sections 14.3.2 and 14.3.3. In the latter I propose a nomological accessibility structure tentatively extending the Lewisian one because partially carrying its structural properties.

Another Lewisian application of his formal semantics in metaphysics is one that produces his counterpart theory of de re modality. The latter is based on a specific interpretation of the accessibility relation in terms of similarity order (also often

---

[2] Saying that world w "obeys" some physical laws means saying that the latter "are laws of" that world and that they "are true" in that world.

called the counterpart relation). The interpreted accessibility relation is symmetrical and not transitive.[3] In Section 14.3.2 I propose a similarity accessibility structure extending the Lewisian one because fully carrying its structural properties.

I want to conclude this section with few introductory remarks about possible worlds. The ways in which philosophers think about the status of possible worlds in modal metaphysics can be mainly divided in two schools of thought.[4] One is inspired by the robust kind of realism proposed by Lewis (1973, 1987) preliminary outlined in the previous section. As I said, Lewis makes clear that all actual and nonactual possible objects are made of only one kind of being: nonactual possibilities exist in just the same way that you and I do (Lewis, 1987). The other school of thought assigns to possible worlds an abstract nature. Possible worlds are ersatz worlds, or world surrogates that are in some sense representational. The two main approaches within this school of thought share the same intent to avoid any realist commitment to nonactuals. On the one side, the Sententialist view of ersatz worlds, whose first articulation was proposed by Carnap (1947); on the other the anti-reductive approach, whose most popular formulation is by Plantinga (1972), republished in Plantinga (2003). This is an anti-reductive strategy because it simply identifies ersatz worlds as "ways the world might have been" without further analysis in terms of sentences.

Now, my extension of Lewisian modal metaphysics does not really settle the dispute between Ersatzers and Lewis, although, as we will see, does not maintain a complete neutrality. This point is unpacked in Section 14.3.2.

14.3 Extending Lewisian Modal Metaphysics

14.3.1 The Topological-Fiber Bundle Structure of String Theory's Moduli Space

In this section I partially summarize parts of the methodology used elsewhere (Vistarini, 2019, ch. 6) to argue for string theory background independence. More precisely, I briefly describe the main algebraic-formal component of that composite argument, although string theory background independence is not among the topics of this essay. The reason I refer to that formal analysis in this essay can be succinctly explained by a couple of points. In the book cited I introduce a "space" whose topology is defined by sets of families of spacetimes (or worlds). These families

---

[3] The symmetrical formal property can be thought quite straightforwardly. The nontransitivity can be thought in the following simplified terms: y is necessarily P just in case all of y's counterparts are P. But being a counterpart of y here means being similar to y. Then, y_0 can be a counterpart of y, and y_1 a counterpart of y_0, without y_1 being a counterpart of y. So, all of y's counterparts y_0 can be P without all the y_0's counterparts y_1 being P.

[4] A detailed introduction to the dialectic between modal realists and Ersatzers is contained in Divers (2002).

contain the sets of physical solutions of quantum string theory. Such topological "space" turns out to be representative of a multiworlds landscape that must not be regarded as a physical megaverse because the worlds (spacetimes) represented in it are completely spatiotemporally isolated from each other—i.e. they are not different spacetime portions of the same megaverse. Rather that topological space should be regarded as a space of possibilities. This space is also equipped with a fiber bundle structure, whose fibers are Hilbert spaces each separately stemming from a single isolated spacetime. Such a topological-fiber bundle structure is proved in Vistarini (2019, ch. 6) to carry dynamical and physical properties of quantum string systems useful to prove string theory background independence.

Now, the rationale behind joining some of the book's trains of thoughts to the different train of thoughts crossing this essay resides in the fact that the very same topological-fiber bundle structure is shown here to also carry structural properties of the Lewisian modal metaphysics. In other words, in this essay I attempt to show that such a topological-fiber bundle structure (connected to quantum string theory) is a plausible accessibility structure extending Lewisian similarity accessibility, as well as tentatively extending Lewisian nomological accessibility.

Now, let's start to unpack the content of this presentation with an intuitive and simplified description of such a "space" and its topological-fiber bundle structure. Such a description is extrapolated from the work cited, so for further details and further context I will refer the reader to it. The thesis of string theory's background independence defended there was a multilayer conclusion of a multilayer argument. Three main argumentative lines contributed to it. One line originated from an analysis of the physical content of perturbative quantum string theory (which is also the most widely known formulation of quantum strings physics).[5] Another argumentative line stemmed from the use of string theory's dualities, which is some cases bring in a nonperturbative formulation of the theory.[6]

The third argumentative line originated from the topological and fiber-bundle structures of the theory's moduli space. Reading through these structures produced two claims of background independence, which will be used in Section 14.3 to extend Lewisian nomological accessibility.[7] The topological-fiber bundle structure I defined in the book does not overlap with the types of topology and bundle structures usually defined on the theory's moduli space by people in string physics circles, even though the underlying set-theoretic structure is the same. This fact is

---

[5] For bibliographic references to past works fully developing this topic and for a brief summary see Appendix 14.A.
[6] For bibliographic references to past works fully developing this topic and for a brief summary see Appendix 14.B.
[7] For formal details and philosophical discussion, for a complete presentation of moduli spaces in general and in this specific context, see Vistarini (2019, ch. 6).

probably due to a mismatch of goals. Moduli spaces are commonly used by string physicists to either explore cosmological applications of the theory[8] or to classify different types of string theory.[9] In this second case the hope is that of finding the "fine moduli space" of the theory, that is, the moduli space that parameterizes and encodes some unique fundamental formulation of string theory, from which every other existing formulation can be gained by means of dualities.[10] Differently from these traditional uses, the topological-fiber bundle structure I defined was supposed to work more like a kind of global "positioning" system. But differently from any ordinary system of that kind, what it is "positioning" is the degree of background independence of string physics.[11]

As I said, the topology I posit on the theory's moduli space is defined by families of spatiotemporal structures, which include the sets of physical solutions of quantum strings theory.[12] The first step of the formal procedure that brings us to the stage of defining this topology is getting these families. The formal process might be concisely described as a formal act of "deforming" in many different ways some initially chosen spatiotemporal structure, followed by a formal act of "gluing together" the many results of these deformations. More precisely, given an initial spacetime geometry (arbitrarily picked within the sets of physical solutions of quantum string theory) if one somehow deforms that geometry, then one produces a collection of different geometries. Then, the collection obtained by means of deformations is also a topological set, or a family equipped with a topology. The topology is induced by the deformations functions and by the fact that the family elements are somehow glued together. The gluing functions equip the collection with neighborhoods for each element. The relations among elements and among neighborhoods satisfy the minimum set of axioms required to regard a set as a topological set. As I show later on, these relations are not spatiotemporal.[13]

---

[8] For example, Novak, Brustein, and de Alwis (2002).

[9] Vafa (1998) and Zwiebach (2009, ch. 25) are among the most popular works on this topic.

[10] The logic used in this application mimics the one that originally introduced moduli spaces in algebraic geometry. In that context, the idea was that the study of intrinsic properties of some type of geometrical objects (for example, properties of elliptic curves not depending on any higher dimensional embedding) might have been made easier by the construction of the finest possible moduli space classifying those objects. The task turned out to be not trivial since not many types of geometrical structures admit the unique fine moduli space. It is much more usual they admit some coarse moduli space potentially improvable in a finer classificatory structure, still far from the finest one. But if the fine moduli space is found for some type of geometrical objects, that structure parameterizes the universal family of that type, that is, one that owns the complete set of intrinsic properties of that type, and for this reason one from which any other incomplete family of that type can be obtained (Vistarini, 2019, ch. 6).

[11] For the topological parts of that structure I heavily relied on the mathematical work of Kodaira (2005).

[12] In this way one gets a topological space of possibilities including the physical ones.

[13] Any type of topological family of spacetimes obtained via deformations of some spatiotemporal structure is formally equivalent to a topological non metrical "manifold." Of course one may always decide to define a metric on it, but one needs not: a metric is not definitional of such a topological manifold. For more details on that see Vistarini (2019, ch. 6).

In technical jargon the initial spatiotemporal structure is also called the central fiber of the family, whereas the ones originating from applying deformation functions to the central fiber are the generic fibers of the family. As I said in Section 14.1, the central structure to which deformations apply is arbitrarily chosen within the sets of quantum string theory's solutions. Then, depending on what central fiber and what type of deformations one chooses, one gets a type of family. Nevertheless, any fiber of any family produced by deformations operators can be promoted to be the central fiber of some other family of deformations. This fact produces a metaphysical landscape of *possibilia* strictly containing the physical ones.

The word *fiber* here is doing a bit of heavy work because it needs some more context to acquire full significance. To this aim, an intuitive account of the type of link between the theory's moduli space M and the families is crucial. The moduli space is said to parameterize all these families of spatiotemporal structures. This formal property of "being a parameter space" means in simple terms that there exist a well-defined map ϕ sending any spatiotemporal structure K_i (for any family of deformations) to some point λ_i of the moduli space.[14] Then, the word *fiber* here refers to the domain of the moduli map ϕ, that is, the word refers to the families of spatiotemporal structures. Another use of the word *fiber* will soon be introduced in connection to the fiber bundle structure on M. The expression that follows visualizes the nature of this link:[15]

$$ M \quad \xleftarrow{\phi} \quad Def(K_0) \quad \xleftarrow{i} \quad K_0 $$
$$ \forall \lambda_i \in M, \lambda_i = \phi(K_i), \forall K_i \in Def(K_0). $$

For any family of deformations $Def(K_0)$, for any arbitrarily chosen K_0, the topological structure of the family (induced by the deformations functions and by the gluing functions) gets mapped onto M through formal maps related to ϕ that I introduce in Section 14.3.2.

Although still incomplete, the presentation made so far can already give an idea of what it means to say that the theory moduli space M gains a topology defined by families of spacetimes. This topology is not equipped with a spatiotemporal metric. And even more, it is not a metrical topology at all. The lack of any type of metric built into such topology since the start is explained by the combination of two facts. First, the atlas of topological neighborhoods covering M is made of

---

[14] The minimal requirement of well-definition does not entail the one-to-one formal character. As we will see in Section 14.3.2, few trivial families of deformations, also called isotrivial, are basically sent to one moduli point.

[15] Sometimes scientific works in deformation theory adopt an alternative, still equivalent, style of notation that uses the inverse of the map ϕ, at least on those neighborhoods of M where ϕ is invertible. Then, on those neighborhoods of M where it is possible to define ϕ^{-1} the link is expressed by $\lambda_0 = \phi^{-1}(K_0)$ and $\lambda_i = \phi^{-1}(K_i), \forall i \neq 0$.

the continuous images (via maps related to ϕ) of the topological neighborhoods composing the topology of the families of deformations. Second, the families' topology is induced by the functions of deformations and by the functions that glue together the many spatiotemporal structures resulting from deformations. Such a topology is not by definition metrical. In particular, the topology of any family of spatiotemporal structures obtained by deformations, does not have a spatiotemporal structure by definition. That explains why the topological relations among points of M are neither spatiotemporal relations, nor more generally metrical.[16]

Now, I introduce the fiber bundle structure $(\overline{H},p)$ defined on top of the moduli space M, which, as I said earlier on, carries any sort of relevant physical and dynamical information. Here I am giving a sketchy description of its feature; for a full analysis I refer the reader to Vistarini (2019, ch. 6). The reason for including in the essay such brief description is that this fiber bundle structure plays a central role in the tentative extension of the Lewisian nomological accessibility relation. The following expression shows the salient features of the fiber bundle:

$$\begin{array}{c} \overline{H} \\ \downarrow{p} \\ M \quad \xleftarrow{\;\phi\;} \quad Def(K_0) \quad \xleftarrow{\;i\;} \quad K_0, \end{array}$$

where $\overline{H}$ is the total space of the fiber bundle and it is defined by the following union:

$$\overline{H} = \coprod_{\forall i,\, \lambda_i \in M} (H_{\lambda_i} \times \mathbb{C}).$$

Although this union is much more than a set-theory union, here I am skipping any details about its extra structure because it is not relevant to the content of this essay.

Each disjunct $H_\lambda \times \mathbb{C}$ is a fiber of the fiber bundle over a point λ_i of the base moduli space M. H_{λ_i} is the Hilbert space identified to some quantum string system Q, whose dynamics are considered to unfold within the spacetime parameterized by λ_i (the spacetime belongs to some family of deformations $Def(K_0)$). The vector space of complex numbers $\mathbb{C}$ represent all the numerical values that the transition functions $f_{\alpha_1 \alpha_2 \ldots \alpha_n}$ of the system Q can take over λ_i. Let me briefly unpack this last claim about transition functions. Assuming that the entire set of physical properties of the system Q—dynamically unfolding in λ_i—is represented by a set of linear operators $\{O_{\alpha_1} O_{\alpha_2} \ldots O_{\alpha_n}\}$, the transition functions of the system are the procedure by which one can calculate the expectation values of all its physical properties:

$$f_{\alpha_1 \alpha_2 \ldots \alpha_n}(\lambda_i) = <O_{\alpha_1} O_{\alpha_2} \ldots O_{\alpha_n}>.$$

---

[16] For more formal details about this see Vistarini (2019, ch. 6).

So, $\overline{H}$ is the fiber bundle resulting from the disjunct union of all the possible Hilbert spaces H_{λ_i}, $\forall i$. Each Hilbert space is identified to a possible counterpart of Q, each dynamically unfolding in an isolated spacetime λ_i. Let's unpack a bit this central point.

Consider Q evolving in a spacetime parameterized in M by λ_0. The corresponding fiber, i.e., $(H_{\lambda_0} \times \mathbb{C})$, with $H_{\lambda_0} \equiv Q$, gives dynamical information about the expectation values that the system's physical properties take as Q evolves in λ_0. Then, following the same logic, the fiber bundle as a whole (the disjunct union) gives the same relevant dynamical information about all counterparts of Q, each counterpart considered to be evolving within a spacetime λ_i parameterizing some spatiotemporal deformation. As I argued, all λ_i are points spatiotemporally isolated because the topological structure of the moduli space M is not equipped with a spatiotemporal metric. Now, the disjointness of the union of the Hilbert spaces composing the fiber bundle formally reflects the isolation of Q from any of its counterparts and the isolation of any Q's counterpart from any other. This formal feature of spatiotemporal isolation is central to both my extensions of the Lewisian metaphysical scheme.

Now, few more things about the projection map p appearing in the diagram, which goes from the total space $\overline{H}$ of the fiber bundle to the moduli space M. The main peculiarity of this map is that its inverse p^{-1} is a continuous function and it is the section of the fiber bundle. More precisely, considering a quantum system Q, as well as all its counterparts, the map p^{-1} acts on all λ_i in the following way:[17]

$$p^{-1}(\lambda_i) = (\psi_{\lambda_i}, f_{\alpha_1 \dots \alpha_n}(\lambda_i)) \in H_{\lambda_i} \times \mathbb{C},$$
$$\forall \lambda_i \in M, \lambda_i = \phi(K_i), \forall K_i \in Def(K_0).$$

Recall that $\forall K_i \in Def(K_0)$ means for all spacetime structures K_i belonging to some family of deformations of some spacetime structure K_0. More importantly, ψ_{λ_i} denotes the quantum state of a system (either Q or any of its counterparts) dynamically unfolding in λ_i, and, as I said earlier, $f_{\alpha_1 \dots \alpha_n}(\lambda_i)$ are the corresponding values taken by the physical properties $\{O_{\alpha_1} O_{\alpha_2} \dots O_{\alpha_n}\}$.

Finally, few more thoughts about how this fiber bundle structure is used in my past work to prove two theses of string theory background independence. As I said, the reason for including these two theses is that they play a role in the tentative extension of Lewisian nomological accessibility proposed in Section 14.3.3.

Through the fiber bundle on the topological structure of the moduli space M it is possible to "compare" the expectation values of all Q's dynamical properties with

---

[17] The general structure of the action of p^{-1}, i.e., that maps a point of the base space to a fiber of the fiber bundle, is built into the definition of a fiber bundle's section.

those of all its counterparts. Those "paths" over M along which Q and its counterparts show to obey the same quantum string laws, are special orbits of moduli points. Each type of orbit turns out to be connected to some type of string duality, because each orbit is made by moduli points parameterizing spacetimes whose spatiotemporal inequivalences, and sometimes even topological ones, do not affect strings dynamics. In other words, different spatiotemporal structures parameterized by the same orbit all have the same strings dynamical laws. The two main types of orbits support two claims of background independence for quantum strings laws.

The first claim is with respect to spacetime geometry, and it says something along this line: some type of local family of different spacetime geometries, still topologically equivalent, can be taken as the data for constructing the same string theory laws without prejudice to choice of a particular member. What type of local family? By looking at the fiber bundle $\overline{H}$, one claims that for any family of topologically equivalent spacetimes, if they are parameterized by a Weyl invariant orbit of points λ_i over the moduli space, then,[18]

$$f_{\alpha_1...\alpha_n}(\lambda_i) = f_{\alpha_1...\alpha_n}(\lambda_j).$$

The second claim is instead with respect to a more radical type of deformations, that is, those producing geometrically inequivalent spacetimes that are also topologically inequivalent: some types of local family of topologically inequivalent spacetimes can be taken as the data for constructing the same string theory laws without any prejudice to choice of a particular member. The types of local family in this case are two. The first is exemplified by any family of geometrically different spacetimes that are topologically inequivalent, but all share one topological invariant, i.e., vanishing Ricci curvature.[19] The second type is constituted by families of topologically inequivalent spacetimes not sharing any topological invariant.[20] Then, the same formula above expressing identities of values of transition functions applies to these cases.

14.3.2 The Topological-Fiber Bundle Structure as a Plausible Accessibility Structure: Extending Lewisian Similarity Accessibility

Lewis wrote about modal metaphysics in many places and in many different ways. Lewisian modal metaphysics is actually an umbrella term to indicate a collection of theses, each arising from different applications of the formal possible worlds

---

[18] For technical details and philosophical discussion on this first claim of background independence, see Vistarini (2019, ch. 6). T-duality is a string duality connected to this type of orbits on the moduli space M.
[19] Vistarini (2019, ch. 6). Mirror symmetries not preserving topological invariants (for example, switching Euler characteristics of the compact extra dimensions) are connected to this kind of orbits on the moduli space M.
[20] The AdS/CFT duality is an instantiation of this form of background independence.

semantics. In this section I consider the thesis obtained via the metaphysical application producing the similarity accessibility structure (or counterpart accessibility structure). As far as I know, the earliest formulation of this thesis can be found in "Counterfactuals" and it is succinctly described by this famous passage:

> I believe, and so do you, that things could have been different in countless ways. But what does this mean? Ordinary language permits the paraphrase: there are many ways things could have been besides the way they actually are. I believe that things could have been different in countless ways; ... I therefore believe in the existence of entities that might be called ways things could have been. I prefer to call them possible worlds.
>
> Lewis, 1973, p. 84

Reality is conceived by Lewis as the sum of the imaginable rather than a sum of what actually exists. But the imaginable exists as well, just not in the form of our actuality. As I said in Section 14.1, Lewisian notion of actuality can be read as an indexical notion whose reference varies with the speaker (Lewis, 1973, pp. 84–91).

A later formulation and defense of the counterpart thesis, regarded by many as the canonical ones, are in *On the Plurality of Worlds*. After arguing that we would be better off in many areas of philosophy if we accept unactualized possibilities, Lewis describes there what he takes to be the best theory of possible worlds. This theory must meet a necessary requirement, that is, any part of any world cannot be spatiotemporally connected to any other part of any other world. The lack of spatiotemporal relations among worlds is what tags individuals as belonging to different worlds. Moreover, a further characterization of the theory is that there is a plenitude of worlds, that is, there is a world for every way things could be (according to some suitably restricted principle of recombination). In this metaphysical context, every way things could be refers to every possible situation more or less similar to the actual state of affairs, in some sense of similarity. Finally, the worlds are all concrete.

Although spelling out what the abstract-concrete distinction comes to in this context is a difficult task, the indexical nature of actuality formulated by Lewis (1973) contributes to explain the Lewisian notion of concreteness by unveiling what this notion is not: concreteness is not a prerogative of the actual world. The concrete nature of possible worlds is a main component of Lewis's realism about them and grounds his many arguments against many forms of *ersatzism*.[21]

As I said in Section 14.2, my extension of both similarity and nomologically accessibility structures does not settle the dispute between Lewisian realism and

---

[21] Indeed, nowhere in his works one can find a unique knock-down argument against any form of ersatzism about the worlds. More precisely, when Lewis looks at alternatives to his own kind of modal realism (1986, ch. 3) by temporarily dropping the requirement that the possible worlds are concrete, he does not achieve a single conclusive argument in favor of concreteness of the worlds, and launches different attacks against different forms of ersatzism.

ersatzism about possible worlds. However, the methodology used to realize this extension inevitably perturbs this neutrality, even though slightly. What I mean with that can be explained in the following way: possible worlds in my extension are obtained as deformations of physical solutions of quantum string theory. It is not always the case that a deformation of the theory's physical solution produces another physical solution. Sometimes it does, sometimes it does not. Let's briefly say that the deformation procedures applied in this context create a metaphysical landscape in which physical *possibilia* (predicted by quantum string theory) are a subset of a broader set of mathematical *possibilia*, where the mathematical ones are still connected to quantum string theory because originating from its formal articulation. Now, this methodology should be also read in light of my broader metaphysical commitment to the idea that the physical content and the formal articulation of an empirically adequate physical theory—like quantum string theory—carry some important weight in defining what it is real, or at least in defining aspects or layers of reality that can be connected to the theory's domain of application.[22] So, based on this enlargement of perspective, one may say that in my extended metaphysical landscape at least a small group of possible worlds are real. And this is definitely not a metaphysical scenario that an Ersatzer about worlds would support. However, settling the dispute between Lewis's realism and ersatzism about possible worlds is not a topic of this essay.

Now, in the metaphysical setting equipped with the counterpart relation, possible worlds are concrete worlds more or less similar to the actual one, in some sense of similarity, and to different extents. Then, Lewisian similarity is the main core of the notion of metaphysical possibility residing in this setting. For this reason, the analysis of Lewisian similarity deeply correlates with the analysis of metaphysical possibility built into this setting. Lewis's criterion for admitting worlds into the possible worlds landscape equipped with similarity branches off transversally to many chapters of *Plurality*. However, the part of the book in which his analysis appears to me as coming across most clearly, is chapter two, even though it is not the main topic of that chapter and needs some work of extrapolation to come to the surface. There Lewis is answering to several objections that try to prove his modal realism leads to paradoxes. Here I want to spend few words on his argumentative line against Martin Davies's (1981, p. 262) objection since it is exactly from that Lewisian anti-paradox strategy that I extrapolate a lesson about Lewis's notion of metaphysical possibility within the similarity accessibility setting.[23]

---

[22] The property of being empirically adequate in this context refers to the fact that quantum string theory admits emergent general relativistic spacetime. For an extensive analysis of this criterion in the general context of quantum gravity theories see Huggett and Wüthrich (2013), for an analysis in the specific context of quantum string theory see Vistarini (2019, ch. 3).

[23] The objection was originally formulated by David Kaplan, but that version was never published. The version of the objection to which Lewis answers is that presented by Davies (1981, p. 261).

As far as I understand, two combined facts mainly conspire to produce the objection. The first amounts to be an unrestricted reading of the Lewisian claim about the existence of a one-to-one correspondence between sets of worlds and propositions; the second fact is the traditional identification between a proposition and the content of some *intentional* mental state, widely supported within the philosophy of language circles.[24]

Now, the paradox described by Davies arises from the following argumentative line: on the one side any proposition corresponds to some set of worlds, on the other side a proposition might be the one being thought by some individual in one of the worlds composing the set identified by the proposition. Equivalently, for each proposition there might be one world in the corresponding set of worlds in which the proposition is thought by someone. And that means, for any set of worlds, there might be a one-to-one correspondence between the set and only one of its elements. This conclusion is the paradox pointed by Davies. Lewis's reply apparently tries to bypass the paradox by weakening the link between propositions and mental states' content. In other words, he rejects the idea that every proposition can be thought. His main strategy to support this rejection would be that of combining a form of functionalism about beliefs (loosely speaking one defining beliefs in terms of their causal role rather than in terms of their mental content) with the requirement that beliefs should attach to natural properties. This combination would produce the conclusion that most propositions cannot be thought, hence, according to Lewis, bypassing the paradox.

Now, whether or not this strategy conclusively defeats the paradox is not a topic of this essay. Recall that the rationale behind this brief presentation of Davies–Lewis debate is that of extrapolating from the Lewisian anti-paradox strategy a lesson about his construction (within the similarity setting) of a notion of metaphysical possibility, one that turns out to be informed by ordinary language, but without ties to the mental states riding that linguistic vehicle. So, paraphrasing the Lewisian passage quoted earlier, ordinary language provides a "*countless*" multiplicity of metaphysical possibilities by providing the "*countless*" spectrum of ways in which worlds can be similar. The use of quotation marks here emphasizes that the notion of "*countless*" multiplicity in this context is not a quantitative notion, one that refers to the cardinality of real numbers. Rather, in virtue of its connection to ordinary language, the notion should be here read as meaning some form of qualitative multiplicity, one that is not equipped with cardinality. In other words, as far as I understand, the Lewisian use of the word *countless* here, intended to characterize

---

[24] The dominant philosophical tradition on this identification is that originating in Brentano's works, where beliefs are mainly characterized in terms of *intentional* mental states, later developed by Russell who coins for the first time the term *propositional attitudes* to denote states of belief.

the many ways in which things can be different, does not really overlap with the traditional use, one comparing what the word denotes with what it is denoted by the word *countable*. Rather, a "*countless*" multiplicity here appears to be incomparably different from a *countable* one.

Now, in my reading, ordinary language equips the Lewisian spectrum of metaphysical possibilities with some minimal "formal" framework harnessing the countless variety of ways in which similarity can be conceived. Although it has been thought for a long time that "ordinary language has not exact logic" (Strawson, 1950) the work made by Chomsky (1957) to develop generative grammar initiated a robust tradition of studies in linguistics about the possibility of developing a logic of ordinary language, or at least about the plausibility that some parts of traditional logic are reflected in ordinary language and its ordinary practice. Then, endorsing this tradition, one can say that ordinary language has a "formal" framework and that the latter can be found at the intersections of grammar and logical forms. And it is in virtue of this property that ordinary language would provide some formal guidelines to conceive Lewisian similarity relation. Whether the last claim would be explicitly endorsed by Lewis is an issue that goes beyond the goal of this essay. In any case, the bottom line for me is that the centrality assigned by Lewisian metaphysics to ordinary language, produces a characterization of Lewisian similarity that is sharper and more rigorous than it would be if grounded on the mental content of ordinary language. Ordinary linguistic games produce some structural distinction between legitimate and nonlegitimate uses of the notion. Yet, ordinary language guidelines produce a qualitative spectrum of legitimate uses, hence a qualitative characterization of similarity.

Now, my extension of the Lewisian similarity relation tries to achieve an accessibility structure characterizable in quantitative terms. Indeed a main feature of this extension is that it provides a formal methodology for quantitatively evaluating similarity, for some composite sense of similarity that I will soon describe and call S^*-similarity. As we will see, the quantitative evaluation introduced in this context does not require setting a metric, so it is perfectly consistent with the nonmetrical nature of the relation. Indeed S^*-similarity will be shown to be an ordering and not a metrical relation. Moreover, it will also turn out that the spectrum of variation of S^*-similarity is a countable one. Then, in the extended scheme things could have been different only in countable ways. We'll see the deep philosophical significance of all this.

Now, to unpack this initial presentation of the extended scheme, I start with explaining what sense of similarity is denoted by S^*. That bring us back to the earlier construction of a topology for the moduli space of string theory.

S^*-similarity can be mainly described by two types of spatiotemporal similarity: a type of spatiotemporal similarity relations between pairs of topologically equivalent possible worlds, and another type of spatiotemporal similarity relations between pairs of topologically inequivalent possible worlds. Possible worlds in the S^*-extended metaphysical scheme are interpreted as spatiotemporal structures, but the moduli space's points parameterizing those structures carry the label of possible worlds.[25] This distinction in two types of spatiotemporal similarity between pairs of worlds is explainable in terms of the topology of the families of spatiotemporal deformations. More precisely, as I said in the previous section, string theory moduli space M parameterizes families of spatiotemporal deformations via the continuous map ϕ:

$$M \quad \xleftarrow{\phi} \quad \bigcup_{K_j} Def(K_j).$$

The moduli space gets from the families a topological cover made of local neighborhoods, each neighborhood being image, via ϕ-related maps, of the topology of some local family of deformations $Def(K_j)$, for some central family-fiber K_j. And, as I said earlier, families of spatiotemporal deformations can be basically of two types: those made of spatiotemporal deformations that are topologically equivalent, and those made of spatiotemporal deformations that are topologically inequivalent.

Also, recall that any point λ_i on the moduli space M is spatiotemporally isolated by any other point because M's topology is not equipped with a spacetime metric and that this property is inherited from the topology of the families: any topological family of spatiotemporal structures obtained via deformation functions does not have by definition a spatiotemporal structure; in fact it does not have by definition any metric at all.

Then, it appears clearly that in order to analyze the extended similarity, one needs to dig into the topologies of both sides, as well as into their connection.

When one deforms any spatiotemporal structure K_j (parameterized by λ_j on the moduli space M) one gets a different spatiotemporal structure K_i (parameterized by a different moduli point λ_i). The resulting structure is related to the original one by some type of S^*-similarity, which is produced by the type of deformation functions applied. Then, one can say that in general the S^*-similarity between pairs of spatiotemporal structures, one deformation of the other, "is" the nonmetrical topology of the families, which is produced by the deformation functions.

The S^*-similarity between the pair (K_j, K_i) gets transferred onto M as S^*-similarity ordering between the two corresponding moduli points λ_j and λ_i

---

[25] Lewisian similarity order among worlds does not explicitly refer to forms of spatiotemporal similarity, although for Lewis any collection of things can be regarded as a world just in case it has some spatiotemporal structure. To be precise the Lewisian accessibility structure in which possible worlds are explicitly interpreted as spacetime wholes is not the similarity one, rather the nomological accessibility. My extension of the latter (that I will denote with N^*) will preserve this interpretation. So, in my extended scheme, both S^*-similarity and N^*-nomological accessibility structures share the same interpretation of possible worlds.

(via maps related to ϕ that I'm going to introduce).[26] In general, the transferred $S^\star$-similarity among moduli points "is" the nonmetrical topology inherited by M from the families of deformations on the other side of ϕ. In particular, the $S^\star$-similarity between λ_j and λ_i (parameterizing the pair (K_j, K_i)) is understood in light of the notion of "closeness" brought in by the inherited nonmetrical topology. Here the quotation marks are used to emphasize that being close on the moduli space with respect to its topology has nothing to do with being close in ordinary spacetime, or in any other metrical sense. Indeed the two moduli points represents distinct and spatiotemporally isolated spacetimes, and if one looks at the points' fibers in the fiber bundle $\overline{H}$ on top of M, one finds two different Hilbert spaces.

Then, let's unpack the mechanism of inheritance of $S^\star$-similarity. This analysis also allows us to dig into the notion of "closeness" among moduli points.

Given a spatiotemporal structure K_0, in how many ways can one deform it to produce families of $S^\star$-similar structures? Preliminary speaking, all the ways in which one can deform that structure produce an ordering having two opposite sides: on one side, a family that varies "as much as possible," on the other, a family that varies "as little as possible." Both sides, as well as the intermediate sector of the ordering, are mapped onto different types of moduli points' neighborhoods. The formal maps in charge of these mappings are those explaining the mechanism by which M inherits the $S^\star$-similarity from the families of deformations. Also, once the mechanism is explained, it will appear more clearly in what sense the spectrum of variation of $S^\star$-similarity among moduli points is an ordering and not a metrical structure.

As I said, to describe how the nonmetrical topology relating families' fibers get transformed into the nonmetrical topology relating moduli points, one needs to introduce some additional formal maps. The first map is obtained in the following way: given ϕ (mapping a family-fiber to a moduli point) it is possible to define its differential, namely, another map that act on ϕ like a functional. Taking the differential of ϕ in this context means taking its derivative with respect to the variations of the family-fibers in $Def(K_0)$, for some central fiber K_0. That is equivalent to say that the derivative is taken with respect to spatiotemporal variations of the central family fiber.[27]

In general, the differential of a function (or map) from space A to space B is a contravariant function (or map) from B's tangent bundle to A's tangent bundle.[28]

---

[26] Note that the moduli map ϕ only maps each fiber of any family onto some moduli point. ϕ does not map relations among fibers. As we will see, in order to transfer the topological structure relating a group of fibers to the group of moduli points corresponding to those fibers, one needs to use maps related to ϕ by reiterated applications of the differential operator.

[27] In general, the derivative of a function (or a map) always introduces a notion of degree of change of that function with respect to the variation of its variables. In this particular case the variables of the function are spatiotemporal deformations of some spatiotemporal structure.

[28] For readers unfamiliar with this formal notion: a tangent bundle of a space A is the union of all tangent spaces A_p at each point p of A.

So, in this particular case the differential of the moduli map ϕ is the contravariant map $\phi^\star$ from the tangent bundle of the moduli space M to the tangent bundle of the family of all possible deformations $\bigcup_{K_j} Def(K_j)$. By restricting ϕ to some local family of infinitesimal deformation $Def(K_0)$ of some structure K_0, one gets that the corresponding controvariant map $\phi^\star$ is defined over the tangent space to M at the point λ_0 (parameterizing K_0) and it has as codomain the tangent space to $Def(K_0)$ at K_0. This contravariant map is also called the Kodaira–Spencer map, based on Vistarini (2019, ch. 6) and Kodaira (2005, ch. 2):

$$\phi^\star : T_{M,\lambda_0} \longrightarrow H^1(K_0, T_{K_0}), \qquad (14.1)$$

where $H^1(K_0, T_{K_0})$ is the tangent space (at the central family-fiber K_0) of the local family of deformation $Def(K_0)$ and T_{M,λ_0} is the tangent space to M at the point λ_0 (parameterizing K_0).[29]

Now, the Kodaira–Spencer map is one of the formal maps in charge of the mechanism of topology's inheritance: a specific type of $S^\star$-similarity among spatiotemporal structures is transformed into a specific type of $S^\star$-similarity among moduli points via the pull-back of the Kodaira–Spencer map. What specific types of $S^\star$-similarity are involved here?

$H^1(K_0, T_{K_0})$ turns out to be the vector space containing all the first-order infinitesimal deformations of K_0 (see Kodaira [2005]). An element in this vector space is a spatiotemporal structure (an isomorphism class) that is spatiotemporally isolated from any other, still any element is comparable to any other with respect to any arbitrarily chosen base.

In general vector spaces do not have spatiotemporal metrics, or any metric, by definition. Having bases and coordinate systems does not mean to be equipped with a metric.[30] One can define on a vector space a relational structure that is weaker than a metrical one, that is, a vector space can be equipped with a topology. Topological vector spaces (also called linear topological spaces) are vector spaces that are also topological spaces, thereby they admit notions of convergence, of sets' closure with respect to the defined topology, and so on. They are mainly investigated in functional analysis.

Now, it should appear clearly that in this context the vector space $H^1(K_0, T_{K_0})$ is equipped with a topology, one identifiable to the network of $S^\star$-similarity relations (among spatiotemporal deformations) produced by first-order infinitesimal deformation functions. Then, the pull-back of $\phi^\star$ maps such a relational network among

---

[29] For detail about Kodaira–Spencer map and the tangent spaces involved I refer the reader to Vistarini (2019, ch. 6); there the simplified presentation of the formal procedures heavily relies on Kodaira (2005, ch. 4).

[30] Parenthetically, that does not entail that we cannot define a metric on a vector space. We can always define an inner product among vectors that induces a metric and produces a mathematical space with more internal structure, but a metric is not definitional of a vector space.

spatiotemporal structures onto local neighborhoods of moduli points that, in virtue of this pull-back, turn out to be linear topological neighborhoods characterized by a specific type of "closeness." Then, the question is: "what type of 'closeness'?"

Before answering let me remind that we are exploring how M inherits the S^*-similarity ordering's sector coming from families of first-order spatiotemporal deformations. Soon I explore other sectors of this topological ordering by introducing new maps (still related to ϕ and to the Kodaira–Spencer map) as well as families of deformations of order higher than one.

Now, an explicit description of the type of "closeness" characterizing the sector connected to $H^1(K_0, T_{K_0})$ requires me to say a few general facts about ordering relations in topological vector spaces. In general the latter can be equipped with forms of ordering among their points (or vectors). In this section I propose a hybrid form of ordering, something at the intersection between general topological tools used in algebraic geometry and the specific topological structure of the theory's moduli space presented in this essay. The hope is that of proposing a topological ordering among moduli points that extends the Lewisian notion of closeness among worlds.

Now, on any topological vector space it is possible to define the specialization order $(\ldots \leq \ldots)$ (or $\geq$) relating two points when one lies in the closure of the other (see Hartshorne [1977]). Sticking to Hartshorne's style of notation, one can say that $x \leq y$ just in case $cl(x)$ is contained in $cl(y)$, that is, just in case the closure of the singleton set $\{x\}$ belongs to the closure of the singleton set $\{y\}$. The notion of closure is purely topological. The closure of a singleton set $\{P\}$ is defined as the intersection of all closed sets containing the singleton set $\{P\}$—the notion of close (or open) sets refers to whatever topology is defined on the space. An equivalent formulation is the following: $x \leq y$ if and only if x is contained in all closed sets that contain y. In this sense one may say that x is a specialization of y. In this general context one speaks of specialization because the more closed sets contain a point, the more properties the point has, the more special it is (see Hartshorne [1977]).[31]

This general characterization of the specialization ordering $(\ldots \leq \ldots)$ is here calibrated to the specific setting of my topological spaces to produce a suitable topological ordering, one I denote with $(\ldots \prec^p \ldots)$. I preliminary define this order in the following way: for each integer p, the ordering characterizes the type of

---

[31] This type of ordering in algebraic geometry is traditionally used to study properties of topological or algebraic manifolds: studying the specialization of the manifold's generic points to closed ones carries relevant information about the manifold's properties. Generic points of a manifold are those contained in every nonempty open set of the manifold topology, whereas the closed points are the most specific ones because contained in every nonempty closed set. For this reason, one can equivalently say that a generic point is a point at that all generic properties of the manifold are true, a generic property being a property which is true at almost every manifold's point (see Hartshorne [1977]).

"closeness" between any pairs of moduli points (λ_i, λ_j) such that λ_i parameterizes a deformation of order p of the spatiotemporal structure parameterized by λ_j. The topological nature of this ordering is straightforwardly shown by the fact that, at the end of the day, it is produced by deformations functions. Such topological ordering defined on string theory moduli space is shown to preserve specialization's features, even though the reading of what specialization means in this context is different from the original meaning.

Now, before digging into this preliminary characterization, recall that at the moment I am looking at the sector of the moduli space ordering gained via the pull-back of the Kodaira–Spencer map. So, for now the deformation order p is set to 1. Let's consider a moduli point λ_0 parameterizing the central fiber K_0 of a family of first-order spatiotemporal deformations $Def(K_0)$ contained in the topological vector space $H^1(K_0, T_{K_0})$. Moreover, let's $\{\lambda_i\}_i$ be the set of moduli points parameterizing the first-order spatiotemporal deformations $\{K_i\}_i$ of the family $Def(K_0)$. The type of "closeness" inherited by this set of moduli points via the Kodaira–Spencer map can be formally described by saying that, for all λ_i in the set,

$$cl^{h1}(\lambda_0) \subseteq cl^{h1}(\lambda_i).$$

Here the closure cl^{h1} is taken with respect to the topology pulled back from $H^1(K_0, T_{K_0})$.[32]

The moduli point λ_0 parameterizing the central fiber K_0 of any family in $H^1(K_0, T_{K_0})$ is a sort of specialization point for any moduli point λ_i parameterizing some deformation in those families. The justification for this claim is in what we already know: by construction the topological structure relating the moduli points of the neighborhood $\phi(Def(K_0))$ reflects the relational structure of the family $Def(K_0)$. As we saw in Section 14.3.1 any generic fiber of a family of deformations is obtained by applying deformation functions to the central fiber K_0. So, one can deduce that for each generic fiber K_i of the family, the central fiber K_0 shares by construction all closed neighborhoods containing K_i. In this sense one can say that the central fiber K_0 is a specialization of K_i, for any generic fiber K_i in the family—for an exhaustive and general analysis on the deformation procedures see Kodaira (2005) and for a simplified analysis confined to spatiotemporal deformations see Vistarini (2019).

Then, consider a moduli point λ_0 parameterizing a spatiotemporal structure K_0. Also consider all the families of first order deformations having K_0 as central fiber, i.e., the whole topological vector space $H^1(K_0, T_{K_0})$. Now, given any moduli point λ_i, one can say that λ_i is "close" to λ_0 according to the $S^\star$-similarity induced by first order deformation functions, that is,

---

[32] Technically speaking, the closure cl^{h1} is taken with respect to the inverse image topology $(\phi^\star)^{-1}(T_{H^1})$.

$$\lambda_0 \prec^1 \lambda_i,$$

just in case

$$cl^{h1}(\lambda_0) \subseteq cl^{h1}(\lambda_i).$$

This is the formal description of the sector of the moduli points ordering corresponding to first-order infinitesimal deformations, as well as it is the formal description of the type of "closeness" characterizing topological neighborhoods of moduli points in that sector.

Now, I want to explore other sectors of the $S^\star$-similarity topological ordering by introducing new maps and new families of deformations of order higher than one.

Given a spatiotemporal structure K_0 admitting a nontrivial $H^1(K_0, T_{K_0})$, such a structure might be deformed by second-order deformations. In this case, to find out how the corresponding ordering sector looks like on the moduli space, one needs to introduce the first derivative of the Kodaira–Spencer map (i.e., second derivative of the moduli map ϕ). Broadly speaking, a different topological vector space is involved, one containing all the second-order deformations' families of K_0, i.e., $H^2(K_0, T_{K_0})$. The vector space's topology in this case "is" the $S^\star$-similarity relation produced by deformation functions of second order. Whether the topological vector space $H^2(K_0, T_{K_0})$ is different from the empty set is not a trivial fact. Not any spatiotemporal structure admitting first-order deformations always admits the existence of second order ones. But skipping any technicality here, which can be found in Kodaira (2005), and by mimicking more or less the same logic describing $S^\star$-similarity produced by first-order deformations, one can more or less repeat the process with the second-order ones.

In this case the type of $S^\star$-similarity differs from the type analyzed earlier, because second-order deformation functions, although usually preserving topological equivalence, produce families of structures less spatiotemporally similar than those produced by first-order deformations. Then, repeating the same reasoning developed in the first-order case, the $S^\star$-similarity produced by second-order deformations of the central fiber K_0 is a topological relation between the two members of any pairs $\{(K_i, K_0)\}_i$, where K_i is any generic fiber of any family of second-order deformations having K_0 as central fiber. Through the new map, i.e., the first derivative of the Kodaira–Spencer map, the moduli space inherits such a topological network. The latter on the moduli space gets transformed into a patch of topological neighborhoods containing moduli points $\{\lambda_i\}_i$, each parameterizing a distinct generic fiber in $\{K_i\}_i$, and each having the moduli point λ_0 (parameterizing K_0) as specialization point.

Then, given the moduli point λ_0 parameterizing a spatiotemporal structure K_0, one can say that λ_k is "close" to λ_0 in relation to the $S^\star$-similarity induced by second-order deformation functions applied to K_0, that is,

$$\lambda_0 \prec^2 \lambda_k,$$

just in case

$$cl^{h2}(\lambda_0) \subseteq cl^{h2}(\lambda_k).$$

Virtually this can be repeated at any order of deformation p (for all positive integer number p). The higher the order of deformation, the lower the $S^\star$-similarity of the possible worlds found in the process. Then a world lying in $cl^{h2}(\lambda_0)$ is less "close" to λ_0 than any other world lying in $cl^{h1}(\lambda_0)$.

Now, before giving a comprehensive formulation of $S^\star$-similarity accessibility for then moving to the issue of regaining Lewisian similarity, I would like to introduce a last crucial feature of this extended accessibility structure. This brings us back to the Kodaira–Spencer maps and to all the other maps obtained as its derivatives of any order. Recall that each map is connected to a different sector of the topological ordering on the moduli space, since each map is connected to a different order of spatiotemporal deformations. Also, in general, different maps have different ranks. A formal property shared by these maps turns out to be that of providing a quantitative description of any sector of the $S^\star$-similarity order on the moduli space: the rank of each map either *computes* or *counts* in "how many non-isomorphic ways" pairs of "worlds" (λ_j, λ_k) in the corresponding ordering sector can be $S^\star$-similar.

Let me unpack this important point. In general and loosely speaking, the rank of a map between vector spaces gives the maximum number of linearly independent vectors in these vector spaces. In our particular case, each map is connected to an order of deformation, so its rank gives the maximum number of linearly independent deformations of that order applicable to some spatiotemporal structure. For the sake of simplicity let's set the order of deformation p to one, and let K_0 be the spatiotemporal structure deformed. Then, the rank of the Kodaira–Spencer map gives the maximum number of linearly independent first order deformations of K_0 in the vector space $H^1(K_0, T_{K_0})$, i.e., it gives the cardinality of any arbitrary base in the vector space. The number of linearly independent deformations of K_0 represents the maximum number of nonisomorphic "ways" in which one can deforms the spatiotemporal structure without redundancies. Let's assume the cardinality is finite and let's fix an arbitrary base. Then, any arbitrary first-order deformation in this vector space is either a vector of the base (a linearly independent deformation) or decomposable in some linear combination of them (a linearly dependent deformation). Under the assumption of finiteness of the vector space's dimension (i.e., of finite cardinality of its bases), the rank of the Kodaira–Spencer map is an integer showing in "how many nonisomorphic ways" one can deform the spatiotemporal structure K_0 to get another one connected to the original by S^*-similarity of the first order. However, the $H^1(K_0, T_{K_0})$'s bases can have infinite cardinality, yet that infinity is countable [see Kodaira [2005]). In this case the rank of the map (represented by an isomorphism class of infinite matrices) *counts* the nonisomorphic "ways" one can deform the spatiotemporal structure K_0 to get another one connected to the original by S^*-similarity of the first order. Importantly, this line of reasoning can be generalized to any other order of deformation p, hence to any vector space $H^p(K_0, T_{K_0})$.

Then, for any order p of spatiotemporal deformation, the spectrum of variation of the corresponding S^*-similarity is *countable*. This formal property turns out to have a deep philosophical significance when one looks at the metaphysical property it calibrates: in the extended metaphysical accessibility structure equipped with S^*-similarity things could have been different in many *countable* ways.

Now, to wrap things up with the construction of S^*-similarity, let's consider a spatiotemporal structure K_0 deformable in a nontrivial way for any order p of deformation. K_0 is parameterized by the moduli point λ_0. Then, for any order of deformation p, λ_0 is the specialization point, with respect to the ordering $(\prec^p)$, of any moduli point parameterizing deformations of K_0 of order p. Recall that in this extended scheme possible worlds are interpreted as spatiotemporal structures, but that the corresponding moduli points carry the labels of possible worlds. Let K_0 be our actual spatiotemporal structure, then the corresponding moduli point λ_0 is labeled as the actual world. Then, for any order p of deformation of K_0, the complete set of families of K_0's deformations of order p is composed by spatiotemporal structures K_i, each parameterized by some moduli point λ_i, each λ_i being a possible world. Then, one can say that

a possible world λ_i is accessible from the actual world λ_0 just in case the former is S^*-similar to the latter for some order p of deformation (i.e., for some type of S^*-similarity). Equivalently, a possible world λ_i is accessible from the actual world λ_0 just in case there exist some integer p such that $\lambda_0 \prec^p \lambda_i$. And that holds just in case $cl^{hp}(\lambda_0) \subseteq cl^{hp}(\lambda_i)$.

Now, interestingly, S^*-similarity has the same basic formal properties as Lewisian similarity: it is nontransitive and non-Euclidean. To briefly show the nontransitivity, let's assume that a spatiotemporal structure K_0 admits deformations of orders one and two at least, that is, it admits the existence of a nontrivial $H^p(K_0, T_{K_0})$ at least for $p = 1, 2$. A first-order deformation K' of K_0 belongs to the $H^1(K_0, T_{K_0})$. Now, any deformation of K_0 in $H^2(K_0, T_{K_0})$ can be seen as first-order deformation of K', but by definition they are second-order deformation of K_0. The same can be shown for any order p of deformation. So, the property of being a deformation of some fixed order p is not transitive. This formal fact produces a nontransitive S^*-similarity that then gets transformed into a nontransitive S^*-similarity on the moduli space.

Also "being a deformation of some fixed order p" cannot be Euclidean: if K' and K'' are two first-order deformations of K_0, they cannot be first-order deformations of each other. This formal fact produces a non-Euclidean S^*-similarity. Also this formal property is straightforwardly transferred to the S^*-similarity on the moduli space.

Now, the connection between S^*-similarity and Lewisian similarity is much stronger than simply satisfying this basic formal features. Indeed I claim that Lewisian similarity order supervenes on S^*-similarity. This is the sense in which the S^*-similarity accessibility structure I propose here is an extension of the Lewisian one. Then, I need to unpack this notion of supervenience to finally describe the nature of this extension.

As I said, Lewisian theory of Humean supervenience develops around the central thesis that there is nothing to reality except the spatiotemporal distribution of local natural properties. And he argued for this thesis by showing how laws, causation, chances, counterfactual dependence, dispositions and so on, could be located within this Humean mosaic.[33]

Now, I want to endorse momentarily this thesis and to consider two possible worlds λ_i and λ_j in the extended metaphysical landscape. These two possible worlds are interpreted as spatiotemporal structures, and let's say they are related by some type of S^*-similarity produced by somehow deforming the spatiotemporal structure of one into the other.

Now, both possible worlds λ_i and λ_j can be seen as two Humean mosaics, since both fit the Lewisian identikit of such structure. These two Humean mosaics are different but S^*-similar because obtained by somehow deforming one into the other. Still, because of Humean supervenience, each mosaic is basis of supervenience for laws, causation and so on. In other words each mosaic is basis of supervenience for a Lewisian world.

The two Lewisian worlds supervening on the two Humean mosaics λ_i and λ_j are similar in the Lewisian sense of closeness in virtue of the S^*-similarity order of their respective underlying Humean mosaics. Speaking of supervenient features like laws for example, the laws supervening on one mosaic λ_i may be the same as those supervening on the mosaic λ_j, in which case this supervenient feature would contribute to a strong overall similarity between the two Lewisian worlds. Nevertheless, the laws supervenient on one mosaic λ_i may not be the same as those supervening on the mosaic λ_j. In this case the spatiotemporal deformations of the

---

[33] Nevertheless, for the sake of accuracy, I would like to say that this statement might be understood in the wrong way, i.e., as if Lewis would confine the notion of reality to a world. But this is not what I mean, as well as it's not what Lewis does. Reality for Lewis is far more extensive than a world. It contains all worlds, and that statement applies to each of them.

underlying Humean mosaics would produce deformations of the supervening laws, hence an overall weaker similarity between the two Lewisian worlds.

The bottom line here is that any qualitative "amount" of Lewisian closeness between pairs of Lewisian worlds supervenes the "closeness" of their Humean mosaics, which is associated to some order p of deformation. Then, in this sense the Lewisian similarity accessibility structure, with its *countless* variation of worlds' closeness, can be regained by the S^*-similarity one. And it can be regained in virtue of Lewisian Humean supervenience. Then, S^*-similarity order might be regarded as a plausible extension of the Lewisian one.

Now, at a first glance the previous paragraph might be considered to be self-defeating. The extension of Lewisian similarity accessibility structure has been built by using core aspects of the formal articulation and physical content of quantum string theory, namely, a quantum gravity theory denying the fundamental nature of spatiotemporal relations. Then, to show how to regain Lewisian similarity order from the S^*-similarity extension, I use Lewisian Humean supervenience, i.e., a thesis assigning to spatiotemporal relations a fundamental nature. Should this combination of strategies produce an internal tension?

There are basically two points I can make here to dissolve this apparent tension. The first is about the special status of quantum string theory. As any other existing approach to quantum gravity, quantum string theory delivers a description of reality at the Planck scale, one in which gravity, the only fundamental force left not quantized by quantum field theory, is quantized. Quantizing gravity is quantizing or discretizing the world's geometry. Differently from any other quantum gravity approach, in quantum string theory quantization of geometry is produced by laws that once applied far from the Planck scale (by taking some low energy limits of their structural equations) morph into many of the empirically tested low-energy laws. For example, some low-energy limit of quantum strings laws reproduces general relativity laws governing classical gravity, i.e., governing the collective behaviors of the gravity's quanta found in the quantum string spectrum of states at the Planck scale.[34] Now, physical laws far from the Planck scale, like general relativity, quantum field theory, the Standard model, and so on, do have the property of supervening on the Humean mosaic of our world. That means, quantum string theory laws far from the Planck scale morph into the physical laws supervening on the Humean mosaic of our world. However, because of their special status of being fundamental laws of our world, they do not supervene the Humean mosaic. Yet this fact would not require a total rejection of Humean supervenience thesis, rather a limitation of its domain of application. Indeed physical laws at the Planck scale have a special metaphysical status not shared with

---

[34] See Polchinski (2005), Witten (1996), and Huggett and Vistarini (2015)

physical laws far from the Planck scale: quantum string laws do not fit any Humean mosaic.

Then, to unpack this last claim, I need the second point. At the Planck scale quantum string physics is background independent. Quantum string laws are more fundamental than any spatiotemporal structure (any geometry with positive, negative, and zero curvatures). Equivalently, the string physical lesson coming from the Planck scale tells that any Humean mosaic (any spatiotemporal structure) is itself an emergent structure. Now, the fact that any Humean mosaic in any Lewisian possible world is more fundamental than laws governing those worlds far from the Planck scale, does not entail that Humean mosaics cannot be emergent. Quantum string laws cannot fit any Humean mosaic because they are the special status laws in charge of predicting and explaining those mosaics, as well as in charge of predicting and explaining their supervenient physical laws. In this context, Humean supervenience is regarded as a metaphysical thesis about an emergent reality.

Now, although the nonapplicability of Lewisian Humean supervenience to quantum strings laws, the particular features of quantum string theory background independence partially support a tentative formulation of a different thesis of supervenience applicable to quantum strings laws in some case. This analysis is part of my tentative extension of the Lewisian metaphysical scheme in which the interpreted accessibility structure is the nomological one. This is the topic of next section.

14.3.3 The Topological-Fiber Bundle Structure as a Plausible Accessibility Structure: Extending Lewisian Nomological Accessibility

Lewisian nomological accessibility among worlds was briefly described in Section 14.2. Let's recall that in this metaphysical setting possible worlds are mereological spacetimes wholes. Moreover, we saw that saying that world u is accessible from world w means saying that u obeys the physical laws of w. As I said, the property of "being symmetric" is not definitional of this relation because the fact that u obeys the physical laws of w does not necessarily entail the converse. For example, world w might be characterized by a Humean mosaic strictly containing the one characterizing u. Then, because of Humean supervenience, there might be some physical laws governing w and supervening the nonshared portion of its mosaic, that are not physical laws of u.

Now, in this section I attempt to extend Lewisian nomological accessibility by using the topological-fiber bundle structure defined to extend Lewisian similarity to S^*-similarity. But this time it is the fiber bundle component read through the lens of string dualities that is playing a central role.

In Section 14.3.1 I introduced the total space of the fiber bundle $\overline{H}$ formally defined by the following union:

$$\overline{H} = \coprod_{\forall i, \lambda_i \in M} (H_{\lambda_i} \times \mathbb{C}).$$

Let's recall that the space base of the fiber bundle is the string theory moduli space M equipped with a nonmetrical topology, one induced by mainly two types of deformations of spatiotemporal structures: spatiotemporal deformations preserving topological equivalence among the structures, and spatiotemporal deformations not preserving topological equivalence:

$$
\begin{array}{l}
\overline{H} \\
\downarrow{\scriptstyle p} \\
M \quad\quad \overset{\phi}{\longleftarrow} \quad \bigcup_{K_j} Def(K_j),
\end{array}
$$

where for all $\lambda_i \in M$,

$$p^{-1}(\lambda_i) = (\psi_{\lambda_i}, f_{\alpha_1 \ldots \alpha_n}(\lambda_i)) \in H_{\lambda_i} \times \mathbb{C},$$

This diagram more or less combines those presented separately in Section 14.3.1. Here I am partially relying on the content of that section. As I said there, given a quantum string system Q, one can choose some set of physical properties of the system, for example the set $\{O_{\alpha_1} O_{\alpha_2} \ldots O_{\alpha_n}\}$, and one can also decide that the spatiotemporal structure in which the system Q dynamically evolves is the one parameterized by the moduli point λ_0. Then, as we saw, the fiber bundle's fiber on top of the moduli point λ_0, i.e., the fiber $p^{-1}(\lambda_0) = (\psi_{\lambda_0}, f_{\alpha_1 \ldots \alpha_n}(\lambda_0)) \in H_{\lambda_0} \times \mathbb{C}$, carries information about the values taken by the chosen physical properties of the system Q as it dynamically unfolds in the spacetime parameterized by λ_0.

Any other fiber of the fiber bundle, each on top of some deformation λ_i of λ_0, is a separate Hilbert space. Then any other fiber carries information about the values taken by the chosen physical properties $\{O_{\alpha_1} O_{\alpha_2} \ldots O_{\alpha_n}\}$ of some counterpart of Q dynamically evolving within the corresponding spatiotemporal deformation of the arena in which Q evolves.

Now, we saw in Section 14.3.1 that through the fiber bundle structure it is possible to "compare" the values taken by Q's physical properties with those of all its counterparts. And this comparison because of string dualities produces the interesting result of dividing the moduli space M in sets of orbits made by moduli points. Each orbit is made of some subset of moduli points λ_i parameterizing some family of spatiotemporal deformations of the spacetime in which Q evolves. Each orbit identifies a family of spacetime deformations in which the corresponding counterparts of Q dynamically evolve in the exact same way as Q does. In other words, all $Q's$ counterparts along an orbit obey the quantum string laws obeyed by Q, despite being in different spatiotemporal structures. The following diagram illustrate what I said so far:

$$\begin{array}{c} \overline{H} \\ \downarrow^{p} \\ M \end{array} \xrightarrow{\ \ \pi\ \ } \frac{M}{\{[\Theta_d]\}_d}.$$

The map π is the canonical projection onto the quotient space $\frac{M}{\{[\Theta_d]\}_d}$ obtained by dividing the topological space M into the orbits $\{[\Theta_d]\}_d$ produced by string dualities and revealed by the fiber bundle structure. As we saw in Section 14.3.1, these orbits are equivalence's classes, where the equivalence relation is defined as dynamical equivalence among different strings models. Each type of orbit is connected to some type of string duality, which in turn points to some degree of strings laws insensitivity to spatiotemporal differences.

Parenthetically, one can in principle imagine nondual worlds obeying the same quantum string laws. In this case such a set would not be an orbit in the quotient space above, because not identifiable by a string duality. So, including such a possibility in this formal analysis would probably require adding some new formal features to the quotient space. It is hard to tell whether these presumed extra features can be imported from the formal articulation of quantum string theory because, as a matter of fact, such a conceptual possibility does not seem to be formally traceable into the physical content of the theory as it stands. Since in this essay the extension of the Lewisian nomological accessibility is informed by quantum string theory, I would like to give priority to orbits of worlds produced by string dualities.

Now, the topological space $\frac{M}{\{[\Theta_d]\}_d}$ is basically the disjunct union of orbits Θ_d. Its topology (the quotient topology) is inherited by that of M via the projection π that basically glues together moduli points connected by the duality relations.[35]

The quotient space might be the appropriate setting in which to extend Lewisian nomological accessibility. Let's unpack this point. Recall that, like in the case of $S^\star$-similarity, possible worlds in this extension are interpreted as spatiotemporal structures parameterized by moduli points λ_i. The extension of Lewisian nomological accessibility is called $N^\star$-nomological accessibility. The following claim is the first of the two claims characterizing this extension:

world λ_i is $N^\star$-nomologically accessible from word λ_0 just in case λ_i obeys the quantum string laws of λ_0.

Because string dualities are the central component around which $N^\star$-nomological accessibility is constructed, the latter is a symmetric relation by definition. Its property of being symmetric is inherited from the symmetric nature of string

---

[35] For more details on quotient topologies, see Kodaira (2005).

dualities. The latter are symmetric because the dynamical equivalence among string models they establish is an exact one. So, the second claim characterizing $N^\star$-nomological accessibility is the following:

if world λ_i is $N^\star$-nomologically accessible from word λ_0, then the converse holds true because they both obey the same quantum string laws in virtue of being dual, for some type of string duality.

Now, from the fact that there are mainly two types of spatiotemporal deformations producing the topology of M, one infers that there must be two types of orbits covering $\frac{M}{\{[\Theta_d]\}_d}$. The first type is made of possible worlds not sharing the same spatiotemporal structure, yet sharing the topological one, and all obeying the same quantum strings laws. This type of orbit is often connected to T-duality. Such dynamical equivalence "identifies" possible worlds obeying the same strings laws, but spatiotemporally inequivalent because produced by deformations smoothly shrinking (or dilating) spatial dimensions of some initial world. However, this type of deformation does not change the topological invariants of the initial world, so the worlds produced all share the same topological invariants.

Fixing as usual some spatiotemporal structure K_0 parameterized by a moduli point λ_0, orbits of this type originating from such moduli points can be regarded as possible worlds $N^\star$-nomologically accessible from world λ_0. Although preserving topological invariants, the large variety of spatiotemporal differences along an orbit constitutes an obstacle to a straight application of Lewisian Humean supervenience to the set of string laws obeyed by the worlds in the orbit. According to this thesis of supervenience, spatiotemporal deformations in general change the supervenient laws, even though in some cases they don't. There might be cases in which the same physical law supervenes on more than one spatiotemporal "token." In other words, Humean supervenience, as many other theses of supervenience formulated in different philosophical contexts, admits its own form of "multiple realizability." But as in many other theses of supervenience, in the Humean one there are implicit constraints built into the notion of multiple realizability in play. In general, such constraints prevent the relation between the base and the presumed supervenient property from becoming too weak to qualify as supervenience. For example, a claim that mental properties can supervene any physical substratum, including blue cheese, endorses a notion of supervenience that is self-defeating, because the multiple realizability in play here seems actually to set a dualistic agenda. *Mutatis mutandis*, within the context of Humean supervenience, a claim that the same physical laws supervene many varieties of spatiotemporally inequivalent structures seems to undermine the very notion of Humean supervenience, because it appears to set a scenario in which those laws are more fundamental than spatiotemporal relations. So, the bottom line here is that Humean supervenience is not applicable to quantum string laws, at least in its original formulation, because the kind of insensitivity shown by these laws to spatiotemporal deformations has nothing to do with the kind of multiple realizability compatible with Humean superveninece. Indeed that insensitivity sets a scenario in which quantum string laws are more fundamental than spatiotemporal structures.

However, as I said, the type of deformations applied in this first case (mainly connected to T-duality) does not change the underlying topology of the original spatiotemporal structure K_0. Even more interestingly, modifying even just one topological invariant of one world residing in the orbit would break the $N^\star$-nomological accessibility to that world. Then, some notion of supervenience applicable to string laws might be formulated in this case. The mosaic would be defined in terms of a topological distribution of local properties and the set of fundamental strings laws governing those worlds would supervene on it.

The second type of orbit made of worlds $N^\star$-nomologically accessible from world λ_0 is instead made of possible worlds not sharing the same spatiotemporal structure, but also topologically inequivalent. Two subtypes are included here. The first subtype contains those orbits parameterizing spatiotemporal structures geometrically inequivalent, as well as topologically inequivalent, but all having vanishing Ricci curvature. Without unpacking technical details let's just say that constant Ricci curvature is basically a property characterizing a specific set of topological structures, those which are differentiable.[36] Then for this set of topological structures, the constant value of the Ricci curvature is a topological invariant. So, the fact that the worlds in the orbit all share the same vanishing Ricci curvature means that the deformations of the central structure K_0 change all topological invariants except one. The string dualities connected to this subtype of orbits are called string mirror symmetries.

Again, a deformation of any of the orbit's worlds done without preserving vanishing Ricci curvature would produce a world not nomologically accessible from world λ_0, hence an outsider. In this case trying to formulate a thesis of supervenience applicable to the fundamental strings laws governing those worlds is much more difficult. It is highly non trivial to figure out what a mosaic would be in this case. One cannot use the same strategy used earlier, one that replaces the notion of geometrical distribution of local facts with that of purely topological distribution. All one has here is a single shared topological property allowing for $N^\star$-nomological accessibility from world λ_0 to the worlds in this subtype of orbits.

Things get even worse with the second subtype, where the worlds $N^\star$-nomological accessible from λ_0 are produced by deformations not preserving

---

[36] A differentiable manifold is not equipped with a metrical (spatiotemporal) structure by definition. It is instead a topological manifold. However, not every topological manifold is a differentiable one.

any topological invariant. Orbits of this type are connected to dualities like anti–de Sitter spacetime conformal field theory (AdS/CFT) correspondence. In this case there is no room for formulating any thesis of supervenience applicable to quantum string laws involving an ad hoc notion of mosaic.

Now, if λ_i is $N^\star$-nomologically accessible from λ_0, then the Lewisian world w supervening world λ_i might be nomologically accessible from the Lewisian world v supervening word λ_0. Let's unpack this claim.

By using the same line of reasoning developed in the previous section, worlds λ_i and λ_0 satisfy by definition all the Lewisian requirements for being Humean mosaics—indeed they are interpreted in the extended accessibility relation as spatiotemporal structures. Then the notion of worlds-supervenience mentioned here is that of Humean supervenience. A Lewisian world is not just a Humean mosaic, but also everything else that supervenes on that mosaic. Let's be focused only on the supervenient feature represented by physical laws far from the Planck scale.

Now, λ_i and λ_0 are two different spatiotemporal structures, and λ_i is $N^\star$-nomologically accessible from λ_0 because obeying the quantum string laws of the latter. Also, the extended accessibility relation is symmetric by definition. For example, it cannot be the case that λ_i obeys a set of quantum string laws strictly containing that obeyed by λ_0. The two Humean mosaics λ_i and λ_0 obey the same quantum string laws. Then, as I said in the previous section, such quantum string laws are more fundamental than the two mosaics and in charge of predicting them.

Now, when applying those quantum string laws far from the Planck scale (by taking suitable low-energy limits of their structural equations) they reproduce a set of physical laws at low energy supervening on the Humean mosaics λ_0 and λ_i. Since Humean supervenience applies to these physical laws, the two supervening Lewisian worlds in general would not share all of them, even though in some cases they would.

In the case that two supervening Lewisian worlds share all of the supervening physical laws, Lewisian nomological accessibility is straightforwardly regained from the $N^\star$-nomologically accessibility connecting λ_i and λ_0. In the more general case in which the two Lewisian worlds share instead only some of the supervening physical laws, Lewisian nomological accessibility might be regained. One gets the Lewisian accessibility back at the supervenient level in all those cases in which one of the two supervenient worlds obeys a subset of the low-energy laws obeyed by the other. For example, let $\{L_1, L_2\}$ be the complete set of supervenient physical laws, if one of the two Lewisian worlds, say, w supervening λ_i, obeys all of them and the other obeys at least L_1 or L_2, one can claim that the Lewisian world w supervening λ_i is nomologically accessible from the Lewisian world v supervening λ_0.

However, Lewisian nomological accessibility might not be always regained from $N^\star$-nomologically accessibility. For example, let $\{L_1, L_2, L_3\}$ be the complete set of supervenient physical laws. The Lewisian world w obeys (L_1, L_2), whereas the Lewisian world v obeys (L_2, L_3). The two Lewisian worlds only share one law. One cannot say that world w is nomologically accessible from world v because world w does not obey L_3. In the same way one cannot say that world v is nomologically accessible from world w because world v does not obey L_1. So, in this setting Lewisian nomological accessibility between the two supervenient worlds cannot be regained from the $N^\star$-nomologically accessibility connecting their underlying mosaics.

14.4 Conclusion

In this essay I propose a metaphysical model of extension of Lewisian modal metaphysics in light of the physical content and formal articulation of quantum string theory. The broad rationale behind the project is that of showing that, although endorsing the metaphysical commitment of quantum string theory to the nonfundamentality of space and time, an attempt of preserving crucial features of Lewisian modal metaphysics (assigning a fundamental role to spatiotemporal relations) is not only possible but also has interesting philosophical implications. My extension applies to two Lewisian accessibility structures, the similarity accessibility and the nomological one. As it turned out in the essay, the formal scheme, constructed by extrapolating parts of the theory's formal articulation and physical content, calibrates a metaphysical accessibility structure extending Lewisian similarity, because of fully carrying its structural properties. It also calibrates a metaphysical accessibility structure tentatively extending Lewisian nomological accessibility, because of only partially carrying its structural properties. So, the two Lewisian accessibility structures can be regained from the two extensions, even though the nomological one only partially. Then, the bottom line is that the extended metaphysical accessibility structures (especially the one extending Lewisian similarity) appear to be somehow "more fundamental" than the Lewisian ones in virtue of the fact that some of their crucial properties smoothly reflect the discreteness of the fundamental quantum nature of reality.

Appendix 14.A

For details about general relativistic spacetime independence I refer the reader to Vistarini (2019, ch. 3). Perturbative string theory arises from the quantization of classical string theory performed via perturbative techniques. Classical string theory is a toy model, that is, it is an unphysical representation of fictional dynamics: there are no such things like classical strings out there. It is widely held that the perturbative formulation of quantum string theory is background dependent, because the perturbative procedure applied to the classical action would put geometry by hand. I argue that this is not the case: some contingent methodological features applied to the toy classical string action are wrongly regarded as revealing the metaphysical commitment of the quantum string theory eventually obtained through that methodology of quantization. Indeed, the "geometry" appearing in the classical action is an arbitrary, free parameter, whose arbitrariness expresses its unphysical nature, and the unphysical nature of the classical string theory in which formally shows up. A uniquely determined class of physical geometries is the result of the quantization of the classical toy action, and the underlying physical quantum string theory dictates the derived, emergent nature of that physical geometry. Perturbative quantum string theory does not posit any geometry at the quantum string scale, rather it admits general relativistic spacetime emergence.

Appendix 14.B

For details about my argument in favor of the emergent nature of the compact extra dimensions I refer the reader to chapters four and five of Vistarini (2019). It was in 1997 when Edward Witten produced strong evidence in support of the idea that physicists could explore the string land beyond the small coupling constants regime. Witten (1996) argued that whenever the coupling constants in any one of the many formulations of string theory become very large, that formulation turns into one of those having small coupling constants. This physical equivalence exchanging different perturbative regimes also transforms their related geometries. This simple and powerful set of findings introduced the notion of string dualities. Over time, the increasingly sophisticated control over different string physical scenarios morphing into each other, produced theoretical findings about dualities pointing to the emergent nature of the compact extra dimensions. The emergence of the compact extra dimensions is a thesis of background independence complementary to that of general relativistic spacetime emergence.

References

Carnap, R. 1947. *Meaning and Necessity*. Chicago: University of Chicago Press.

Chomsky, N. 1957. Logical structures in language. *Journal of the Association for Information Science and Technology*, **8**, 243–333.

Davies, M. 1981. *Meaning, Quantification, Necessity: Themes in Philosophical Logic*. London: Routledge.

Divers, J. 2002. *Possible Worlds*. London: Routledge.

Hartshorne, R. 1977. *Algebraic Geometry*. New York: Springer.

Huggett, N., and Vistarini, T. 2015. Deriving general relativity from string theory. *Philosophy of Science*, **82**(5), 1163–1174.

Huggett, N., and Wüthrich, C. 2013. Emergent spacetime and empirical (in)coherence. *Studies in History and Philosophy of Modern Physics*, **44**(3), 276–285.

Kodaira, K. 2005. *Complex Manifolds and Deformation of Complex Structures*. Berlin: Springer-Verlag.

Lewis, D. 1973. Causation. *The Journal of Philosophy*, **70**(17), 556–567.

Lewis, D. 1987. *On the Plurality of Worlds*. Oxford: Blackwell.

Novak, E., Brustein, R., and de Alwis, S. 2002. M-theory moduli space and cosmology. *Physical Review D: Particles and Fields*, **68**(4).

Plantinga, A. 2003. *Essays on the Metaphysics of Modality*. Oxford: Oxford University Press.

Polchinski, J. 2005. *String Theory, Superstring Theory and Beyond, Vol I*. Cambridge: Cambridge University Press.

Strawson, P. 1950. On referring. *Mind*, **59**(235), 320–344.

Vafa, C. 1998. Geometric physics. *Proceedings of the International Congress of Mathematics*, **1**, 537–556.

Vistarini, T. 2019. *The Emergence of Spacetime in String Theory*. London: Routledge.

Witten, E. 1996. Reflections on the fate of space-time. *Physics Today*, **49N4**, 24–30.

Zwiebach, B. 2009. *A First Course in String Theory*. Cambridge: Cambridge University Press.

15

What Can (Mathematical) Categories Tell Us about Spacetime?

KO SANDERS

15.1 Introduction

There are good reasons to believe that the classical structure of spacetime, as it appears in general relativity, breaks down at small length scales of the order of the Planck scale (Doplicher, Fredenhagen, and Roberts, 1995). This poses a problem in particular for any theory of quantum gravity, which should extend to such short length scales. Assuming that the classical concept of spacetime (described as a manifold) is no longer viable as a fundamental concept in such a theory, one needs to explain how it emerges as an approximate concept in the appropriate (long distance) limit.

In order to understand what is required of such an explanation, it is necessary to have a good understanding of the classical structure of spacetime. In this essay I will therefore focus on the concept of spacetime as it appears in the axiomatic framework called locally covariant quantum field theory (LCQFT) (Brunetti, Fredenhagen, and Verch, 2003). This framework provides a language to formulate quantum field theories in the presence of gravitational background fields, and it encompasses some of the most precise physical theories that one can formulate without quantizing gravity. A key aspect of LCQFT is the way that it implements the conditions of locality and general covariance.[1] These conditions have been crucial for a consistent, perturbative treatment of interacting quantum field theories (Hollands and Wald, 2015), and they have even led to a perturbative description of quantum gravity (Brunetti, Fredenhagen, and Rejzner, 2016).

The axiomatic framework of LCQFT is precise, flexible, and quite general. On the one hand it is similar to algebraic quantum field theory (AQFT) (Haag,

---

[1] There are several papers before Brunetti et al. (2003) that combined the locality and covariance of quantum fields in a fruitful way, but mostly without employing the language of category theory (Kay, 1979, 1992; Hollands and Wald, 2001; Verch, 2001).

1991), an axiomatic framework for quantum field theories in Minkowski space, which has also been used by philosophers of physics because of its suitability for studying the foundations of quantum field theory (Halvorson, 2007). On the other hand, it relies heavily on basic tools from category theory, which are used to formulate locality and general covariance. To the best of my knowledge, the use of category theory in this context, to express covariance and especially locality, has not been investigated by philosophers of physics before (see, however, Halvorson and Tsementzis [2017] for the use of categories in general philosophy of science and Weatherall [2017] for related applications in physics). Such an investigation is warranted, because it lies at the basis of some of the most profound results in LCQFT (Fewster and Verch, 2012; Fewster, 2015), which address the question: "How do we know that a theory describes the same physics in all spacetimes?"

I argue in this essay that the language of category theory, as used in LCQFT, carries more significance than a mere bookkeeping tool. It is used to give a precise and explicit statement of how a physical theory gives rise to a kind of model for modal logic, expressing a systems view of the world, and how spacetime acts as an organizing principle that brings structure to our understanding of the world. This aspect of spacetime is tightly related to locality and general covariance. However, it is not so directly related to the manifold structure, which means that this aspect of spacetime might be generalized and need not be lost in a quantum theory of gravity.

The argument that I present is to be viewed as a proposal, coming from a mathematical physicist. I focus on philosophical aspects to the best of my ability. It is my intention that this essay serves as an invitation to the philosophical community to take up a more thorough investigation of the use of category theory to describe spacetime in physical theories like LCQFT.

The structure of this essay is as follows. I first review the basic concepts of LCQFT in Section 15.2. In Section 15.3, I then review the argument that categories can be used in physics as a kind of models for modal logic and how the categorical formulation of locality and general covariance in LCQFT can be viewed as an organizing principle. In Section 15.4, I argue that this organizing principle has some nontrivial content. For that purpose, I review how it reflects on the identity problem ("How do we know that a theory describes the same physics in all spacetimes?"). In Section 15.5, I revisit the structure of the category of localization regions, a key category in LCQFT. In Section 15.6, I formulate two proposals to characterize the notion of spacetime in a general context, which, in a generalized sense, should also apply to theories of (quantum) gravity. Section 15.7 contains some concluding remarks.

15.2 Categories in Locally Covariant Quantum Field Theory

The framework of LCQFT, first introduced by Brunetti et al. (2003), is quite flexible and has since appeared in different variations, adapted to differing circumstances (Sanders, 2010; Fewster and Verch, 2012; Sanders, Dappiaggi, and Hack, 2014; Zahn, 2014; Fewster and Schenkel, 2015). The subject of my argument is the use of category theory to implement locality and general covariance, which is common to all of these variations. In this section I first present the main parts of the original framework, which are relevant for this essay, followed by a discussion of some of the possible variations. I freely make use of basic notions from category theory (see MacLane [1971] for a general reference).

In the spirit of AQFT (Haag, 1991), it is assumed that a quantum system can be described by a C^*-algebra $\mathcal{A}$ with unit, whose self-adjoint operators correspond to observable quantities. A state ω of the system is a normalized positive linear functional $\omega \colon \mathcal{A} \to \mathbb{C}$, and one may recover a Hilbert space formulation by using the Gelfand–Naimark–Segal (GNS)-construction. Instead of focusing on a single system, LCQFT uses many of them: one for each globally hyperbolic Lorentzian manifold. For this reason one introduces a category, Alg, whose objects are C^*-algebras $\mathcal{A}$ with unit, and whose morphisms are injective[2] *-algebra homomorphisms $\alpha \colon \mathcal{A}_1 \to \mathcal{A}_2$ that preserve the unit. The composition of morphisms is simply the composition of maps.

A LCQFT should describe a quantum field theory in every globally hyperbolic Lorentzian manifold M. Here M consists of a smooth, connected manifold $\mathcal{M}$ of dimension 4, together with a smooth Lorentzian metric g, an orientation and a time orientation, and M admits a Cauchy surface. The collection of all globally hyperbolic Lorentzian manifolds, where the theory can be localized, also forms a category, Loc, whose morphisms are isometric embeddings $\psi \colon M_1 \to M_2$, which preserve the orientation, time orientation, and the causal structure. The latter means in particular that any causal curve γ in M_2 that connects points in $\psi(M_1)$ must lie entirely in $\psi(M_1)$.

There are several pragmatic reasons to impose global hyperbolicity on the class of Lorentzian manifolds. First, it ensures the well-posedness of initial value problems for normally hyperbolic partial differential equations, which is helpful when constructing quantum field theories. Indeed, it implies the existence of a diffeomorphism $\mathcal{M} \simeq \mathbb{R} \times \Sigma$, where each hypersurface $\{t\} \times \Sigma$ is a Cauchy surface and t defines a global time coordinate (which is highly nonunique). Second, in the mentioned diffeomorphism, Σ can be any oriented, paracompact manifold, so the class of globally hyperbolic Lorentzian manifolds is rather large. It certainly

---

[2] The injectivity assumption is not very well motivated, and it is known to fail for free electromagnetism, due to Gauss's law (Sanders et al., 2014). For this reason Sanders et al. (2014) suggested to drop it.

includes many physically interesting situations, such as cosmological scenarios and black holes. The question whether there are also more fundamental reasons to impose global hyperbolicity is not fully clarified. It has been pursued especially by Kay (1992, 1997), whose investigations (predating the categorical framework of LCQFT) strongly suggest the necessity of certain constraints on the global causal structure of the Lorentzian manifold, including, e.g., time orientability.

Let me comment at this point on the relation between the category Loc and the notions of locality and general covariance in general relativity. It is often said that the symmetry group of general relativity is the diffeomorphism group of the underlying manifold $\mathcal{M}$. In some sense, the category Loc refines this symmetry statement by including locality. Indeed, a general morphism $\psi\colon M_1 \to M_2$ in Loc can be written as a composition $\psi = \iota \circ \phi$ of morphisms of a special kind. Here $\phi\colon M_1 \to \psi(M_1)$ is a diffeomorphism, as used in the formulation of general covariance. $\iota\colon \psi(M_1) \to M_2$ corresponds to the canonical inclusion of a subset of the Lorentzian manifold M_2, expressing locality.

The fundamental definition of Brunetti et al. (2003) states that a LCQFT is a covariant functor $\mathbf{A}\colon$ Loc $\to$ Alg. This means that it associates to every Lorentzian manifold M in Loc a C^*-algebra $\mathbf{A}(M)$ in Alg, and that it associates to every morphism $\psi\colon M_1 \to M_2$ in Loc a morphism $\mathbf{A}(\psi)\colon \mathbf{A}(M_1) \to \mathbf{A}(M_2)$ in Alg. The algebra $\mathbf{A}(M)$ describes the quantum theory in the localization region M. The morphisms $\mathbf{A}(\psi)$ describe how an embedding of Lorentzian manifolds gives rise to an embedding of the corresponding quantum systems. The requirement that $\mathbf{A}$ is a functor expresses the locality and general covariance of the theory. The simplest example of a LCQFT is the free scalar field with given mass and scalar curvature coupling (Brunetti et al., 2003).

On top of the requirement that a LCQFT is a functor $\mathbf{A}$, one may impose additional axioms. Let me mention here the time-slice axiom, which encodes the existence of a dynamical law, and local commutativity. A morphism $\psi\colon M_1 \to M_2$ is called a Cauchy morphism when its range $\psi(M_1)$ contains a Cauchy surface of M_2. Using a dynamical law it should then be possible to predict the behavior of the system throughout M_2 using information available in $\psi(M_1)$. Accordingly, $\mathbf{A}$ is said to satisfy the time-slice axiom whenever $\mathbf{A}(\psi)$ is an isomorphism for all Cauchy morphisms ψ. For local commutativity one considers two morphisms $\psi_1\colon M_1 \to M$ and $\psi_2\colon M_2 \to M$ into the same target Lorentzian manifold M, such that the ranges $\psi_1(M_1)$ and $\psi_2(M_2)$ are spacelike separated. $\mathbf{A}$ is said to satisfy local commutativity whenever all operators in the range of $\mathbf{A}(\psi_1)$ commute with all operators in the range of $\mathbf{A}(\psi_2)$.

In addition to an algebra of observables, the specification of a quantum system requires the choice of a set of states. For a C^*-algebra one could take the set of all algebraic states, but in physical applications this set often contains many states with bad behavior. This suggests that the theory should also specify a subset of states, $\mathbf{S}(M)$, on the algebra $\mathbf{A}(M)$ for each Lorentzian manifold M. These sets of states are subjected to some conditions as well. E.g., they should be closed under taking mixtures (i.e., convex combinations) and for any morphism $\psi \colon M_1 \to M_2$ the pullback $\mathbf{A}(\psi)^* \omega \colon = \omega \circ \mathbf{A}(\psi)$ of a state in $\mathbf{S}(M_2)$ should be a state on $\mathbf{S}(M_1)$. This means that the criterion that selects the sets of admissible states $\mathbf{S}(M)$ must be local and generally covariant as well. For free fields one typically uses the Hadamard condition (Wald, 1994).

Whereas Minkowski space has a group of symmetries, the structure of the category Loc is much more complicated. This has important consequences for the formulation of LCQFTs. Under quite general circumstances, there is no choice of a well-behaved natural state on all Lorentzian manifolds (Fewster and Verch, 2012), so there is no substitute for the Minkowski vacuum. Nevertheless, it is possible to treat interacting quantum field theories in curved Lorentzian manifolds in a perturbative way (see Hollands and Wald [2015] for a review). This requires the definition of renormalized (Wick) powers and time-ordered products of the quantum fields. The crucial insight here is that the freedom available in choosing a regularization and renormalization procedure is severely restricted by imposing the conditions of locality and general covariance (Hollands and Wald, 2001). Proceeding in this way, mathematical physicists have also formulated perturbative Yang–Mills theories and even a perturbative description of quantum gravity in a categorical framework (Hollands, 2008; Fredenhagen and Rejzner, 2013; Brunetti et al., 2016). Note that this formulation of perturbative general relativity is by construction local and generally covariant.

As I mentioned, the framework of LCQFT is quite flexible and has appeared in different variations. One type of variation is to replace the category Loc by another category BkGrnd, encoding the non-dynamical background structures for the theory. Sanders (2010) uses a category of Lorentzian manifolds with spin structure in order to describe the free Dirac field. External source fields can also be included (Zahn, 2014; Fewster and Schenkel, 2015), leading, e.g., to a quite natural formulation of the Aharonov–Bohm effect in electromagnetism (Sanders et al., 2014). If one can include external source fields into the description of the background structure, one can also imagine removing the metric from the background structure and including it with the dynamical fields. In that case, one would replace the category Loc of globally hyperbolic Lorentzian manifolds by a suitable category Man of manifolds, e.g., in order to obtain a description of general relativity combined with other classical fields. I address this approach, and any conceptual issues surrounding it, in Section 15.6.

In a similar way one can replace the target category without losing any of the core ideas. Brunetti et al. (2003) themselves used a category of more general topological *-algebras in the description of the free scalar field, which is convenient to make contact with the Wightman axioms. In a general analysis of locally covariant theories, Fewster and Verch (2012) go much further and replace the target category by a quite general category Phys, which is restricted only by requiring a few basic properties.[3] The objects of Phys are supposed to give a full description of the physical system. (They could consist, e.g., of a C^*-algebra together with a set of admissible states.) The consequences that Fewster and Verch (2012) derive in this categorical approach rely on almost nothing except locality and general covariance, which lends them a very wide range of validity.

15.3 Categories, Modal Logic and Spacetime

From the review of the basic structure of LCQFT in Section 15.2 it is apparent that the language of category theory plays an important role. At the very least it is a convenient bookkeeping device, which exhibits the structure of locally covariant theories in a conceptually clear language. This is an advantage that has contributed to the recent theoretical development of perturbatively interacting quantum field theories in curved Lorentzian manifolds. However, one may wonder whether the effectiveness of category theory has a deeper cause than mere bookkeeping. In this section I argue that this is the case, by relating the categorical structure to a systems view of the world and the corresponding modal logic. I then conclude that spacetime in LCQFT acts as an organizing principle that brings structure to our understanding of the world.

Let us first follow Fewster and Verch (2012) and consider the general target category, Phys, whose objects are (mathematical) descriptions of physical systems. Each system describes a part of the world. More precisely, we can think of the specification of an object in Phys as tantamount to the specification of (1) certain boundary conditions, like the localization region and external fields as well as (2) the physical operations that can be performed on the system and (3) the states that the system can be in. The morphisms in this category may be thought of as expressing a subsystem relation. The statement that Phys is a category imposes some relations on the morphisms, which can be seen as a consistency criterion that is needed to properly express the notion of subsystem. E.g., when $\alpha \colon P_1 \to P_2$ and $\alpha' \colon P_2 \to P_3$ are morphisms, expressing how P_1 is a subsystem of P_2 and how P_2 is a subsystem of P_3, then $\alpha' \circ \alpha$ expresses the way in which P_1 is a subsystem of P_3. Thus category theory provides a convenient way to express the systems view of the world.

---

[3] To be precise, the morphisms of Phys are required to be monomorphisms (which essentially means they are injective), and the category is required to have equalizers, intersections and unions in the categorical sense.

It is a crucial part of the categorical language of LCQFT, and of theoretical physics in general, that theories can also describe situations that are not actually the case. The branch of logic that studies the truth values of statements and the validity of arguments that involve such situations is called modal logic (Forbes, 1985). Modal logic makes use of models in order to decide whether an argument is valid or not. I now recall how the categorical structure of LCQFT makes it a kind of model in the sense of modal logic (Sanders, 2009).

The most common type of model theory makes use of possible worlds: complete alternatives for how things might be. Objects may or may not exist at a certain possible world and propositions and predicate statements may or may not hold. A typical system as described earlier can be embedded into a larger system, however, so it provides at best only an incomplete description of circumstances. For this reason it is appropriate to use a less-well-known model theory for modal logic, which makes use of incomplete sets of circumstances, so-called possibilities (Forbes, 1985, pp. 18–22). Due to its incompleteness, a possibility may not assign a truth value to every logical sentence. However, the described circumstances do act as truthmakers for some logical sentences. Furthermore, possibilities can be refined by extending the set of circumstances that they specify, thereby extending the set of logical sentences that are assigned a truth value. This refinement is perfectly aligned with the structure of a category like Phys.

To be precise, we think of each object in Phys as a possible system that a given physical theory can describe and that plays the role of a possibility in the sense of modal logic. The morphisms of the category, which express the subsystem relation, then correspond to a refinement of such possibilities. I do not claim that every category must be a viable model for modal logic in this way, but I do claim that this should hold for any category like Phys, which expresses the systems view of the world in a physical theory. The possibility semantics not only fits well with the systems view of the world, it also corresponds with experimental praxis: we would like to be able to make predictions for a certain laboratory experiment, without prescribing a complete set of circumstances for the entire world; making assumptions about the system in question should be enough.

Let me point out here that the extension of a system to a supersystem is usually not the only kind of refinement of possibilities that can occur in physics. Given a possible system, its state also specifies a set of circumstances (e.g., initial conditions). The states should, therefore, describe a model for modal logic in their own right. However, given the choice of a system, the state is supposed to describe a complete set of conditions, so the use of possible worlds is more appropriate in this case. Combing the two kinds of modalities, a possibility in general comprises the choice of a possible system together with the choice of a possible state. A refinement then consists of the embedding into a supersystem, together with an extension of the state.

I return to the distinction between the two modalities in Section 15.6, but let me note here that their interplay should be subject to certain restrictions, which are motivated by physics. It is clear that the refinements of a possibility are typically not unique. Concerning their existence, note that if a system is known to be in a state that admits no extension to any supersystem, then one would expect the system to be the whole universe. Now let us consider the case where a state admits extensions to some of the possible supersystems but not to others. Implicitly, such a state contains information about the nature of any possible supersystems as well as about the system itself. It is desirable to prohibit this kind of behavior, because it indicates that we have made a poor choice in determining what information is needed to specify the system. In the context of LCQFT, the term *local physical equivalence* has been introduced to describe the assumption that essentially every state of a possible system can be extended to every supersystem (Fewster, 2007).

Let me now come to the use of functors in LCQFT. For a general morphism $\alpha: P_1 \rightarrow P_2$, P_1 is a subsystem of P_2, so it describes less degrees of freedom. There may be many ways in which this can occur, e.g., when P_2 describes a system consisting of several fields and P_1 focuses only on one of these. The subsystems appearing in LCQFT are of a very specific type, however, because they are in the image of a functor $\mathbf{A}: \texttt{Loc} \rightarrow \texttt{Phys}$. We have seen in Section 15.2 that this functorial structure expresses the fact that the description of the system $\mathbf{A}(M)$ should depend in a local and covariant way on the localization region M. The systems localized in different Lorentzian manifolds are, therefore, not independent. Because the principles of locality and general covariance are closely tied to the notion of spacetime, the functorial structure is characteristic for how spacetime enters into LCQFT.

In more general terms, if a physical theory gives rise to a category $\texttt{Phys}$ of possible systems, the existence of the functor $\mathbf{A}$ (and the validity of other axioms of LCQFT) may be viewed as a condition on the category $\texttt{Phys}$. It states that a possible system $\mathbf{A}(M)$ can be determined from its localization region M: it describes the degrees of freedom localized in that region. The subsystem relation corresponds to a possible enlargement of the region M (or rather, a morphism in $\texttt{Loc}$). LCQFT, therefore, stipulates in a precise and explicit way that spacetime is to be viewed as an organizing principle, which brings additional structure to the category $\texttt{Phys}$ of systems that a theory describes.

I will not attempt to place this statement in the appropriate philosophical or epistemological context—as a mathematical physicist this lies well outside my expertise. Instead I will address the possible criticism that the conclusions

I have drawn about the notion of spacetime in LCQFT are not very useful, because they rely only on elementary concepts of category theory and they do not contain any deep statements. In response it should be noted that the morphisms of the category Loc have a rich structure, which is much more complicated than the diffeomorphism group of a single manifold. I illustrate in the next section how this structure can be used to analyze deep questions and prove nontrivial results.

15.4 On Doing the Same Physics in All Spacetimes

One of the most successful theories in theoretical physics is the standard model of elementary particles, which describes a wide range of physical phenomena to a very high accuracy. This model is normally formulated in Minkowski space, which models the complete absence of gravity. This is a very good approximation to actual experiments, where the gravitational fields are extremely weak and the detectors are located at large distances compared to the size of the interaction region. However, as a matter of principle it seems desirable to be able to formulate the model also in the presence of (weak) gravitational fields, just to emphasize that the absence of gravity is really an approximation and not an essential assumption. Following this reasoning ultimately leads us to the question how to include gravity as a dynamical (and quantized) theory in the standard model. At an intermediate step, however, where the gravitational fields are weak but not negligible, it leads to the question how to formulate the standard model in a fixed curved Lorentzian manifold. One aspect of this question is the following: given a theory in some Lorentzian manifold, how can we identify it as the standard model, despite the influence of the background gravitational field?

In the general framework of LCQFT, Fewster and Verch (2012) investigated this question, which they formulated as follows: "What does it mean for a theory to describe the same physics in all spacetimes?" They point out that this is a difficult question, which is not adequately answered by specifying a Lagrangian. The question seems similar to a fundamental problem of identity that arises in philosophy: "How do we know that object A is the same as thing as object B, but at a different place and time and under different circumstances?" In everyday life we seem reasonably good at identifying objects, despite changing circumstances, but it is difficult to provide a clear theoretical explanation of why this is so. One of the arguments that may be presented in this context relates to the notion of spacetime: two objects that occupy the same region in spacetime are identical. Such arguments seem problematic in the context of classical field theory and quantum theory, let alone for theories in which spacetime is not a fundamental concept. Nevertheless, it is interesting to note that the analysis of Fewster and Verch (2012) also invokes spacetime, because it makes essential use of the functorial structure of LCQFT, as I now briefly review. (See Fewster [2015] for a more detailed and accessible presentation.)

Let us consider two LCQFTs, **A** and **B**, given as functors from Loc to some category Phys. Fewster and Verch (2012) call **A** a subtheory of **B**, whenever there is a natural transformation $\zeta : \mathbf{A} \to \mathbf{B}$ between the functors. This means that the following holds: for every Lorentzian manifold M in Loc there is a morphism $\zeta_M : \mathbf{A}(M) \to \mathbf{B}(M)$ in Phys, and for every morphism $\psi : M_1 \to M_2$ in Loc there holds:

$$\zeta_{M_2} \circ \mathbf{A}(\psi) = \mathbf{B}(\psi) \circ \zeta_{M_1}. \tag{15.1}$$

The first condition states that on each Lorentzian manifold M, the system $\mathbf{A}(M)$ is a subsystem of **B**, so it describes less degrees of freedom. The second condition essentially requires that the degrees of freedom from **B** that are left out in **A** are independent of the localization region that is being considered. Thus, it makes sense to restrict $\mathbf{B}(M_2)$ to $\mathbf{A}(M_1)$, without the need to prescribe the order in which the localization region and the degrees of freedom are being restricted.

Two theories are said to be equivalent when all the ζ_M are isomorphisms in Phys. In the language of category theory this means that $\zeta : \mathbf{A} \to \mathbf{B}$ is a natural isomorphism. When **B** is the same functor as **A**, the natural isomorphisms $\zeta : \mathbf{A} \to \mathbf{A}$ form a group, which is interpreted as the global gauge transformations of the theory (Fewster, 2013).

It should be emphasized that, from a physical point of view, it is not sufficient to require only the existence of the isomorphisms ζ_M, without imposing Equation (15.1). Fewster (2015) points out that this weaker condition is already satisfied in an analogous situation when $\mathbf{A} : \text{BkGrnd} \to \text{Alg}$ is a free scalar quantum field with an external source term J (which is part of the background structure included in the objects of BkGrnd) and when $\mathbf{B} : \text{BkGrnd} \to \text{Alg}$ is the same field with a source term λJ for any constant λ. It is desirable that a nontrivial rescaling of the coupling to the external sources leads to an inequivalent theory, and this is achieved by also requiring (15.1). In fact, an analogous issue had already appeared in AQFT, where it has been recognized that the information of the theory is encoded in the subsystem relationship of the theory (Haag, 1991).[4] The main difference between AQFT and LCQFT is that the morphisms in the latter case have a much richer structure, because the category includes Lorentzian manifolds with nontrivial gravitational background fields and nontrivial topology.

---

[4] In AQFT, and also for C^*-algebraic LCQFTs, the local algebras are typically expected to be isomorphic to the unique, hyperfinite von Neumann factor of type III_1 (Buchholz et al., 1987; Halvorson, 2007). The algebras by themselves, without their morphisms, therefore carry very little information.

Fewster and Verch now propose the following line of reasoning. Suppose that we had a criterion to decide whether an LCQFT **A** describes the same physics in all spacetimes. Assume that two theories **A** and **B** satisfy this criterion, and that there exists a Lorentzian manifold M such that $\mathbf{A}(M)$ and $\mathbf{B}(M)$ are equivalent. Then both theories should describe the same physics in all Lorentzian manifolds, i.e., **A** should be equivalent to **B**. More technically, a criterion $\mathcal{C}$ on the class of LCQFTs has the same physics in all spacetimes (SPASs) property if and only if for any natural transformation $\zeta : \mathbf{A} \to \mathbf{B}$ between LCQFTs satisfying this criterion, ζ is a natural isomorphism as soon as ζ_M is an isomorphism for some M. The requirement that the isomorphism ζ_M is part of a natural transformation $\zeta : \mathbf{A} \to \mathbf{B}$ prevents the isomorphism at M from being an accidental pathology that does not respect locality and general covariance.

Perhaps surprisingly, Fewster and Verch (2012) show that the axioms of LCQFT by themselves (including the time-slice axiom and local commutativity) are not sufficient to ensure that a theory describes the same physics in all spacetimes. For rather general LCQFTs **A**, Fewster (2015) constructs a natural transformation ζ to another theory **B** such that ζ_M is an isomorphism for some Lorentzian manifolds M, but not for all of them. This signals the need for an additional criterion $\mathcal{C}$ to supplement the axioms of LCQFT.

Fewster and Verch (2012) goes on to propose a criterion $\mathcal{C}$ that does enjoy the SPASs property: dynamical locality. Once again this criterion makes essential use of the categorical structure of the theory, combined with the time-slice axiom and local commutativity. I will not give a full description of this criterion but only indicate its main points, which should suffice to show that the categorical language of LCQFT allowed Fewster and Verch to prove nontrivial statements that address their identity problem .

Dynamical locality is defined for LCQFTs **A** whose target is the category Alg of C^*-algebras and that satisfy local commutativity and the time-slice axiom. Using the time-slice axiom one can study how the system $\mathbf{A}(M)$ responds to a perturbation δg of the metric, or of another external source field, supported in any compact region of the Lorentzian manifold M. This response is encoded in the relative Cauchy evolution,

$$\mathrm{rce}_{M,\delta g} : \mathbf{A}(M) \to \mathbf{A}(M),$$

which is defined in Brunetti et al. (2003). Consider, e.g., a morphism $\psi : M_1 \to M$ and define the kinematic subalgebra:

$$\mathbf{A}^{\mathrm{kin}}(\psi(M_1)) := \mathbf{A}(\psi)(\mathbf{A}(M_1)),$$

of $\mathbf{A}(M)$. For any operator X in $\mathbf{A}^{\text{kin}}(\psi(M_1))$ and any perturbation δg supported in a compact region that lies space-like to $\psi(M_1)$ one may show that:

$$\text{rce}_{M,\delta g}(X) = 0,$$

due to local commutativity. This follows naturally from the localization of X and δg and the fact that the theory has a dynamical law that respects the causal structure of the Lorentzian manifold.

The idea of dynamical locality is that one can also determine the localization region of an operator $X \in \mathbf{A}(M)$ dynamically. The dynamical algebra $\mathbf{A}^{\text{dyn}}(\psi(M_1))$ is essentially generated by operators $X \in \mathbf{A}(M)$ such that $\text{rce}_{M,\delta g}(X) = 0$ for all perturbations δg with compact support space-like to $\psi(M_1)$. Dynamical locality requires that the kinematical and dynamical ways of localizing operators agree:

$$\mathbf{A}^{\text{dyn}}(\psi(M_1)) = \mathbf{A}^{\text{kin}}(\psi(M_1)).$$

Dynamical locality has the SPASs property (Fewster and Verch, 2012). Thus, drawing a parallel to the philosophical identity problem, Fewster and Verch replace the claim that an object may be identified by the spacetime region that it occupies, with a precise mathematical result that suggests that a quantum field theory in different Lorentzian manifolds may be identified when it is dynamically local, a property that depends on locality, general covariance, the causal behavior of the theory and its response to perturbations of the metric (or other external source fields).

This is not the place to try to disentangle the role of the metric and the role of manifold in this study, so let me conclude this section instead with the following remark. It is known that dynamical locality holds for the free scalar field, but it fails, e.g., for free electromagnetism due to Gauss's law (Fewster, 2015; cf. footnote 2). Indeed, Fewster and Verch do not claim that the dynamical locality criterion is a fully satisfactory solution to their identity problem. However, their arguments do show that LCQFT provides a suitable language to address the problem and to prove nontrivial results that can contribute to its clarification.

15.5 The Structure of the Category Loc

In LCQFT, spacetime enters only through a functor $\mathbf{A} \colon \text{Loc} \to \text{Phys}$, which essentially represents the category Loc in the category Phys of systems that a certain physical theory can describe. It is, therefore, worthwhile to have a closer look at the key category Loc.

We have already seen that Loc has sufficiently many objects to describe a wide range of physical situations. Instead of fixing a single Lorentzian manifold to model the universe, LCQFT treats all the objects in Loc on an equal footing. Paraphrasing

Brunetti et al. (2003), fixing a background manifold seems artificial and at variance with the principles of general relativity. It would essentially impose a prejudice on the global shape of the universe, even though such global aspects should be irrelevant for all local physical processes. The use of a category theoretical language to eliminate any such prejudice is analogous to the use of a differential geometric language to eliminate any prejudice for a choice of (local) coordinates.

We have also seen that the morphisms of Loc encode the symmetries of general relativity, comprising both general covariance and locality, and I have mentioned some ramifications, e.g., for the (non-)existence of natural vacuum states (Fewster and Verch, 2012) and for the renormalization ambiguities in perturbatively interacting theories (Hollands and Wald, 2001). Indeed, the importance of Loc can be highlighted even more by an analogy with AQFT. There it is known that the local algebras are almost entirely fixed by a few physically relevant conditions (Buchholz et al., 1987; Halvorson, 2007), so they do not carry much information. Instead, most of the nontrivial information of a QFT is encoded in the embeddings between the local algebras. A similar argument should apply to LCQFT, where the embeddings arise, via the functor **A**, from the category Loc, which has a much richer structure than, say, the Poincaré group.

The category Loc was defined in terms of Lorentzian manifolds. It is interesting to note, however, that much of its structure can probably be characterized directly in terms of objects and morphisms, which lends itself to an interpretation in terms of localized systems and subsystems. Loc has a non-empty class of objects, and all morphisms are monomorphisms (i.e., "injective"). The category also has a notion of subobjects, i.e., a distinguished class of morphisms, denoted by $M_1 \subset M_2$. Every object has a proper subobject (i.e., a noninvertible subobject morphism into the given object), which essentially means that there is no smallest size for systems. Let me also sketch some ideas that lead to less elementary properties. First, it should be possible to define a point in an object M as a suitable infinite filtration of sub-objects, $\ldots \subset M_3 \subset M_2 \subset M_1 \subset M$. Note, however, that a functor **A** need not assign a physical system to a point, in line with the idea that quantum fields are distributional in nature and require an open region to be well defined. Second, for the set of all points in M it should be possible to define a topology, using the subobjects of M to define a topological basis. Proceeding in this way, it should be possible to formulate the necessary topological and differential geometric properties to recover the smooth manifold structure of each object M of Loc, and the fact that morphisms are smooth embeddings.

The Lorentzian metric enters the definition of Loc through the causal structure, which is required to be preserved by all morphisms. It is unclear if the causal structure can be recovered from the structure of the morphisms, but the metric itself seems beyond such an approach, because we can transform it in a conformal rescaling without affecting the causal structure. Of course this is not surprising: the metric should be related to the fields or matter of a physical theory, so it seems much more promising to try to recover the metric and the causal structure from the functor **A**, using the physical properties of the theory under consideration, in the spirit of, e.g., Ehlers, Pirani, and Schild (2012).

In conclusion, starting with an abstract category C to model the localization regions of physical systems, it could be feasible to formulate a list of properties of C, directly in terms of its objects and morphisms, that make it equal (or equivalent) to a category Man of smooth (paracompact) manifolds with smooth embeddings. This already suffices to capture some typical aspects of LCQFT: locality, diffeomorphism invariance and the absence of a prejudice on the global shape of the universe. The metric (and causal) structures in Loc can probably not be fully explained in such abstract terms, but they might be recovered instead from the physical theory at hand via the functor **A**.

15.6 On the Notion of Spacetime

I have argued that the use of category theory in LCQFT has a deeper meaning than mere bookkeeping and that it allows to analyze deep questions. In this section I discuss the question what the categorical point of view can tell us about the notion of spacetime in the context of very general physical theories, including theories of quantum gravity.

In classical general relativity (GR), we use a manifold as the background on which physical fields, including the metric field, are defined. The role of this manifold is somewhat puzzling, due to the following paradox. Although it is commonly understood that the individual points of a manifold in GR do not have any intrinsic physical meaning, the manifold does seem to be indispensable when formulating a field theory. Classically, the resolution is that we can use the values of the fields to identify points in the manifold, in principle with infinite accuracy. Unfortunately this resolution is no longer available in quantum gravity, where we expect all fields (including the metric) to be quantum fields. Instead of sharp field values, we have unsharp measurement results, making the viability of an underlying manifold structure highly questionable and making the paradox above all the more urgent. I think that the categorical perspective may help to clarify this paradox.

Let me first discuss a conceptual issue that already arises in classical GR. In LCQFT we assume the existence of a functor **A**: BkGrnd $\to$ Phys, where the category BkGrnd encodes the available background structure, which always consists of a globally hyperbolic Lorentzian manifold with a fixed metric, possibly with additional source fields. More general theories, such as pure GR or quantum gravity, will not have so much nondynamical background structure. For those theories it seems natural to replace the category Loc by, e.g., a category of manifolds, Man.

Suppose that there were a description of classical GR (in vacuum) as a functor **GR**: Man → Phys for a suitable category of (paracompact) manifolds and a category of physical systems. This means that the choice of a manifold $\mathcal{M}$ in Man determines a physical system. There is certainly no problem in the claim that $\mathcal{M}$ determines a set of solutions to Einstein's equation, or other mathematical, geometric objects. However, in order to do physics, the description of the system should normally include a class of observables, which may be more problematic. If we were to model the observables by test functions (or test tensors), as is usually done in field theory, then we would think of these observables as being localized in the region $\mathcal{M}$. However, the physical interpretation of this model is that we should set up an experimental device that is localized in $\mathcal{M}$. In order to do this, we need to have some way of physically identifying the points of $\mathcal{M}$. This is prohibited by the common understanding that the individual points of a manifold in GR do not have any intrinsic physical meaning.

One viable attempt to circumvent this problem is to include nondynamical background fields in the category Man, while keeping the metric dynamical. The background fields may be used to identify points of the manifold and they give rise to a stress tensor, which need not, however, admit any metric solving Einstein's equations, due to the nonlinearity of these equations. We will not be concerned with the mathematical difficulty of determining when the set of solutions is nonempty, because this approach has a more serious drawback: it cannot capture the dynamics of all fields simultaneously.

A better resolution can be sought in a change of the nature of the category Phys. We saw in Section 15.3 that the systems view of the world is captured by the use of suitable categories Phys associated to physical theories. This immediately raises the question: "Which categories are suitable for this purpose?" Some issues that one may wonder about, and that are worthy of further investigation, are the following: Should it be allowed that systems can be proper subsystems of themselves? Do we insist that the morphisms have to be monic (i.e., injective), or does that exclude Gauss's law (Fewster and Verch, 2012; Sanders et al., 2014)? For the purpose of doing GR or quantum gravity, one may argue that the category Phys should follow the usual strategy of using partial observables, rather than observables, to circumvent the lack of local observables (see, e.g., Rovelli [2008] and Brunetti et al. [2016]). States should then assign (expectation) values to suitable combinations of partial observables. It would be worthwhile to formulate this idea in a categorical setting, consistent with the systems view of the world, and to motivate the structure of the objects of this category Phys in as general setting as possible. I expect that this should lead to a satisfactory formulation of classical GR and other classical Lagrangian field theories as a functor $\mathbf{A}\colon \texttt{Man} \to \texttt{Phys}$.

Let me now come to the notion of spacetime. I propose to mathematically characterize the notion of a classical spacetime by a functor $\mathbf{A}\colon \texttt{Man} \to \texttt{Phys}$. This places the dynamical metric in the category $\texttt{Phys}$, which is the most natural choice in the context of GR and quantum gravity, and it rephrases the manifold aspects only in terms of a functor. In this way, spacetime acts as an organizing principle, which brings structure to the theory with the category of systems $\texttt{Phys}$. Theories that do not admit such a functor, do not have a classical spacetime in this sense. To put this characterization of spacetime into context, note that it suggests a shift in emphasis. The most fundamental aspect of spacetime may not be that the objects of $\texttt{Man}$ are manifolds, consisting of points with local charts etc. Instead, it might be how we can use the spacetime: we use the manifolds to label systems in $\texttt{Phys}$, and we use the shrinking of the spacetime region to label a corresponding subsystem relation. It may be of interest to find a direct characterization of those categories $\texttt{Phys}$ that admit a classical spacetime, or even better, to construct a spacetime functor from a given category $\texttt{Phys}$.

It is likely that quantum gravity theories do not admit a classical spacetime in this sense. Nevertheless, we can still use the categorical framework and replace the category $\texttt{Man}$ by guessing a suitable alternative, e.g., a category $\texttt{Set}$ of discrete sets (in analogy to a category of discrete causal sets, as briefly suggested in Fewster [2015]). Alternatively, if the structure of $\texttt{Man}$ can be recovered from a list of properties on systems and subsystems, we may take the systems view of the world as more fundamental than the manifold structure and eliminate manifolds as fundamental objects. We could then critically analyze the list of properties sketched in Section 15.5 and remove or replace all those properties that seem unfounded, e.g., all those referring to points and perhaps also to subobjects. In this way one may retain a well-motivated, but weaker organizing principle.

To explain spacetime as an emergent notion, one needs to argue that a quantum gravity theory admits at least an approximate notion of spacetime in some sense. One may try to show, e.g., that procedures allowing us to reconstruct the functor $\mathbf{A}\colon \texttt{Man} \to \texttt{Phys}$ for classical field theories may fail in the case of quantum gravity, but that they still lead to a functor $\mathbf{QG}\colon \texttt{C} \to \texttt{Phys}$ with a category $\texttt{C}$ that is close to $\texttt{Man}$, in some sense. More importantly, one will need to show that the functor $\mathbf{QG}$ admits suitable limits, which describe a classical field theory or a LCQFT. These limits should allow us to somehow recover the category $\texttt{Man}$ or $\texttt{Loc}$ from $\texttt{C}$, so they should be properly formulated as the convergence of a family of functors $\mathbf{QG}_{\lambda}\colon \texttt{C} \to \texttt{Phys}$, as the parameter λ goes to 0, to a functor $\mathbf{A}_0\colon \texttt{Man} \to \texttt{Phys}$ or $\mathbf{A}_0\colon \texttt{Loc} \to \texttt{Phys}$.

The key point in these limits is that it is presumably much harder to approximate a functor like $\mathbf{A}\colon \mathtt{Man} \to \ldots$, with the full structure of $\mathtt{Man}$, rather than a single manifold $\mathcal{M}$ or the diffeomorphism group $\mathrm{Diff}(\mathcal{M})$. E.g., if the metric becomes nondynamical, a single system can give rise to many background configurations. Moreover, the morphisms in $\mathtt{Loc}$ need to know about the causal structure of the objects, which may become available only in the limit. All this means that the limiting procedure for functors needs to be carefully considered. In short, including locality (i.e., the subsystem relation) in the picture when requiring spacetime to arise as an emergent concept may impose very stringent constraints on quantum gravity theories.[5]

To conclude this section, let me also propose a characterization of spacetime that is more verbal rather than mathematical. For this purpose I try to rephrase some of the preceding discussion in the terms of modal logic used in Section 15.3. There we saw that the specification of a possibility consists of two parts: the specification of a possible system, which corresponded to a choice of a Lorentzian manifold, together with the choice of a possible state on the system. In LCQFT, it is the interplay between these two types of modalities that seems to be characteristic for the way that spacetime enters. Both the choice of system and the choice of state specify some circumstances (like boundary or initial conditions), but they are not on an equal footing. In LCQFT, the possible system must be chosen first, because it determines not only a set of boundary conditions for the system, but it determines the set of observables, which essentially make up the system. The choice of a state, on the other hand, does not alter the set of observables, but only assigns probability distributions to them. On top of that, the distinction between these two kinds of circumstances should be subject to certain conditions, such as local physical equivalence, as indicated in Section 15.3.

The formulation in terms of modal logic also applies to theories of the form $\mathbf{A}\colon \mathtt{Man} \to \mathtt{Phys}$. The category $\mathtt{Phys}$ may be quite different from the case of LCQFT, e.g., it may specify only partial observables. Nevertheless, the choice of a possibility should still consist of two parts: the choice of a possible manifold and the choice of a state. This discussion leads to a second proposal on the characterization of spacetime (perhaps in a somewhat generalized sense). It is closely related to the mathematical characterization given earlier, but it is formulated in terms of the possibilities semantics: The notion of spacetime (in a generalized sense) can be characterized as those circumstances that need to be specified in order to determine

---

[5] In this regard it is interesting to point out a previous proposal to describe quantum gravity as a topological quantum field theory (Barrett 1995; see also Atiyah [1988]). This framework does make use of an elegant category theoretic formulation, but in a way that does not reflect locality very well, because it does not allow for a comparison of open subsets in different manifolds. Barrett (1995) does not investigate local observables in a systematic way. I thank one of the referees for bringing this proposal to my attention.

a system (e.g., the partial observables) of a theory. A state can be characterized as those circumstances that need to be specified in order to determine expectation values for combinations of partial observables. Theories for which the possibilities admit a precise distinction in these two aspects, admit a (generalized) notion of spacetime.

15.7 Conclusions

The original reason for Brunetti et al. (2003) to introduce the language of category theory in LCQFT, was to express the requirements of locality and general covariance. However, I have argued that this language is appropriate quite generally to express a systems view of the world, in which a physical theory gives rise to a category of possible systems, which can serve as a kind of model for modal logic. The category of Lorentzian manifolds Loc, and the requirement that theories are functors $\mathbf{A}$: Loc $\to$ Phys, are a concrete expression of how spacetime acts as an organizing principle that brings structure to our understanding of the world. I have also argued that the functorial structure of spacetime may be highly nontrivial, because the source category may have many morphisms that lead to a rich structure. To illustrate this point I have used the work of Fewster and Verch (2012), who use the structure of Loc to shed light on the deep question of identifying a quantum field theory in different spacetimes. However, I did not attempt to disentangle the roles of locality, general covariance, the metric, and the causal behavior of the theory in their argument.

I have considered the notion of spacetime without a metric in Section 15.6 as an organizing principle in the general context of the systems view of the world, expressed in the language of category theory. In this context I have formulated two proposals to characterize the notion of spacetime. The first is a mathematical condition: the existence of a functor $\mathbf{A}$: Man $\to$ Phys, where Man is a category of manifolds. This raises the interesting question when such a functor exists and under what conditions we can reconstruct it from the category Phys. Furthermore, if the structure of Man can be formulated entirely in terms of relations between objects and morphisms, as I argued in Section 15.5, this may help clarify how quantum gravity can be about quantized fields without having underlying manifolds as a fundamental notion. The second proposal is formulated in more abstract terms: spacetime can be characterized as the set of circumstances that specify a system of a theory. Although both formulations are somewhat analogous, I do not expect them to be fully equivalent, nor do I expect either of them to be the last word. I do hope, however, that they at least illustrate that the categorical language introduced by Brunetti et al. (2003) is also suitable for the discussion of conceptual and philosophical questions, like those surrounding spacetime and quantum gravity.

Along the way I have mentioned several additional questions. How can we split the possibilities of modal logic into possible systems and possible states? Which categories can act as a category of possible systems Phys? What does a category Phys look like that uses partial observables instead of observables? An investigation of these and of the proposals that I formulated will require input from philosophers of science as well as from mathematicians and physicists. It is the purpose of this essay to raise these questions to the philosophical community, and to promote a dialogue with the mathematics and physics communities.

Acknowledgments

I would like to thank the organizers of the "Space and Time after Quantum Gravity Essay Contest" and the anonymous referees for their careful reading and constructive comments. I am also grateful to the audience of the corresponding presentation at the University of Geneva, and to the participants of the 18th UK and European Conference on Foundations of Physics and of the workshop on "Recent Mathematical Developments in Quantum Field Theory" at the MFO, for their questions and comments.

References

Atiyah, M. F. 1988. Topological quantum field theory. *Inst. Hautes Études Sci. Publ. Math.*, **68**, 175–186.

Barrett, J. W. 1995. Quantum gravity as topological quantum field theory. *J. Math. Phys.*, **36**, 6161–6179.

Brunetti, R., Fredenhagen, K., and Rejzner, K. 2016. Quantum gravity from the point of view of locally covariant quantum field theory. *Commun. Math. Phys.*, **345**, 741–779.

Brunetti, R., Fredenhagen, K., and Verch, R. 2003. The generally covariant locality principle—A new paradigm for local quantum field theory. *Commun. Math. Phys.*, **237**, 31–68.

Buchholz, D., D'Antoni, C., and Fredenhagen, K. 1987. The universal structure of local algebras. *Commun. Math. Phys.*, **111**, 123–135.

Doplicher, S., Fredenhagen, K., and Roberts, J.E. 1995. Space-time quantization induced by classical gravity. *Commun. Math. Phys.*, **172**, 187–220.

Ehlers, J., Pirani, F. A. E., and Schild, A. 2012. Republication of: The geometry of free fall and light propagation. *Gen. Relativ. Gravit.*, **44**, 1587–1609.

Fewster, C. J. 2007. Quantum energy inequalities and local covariance II: Categorical formulation. *Gen. Relativ. Gravit.*, **39**, 1855–1890.

Fewster, C. J. 2013. Endomorphisms and automorphisms of locally covariant quantum field theories. *Rev. Math. Phys.*, **25**, 1350008.

Fewster, C. J. 2015. Locally covariant quantum field theory and the problem of formulating the same physics in all space-times. *Philos. Trans. A*, **373**, 20140238.

Fewster, C. J., and Schenkel, A. 2015. Locally covariant quantum field theory with external sources. *Ann. Henri Poincaré*, **16**, 2303–2365.

Fewster, C. J., and Verch, R. 2012. Dynamical locality and covariance: What makes a physical theory the same in all spacetimes? *Ann. Henri Poincaré*, **13**, 1613–1674.

Forbes, G. 1985. *The Metaphysics of Modality*. Oxford: Clarendon Press.

Fredenhagen, K., and Rejzner, K. 2013. Batalin-Vilkovisky formalism in perturbative algebraic quantum field theory. *Commun. Math. Phys.*, **317**, 697–725.

Haag, R. 1991. *Local Quantum Physics*. Berlin: Springer.

Halvorson, H. 2007. Algebraic quantum field theory. Pages 731–864 of: Butterfield, J. and Earman, J. (eds.). *Philosophy of Physics*. Handbook of the Philosophy of Science. Amsterdam: Elsevier.

Halvorson, H., and Tsementzis, D. 2017. Categories of scientific theories. Pages 402–429 of: Landry, E. (ed.). *Categories for the Working Philosopher*. Oxford: Oxford University Press.

Hollands, S. 2008. Renormalized Yang-Mills fields in curved spacetime. *Rev. Math. Phys.*, **20**, 1033–1172.

Hollands, S., and Wald, R. M. 2001. Local Wick polynomials and time ordered products of quantum fields in curved spacetime. *Commun. Math. Phys.*, **223**, 289–326.

Hollands, S., and Wald, R. M. 2015. Quantum fields in curved spacetime. *Phys. Rep.*, **574**, 1–35.

Kay, B. S. 1979. Casimir effect in quantum field theory. *Phys. Rev. D*, **20**, 3052–3062.

Kay, B. S. 1992. The principle of locality and quantum field theory on (non-globally hyperbolic) curved spacetimes. *Rev. Math. Phys.*, **04**, 167–195.

Kay, B. S. 1997. Quantum fields in curved spacetime: non global hyperbolicity and locality. Pages 578–588 of: Doplicher, S., Longo, R., Roberts, J. E., and Zsido, L. (eds.). *Operator Algebras and Quantum Field Theory*. Boston: International Press.

MacLane, S. 1971. *Categories for the Working Mathematician*. New York: Springer.

Rovelli, C. 2008. *Quantum Gravity*. Cambridge: Cambridge University Press.

Sanders, K. 2009. *Aspects of Locally Covariant Quantum Field Theory*. PhD thesis, University of York.

Sanders, K. 2010. The locally covariant Dirac field. *Rev. Math. Phys.*, **22**, 381–430.

Sanders, K., Dappiaggi, C., and Hack, T.-P. 2014. Electromagnetism, local covariance, the Aharonov-Bohm effect and Gauss' law. *Commun. Math. Phys.*, **328**, 625–667.

Verch, R. 2001. A spin-statistics theorem for quantum fields on curved spacetime manifolds in a generally covariant framework. *Commun. Math. Phys.*, **223**, 261–288.

Wald, R. M. 1994. *Quantum Field Theory in Curved Spacetime and Black Hole Thermodynamics*. Chicago and London: University of Chicago Press.

Weatherall, J. O. 2017. Categories and the foundations of classical field theories. Pages 329–348 of: Landry, E. (ed.). *Categories for the Working Philosopher*. Oxford: Oxford University Press.

Zahn, J. 2014. The renormalized locally covariant Dirac field. *Rev. Math. Phys.*, **26**, 1330012.

Index

358